Paths to a Green World

Paths to a Green World

The Political Economy of the Global Environment

Second edition

Jennifer Clapp and Peter Dauvergne

The MIT Press
Cambridge, Massachusetts
London, England

First published 2005
Second edition 2011

For information about special quantity discounts, please email special_sales@mitpress.mit.edu

This book was set in Sabon by Toppan Best-set Premedia Limited. Printed and bound in the United States of America.

Library of Congress Cataloging-in-Publication Data

Clapp, Jennifer, 1963–
Paths to a green world : the political economy of the global environment / Jennifer Clapp and Peter Dauvergne.—2nd ed.
p. cm.
Includes bibliographical references and index.
ISBN 978-0-262-51582-5 (pbk. : alk. paper)
1. Environmental economics. 2. Environmental policy. 3. Global environmental change. 4. Globalization—Economic aspects. I. Dauvergne, Peter. II. Title.
HC79.E5.C557C53 2011
333.7—dc22

2010032798

10 9 8 7 6 5 4 3

For our families.

Contents

Illustrations

Tables

Figures

Boxes

Preface

This second edition of *Paths to a Green World* is a highly ambitious book. Thoroughly updated, this revised edition remains the only book to concentrate exclusively on the political economy of the global environment, striving to integrate the debates within the "real world" of global policy and the "academic world" of theory. It moves well beyond the traditional academic focus on international agreements and institutions in an effort to capture the views on politics, economics, and the environment within the halls of global conferences, on the streets during antiglobalization protests, and in the boardrooms of international agencies, nongovernmental organizations, and industry associations. In doing so, it investigates the debates over globalization, environmentalism, economic growth, poverty, consumption, trade, corporate investment, and international finance. It does so from a variety of angles—economic, political, ecological, and social.

The book explicitly does not advocate for a particular perspective on how politics and economics relate to the health of the global environment. Instead, it offers an original typology of worldviews—what we call market liberal, institutionalist, bioenvironmentalist, and social green—to classify the various debates present in political and academic arenas. This typology is, we believe, parsimonious enough for readers to grasp the key threads with ease, yet nuanced enough to rouse vigorous debate. The book fills, in our view, a critical gap in the literature on global environmental change. It meets an immediate need in the field of global environmental politics, by providing comprehensive coverage of the political economy of the global environment that includes policy and corporate views that academics often downplay or ignore. The typology we propose in the book, we hope, will also meet a much more imposing need: to help scholars, bureaucrats, industrialists, and activists communicate in a common language. This latter goal is perhaps too ambitious, perhaps even naive. But given the enthusiastic response to the 2005 edition of this book, we are encouraged that a second edition will facilitate such dialog even further.

We have tried our best to explain the complexities of the political economy of global environmental change without disciplinary jargon. Naturally, the book uses terminology; otherwise, it could only skim the surface of the core debates. Yet at every turn we strive to explain debates and define terms in ways that transcend disciplines. Our hope is that those from a range of educational backgrounds—including development studies, economics, environmental studies, geography, human ecology, international law, philosophy, political science, and sociology—can use this book for a big-picture snapshot of the core debates.

Like the first edition of *Paths to a Green World*, this one will also function well as a university textbook to introduce the debates on the interface between political economy and global environmental change. Instructors using it as a textbook may want to add case studies of particular global environmental problems. In our own teaching, for example, we add lectures and readings on the political economy of climate change, deforestation, food security, nonrenewable resource extraction, ozone depletion, persistent organic pollutants, and trade in hazardous waste. But other global environmental issues—like acid rain, biodiversity loss, desertification, energy use, overfishing, genetically modified organisms, trade in endangered species, transboundary pollution, and whaling, as well as many others—would work equally well.

Instructors may also want to integrate some literature with more of a disciplinary focus to expose students to the particular terminology and research methods that their discipline uses to analyze the political economy of global environmental change. One of us, for example, teaches in a department of political science and supplements this book with readings that reflect the language and debates in the fields of international relations and global environmental politics. The other teaches both environmental and international studies and supplements the book with readings that reflect the learning of the students in these programs. It is, we believe, worthwhile to encourage students to think beyond disciplinary boundaries. Yet it is often just as valuable to embed some learning within one or two disciplines, because this can allow for a more erudite analysis of the core questions in a particular discipline.

We trust that all who choose to continue—regardless of the reason for beginning—will read with the curiosity of a true student, so each of the worldviews can spring equally to life in the analysis in the rest of the book.

Acknowledgments

We would like to thank all who have inspired us in one way or another with the first and second editions of this book. The feedback from the many instructors using the first edition as a textbook has been especially encouraging and rewarding. For invaluable assistance, special thanks go to Sam Grey, Ryan Pollice, and Linda Swanston for their exceptional support on the research, graphics, and data. Sharon Goad, Joshua Gordon, Jeca Glor-Bell, Sanushka Mudaliar, Christopher Rompré, Michael Stevenson, and Jason Thistlethwaite also helped immeasurably with the research. We are grateful to the Australian Research Council and the Social Sciences and Humanities Research Council of Canada for financial support for this research. For perceptive comments and valuable guidance, we would like to thank Ben Cashore, Herman Daly, Catherine Dauvergne, Torben Drewes, Robert Falkner, Derek Hall, Eric Helleiner, Osamu Nakano, Matthew Paterson, Rodrigo Pinto, and three anonymous reviewers for The MIT Press. We also owe Osamu Nakano additional thanks for translating the first edition into Japanese. We are grateful to Clay Morgan of the MIT Press for shepherding both editions through the publication process and Kathleen Caruso and her team for their careful editing of manuscript. And we would like to thank our friends and families for their moral support and patience throughout this project. We accept full responsibility for any errors or omissions that are bound to arise in a book of this scope, and as with our first edition, we would appreciate any feedback from our readers on the second one.

Paths to a Green World

Acronyms

ADB	Asian Development Bank
AfDB	African Development Bank
AIDS	Acquired immune deficiency syndrome
AoA	Agreement on Agriculture
ASEAN	Association of Southeast Asian Nations
BASD	Business Action on Sustainable Development
BECC	Border Environmental Cooperation Committee
BIS	Bank for International Settlements
CBD	Convention on Biological Diversity
CCX	Chicago Climate Exchange
CDM	Clean Development Mechanism
CDP	Carbon Disclosure Project
CEC	Commission on Environmental Cooperation
CEO	Chief executive officer
CFCs	Chlorofluorocarbons
CIDA	Canadian International Development Agency
CIS	Commonwealth of Independent States
CITES	Convention on International Trade in Endangered Species of Wild Fauna and Flora
CO_2	Carbon dioxide
COP	Conference of the Parties
CPRs	Common property regimes
CSD	Commission on Sustainable Development (UN)
CSR	Corporate social responsibility

CTE	Committee on Trade and Environment
DAC	Development Assistance Committee (OECD)
DCSD	Danish Committee on Scientific Dishonesty
DDT	Dichlorodiphenyltrichloroethane
EBRD	European Bank for Reconstruction and Development
ECAs	Export credit agencies
ECGD	Export Credit Guarantee Department (UK)
ECOSOC	Economic and Social Council (UN)
EFIC	Australian Export Finance and Insurance Corporation
EIA	Energy Information Administration
EIR	Extractive Industries Review
EKC	Environmental Kuznets curve
EMAS	Eco-Management and Audit Scheme
EPA	Environmental Protection Agency (U.S.)
EU	European Union
EU ETS	European Union Emissions Trading Scheme
FAO	Food and Agriculture Organization of the United Nations
FDA	Food and Drug Administration (U.S.)
FDI	Foreign direct investment
FOE	Friends of the Earth
FOEI	Friends of the Earth International
FSC	Forest Stewardship Council
G-77	Group of 77
GAST	General Agreement on Sustainable Trade
GATS	General Agreement on Trade in Services
GATT	General Agreement on Tariffs and Trade
GDP	Gross domestic product
GEF	Global Environment Facility
GEMI	Global Environmental Management Initiative
GHG	Greenhouse gas
GMOs	Genetically modified organisms
GNI	Gross national income
GNP	Gross national product

GPI	Genuine progress indicator
HDI	Human development index
HIPC	Heavily indebted poor countries
HIV	Human immunodeficiency virus
IBRD	International Bank for Reconstruction and Development
ICC	International Chamber of Commerce
ICSID	International Center for the Settlement of Investment Disputes
IDA	International Development Association
IDB	Inter-American Development Bank
IFAD	International Fund for Agricultural Development
IFC	International Finance Corporation
IFG	International Forum on Globalization
IIC	International Insolvency Court
IISD	International Institute for Sustainable Development
IMF	International Monetary Fund
ISEW	Index of sustainable economic welfare
ISO	International Organization for Standardization
ITU	International Telecommunications Union
IUCN	International Union for Conservation of Nature and Natural Resources
JBIC	Japan Bank for International Cooperation
LETS	Local exchange trading systems
LPI	Living planet index
MAI	Multilateral Agreement on Investment
MARPOL	Convention for the Prevention of Pollution by Ships
MEAs	Multilateral environmental agreements
MFN	Most favored nation
MIGA	Multilateral Investment Guarantee Agency
MMPA	Marine Mammal Protection Act (U.S.)
NAAEC	North American Agreement on Environmental Cooperation
NAFTA	North American Free Trade Agreement
NASA	National Aeronautics and Space Administration

NGOs	Nongovernmental organizations
NIEO	New International Economic Order
ODA	Official development assistance
OECD	Organization for Economic Cooperation and Development
OPEC	Organization of the Petroleum Exporting Countries
OPIC	Overseas Private Investment Corporation (U.S.)
PANNA	Pesticide Action Network North America
PCBs	Polychlorinated biphenyls
POPs	Persistent organic pollutants
PPMs	Production and processing methods
PPPs	Public-private partnerships
PRI	Principles for Responsible Investment
PVC	Polyvinyl chloride
SAL	Structural adjustment loan
SAPs	Structural adjustment programs
SPS	Agreement on the Application of Sanitary and Phytosanitary Measures
SUV	Sport utility vehicle
TBT	Agreement on Technical Barriers to Trade
TNC	Transnational corporation
TRIPS	Trade-Related Intellectual Property Rights Agreement
TWN	Third World Network
UNCED	United Nations Conference on Environment and Development
UNCTAD	United Nations Conference on Trade and Development
UNCTC	United Nations Center for Transnational Corporations
UNDP	United Nations Development Programme
UNEP	United Nations Environment Programme
UNEP FI	UNEP Finance Initiative
UNFCCC	United Nations Framework Convention on Climate Change
UNFPA	United Nations Fund for Population Activities
UNIDO	United Nations Industrial Development Organization
USAID	United States Agency for International Development

WBCSD	World Business Council for Sustainable Development
WCED	World Commission on Environment and Development
WEO	World Environment Organization
WHO	World Health Organization
WICE	World Industry Council on the Environment
WLO	World Localization Organization
WRI	World Resources Institute
WWF	WWF Network (formerly World Wildlife Fund/World Wide Fund for Nature)

1

Peril or Prosperity? Mapping Worldviews of Global Environmental Change

The sun could well engulf the earth in about seven or eight billion years. "So what," you might shrug. "The extinction of earth, beyond the horizon of human time—ridiculous, not worth imagining." Yet some environmentalists believe that waves of smaller disasters—like climate change, deforestation, toxic pollution, and biodiversity loss—are already destroying the planet. Without doubt, many of the world's poorest people have already collided with their sun, dying from disease, starvation, war, and abuse. The beginning of the end, these environmentalists lament, is already upon us. We, as a species, are now beyond the earth's carrying capacity, a trend accelerating in the era of globalization. Unless we act immediately with resolve and sacrifice, in a mere hundred years or so, humanity itself will engulf the earth. The future is one of peril.

Many environmentalists rebel against such catastrophic visions. Yes, there are undeniable ecological problems—like the changing global climate, the pollution of rivers and lakes, and the collapse of some fish stocks—but some ecological disturbance is inevitable, and much is correctable through goodwill and cooperation. There is no crisis or looming crisis: to think so is to misread the history of human progress. This history shows the value of positive thinking, of relying on human ingenuity to overcome obstacles and create ever-greater freedom and wealth with which we can ensure a better natural environment. Globalization is merely the latest, though perhaps the most potent, engine of human progress. The future is one of prosperity.

Who is correct? Do the pessimists need antidepressants? Do the optimists need a stroll through a toxic waste dump in the developing world? Less flippantly, what is the middle ground between these two extremes? What are the causes and consequences of global environmental change?

Are ecological problems really as severe as some claim? Does the cumulative impact of these problems constitute a crisis? How is the global community handling them? Why are the efforts to resolve some problems more successful than others? Why are environmental problems worse in some parts of the world? And what is the relationship to global political and economic activity? These are tough questions, and we do not pretend to know the answers with absolute certainty. A quick survey of the typical answers to these questions reveals an almost endless stream of contradictory explanations and evidence. Each answer can seem remarkably logical and persuasive. The result for the thoughtful and "objective" observer is often dismay or confusion.

Given this, how does one even begin to understand global environmental change? It helps, we believe, to begin with the big picture, rather than delving immediately into in-depth studies of particular environmental issues. Understanding this big picture is, in our view, necessary *before* we can fully understand the various interpretations of the *specific* causes and consequences of environmental problems. In the quest for knowledge and a role in a world overloaded with information and experts, far too often this larger picture is ignored—or at least poorly understood. For problems as intricate as global environmental ones, this can lead to muddled analysis and poorly formulated recommendations. Without this broad perspective, for example, "solving" one problem can ignore other related problems, or create even greater problems elsewhere.

How polities and societies allocate financial, human, and natural resources directly influences how we manage local, national, and ultimately global environments. The issues that shape the relationship between the global political economy and the environment are, of course, often technical and scientific. But they are frequently also socioeconomic and political. Our hope is that by sketching the arguments and assumptions about socioeconomic and political causes with the broadest possible strokes, we will assist readers in a lifelong journey of understanding the causes and consequences of global environmental change, as well as the controversies that surround it. This is a small yet essential step to eventually solving, or at least slowing, some of these problems.[1] To introduce these topics, we map out a new typology of worldviews on the political economy of global environmental change.[2]

Four Environmental Worldviews

We present four main worldviews on global environmental change and its relationship to the global political economy: those of *market liberals*, *institutionalists*, *bioenvironmentalists*, and *social greens*. These labels are intentionally transdisciplinary. Many books on the global environment confine the analysis to one disciplinary box—by limiting it, say, to political science theories or to economic models. This leaves far too many questions badly answered and far too many questions unasked. But we have had to make some choices. It is, of course, impossible to cover all disciplinary perspectives in one book. In our case, we have chosen to rely mostly on the tools of political science, economics, development studies, environmental studies, political geography, and sociology. This focus, we believe, is narrow enough to do justice to the literature in these disciplines, while still broad enough to provide new insights into the sources of environmental change and the possible options—both theoretical and practical—for managing it.

These are ideal categories, somewhat exaggerated to help differentiate between them, although certainly there are some champions of each of these worldviews in the real world that do adhere to the extreme end of each of these viewpoints. By mapping out these four very different worldviews in their extremes, we aim to help students navigate a seemingly unmanageable avalanche of conflicting information and analysis. Within each category, we have tried to group the ideas of thinkers—not just academics, but equally policy makers and activists—with broadly common assumptions and conclusions. This we hope provides a sense of the debates in the "real" world—that is, within bureaucracies, cabinet meetings, international negotiations, activist campaigns, and corporate boardrooms, as well as in classrooms. Our approach, in a sense, tries to capture the broader societal debates about environment and political economy, rather than just the academic debates over the theories of the political economy of the environment, which often cover a more narrow range of viewpoints.

Naturally, given the breadth of our labels, many disagreements exist among those in each category. We have tried to show the range of views subsumed under each of the four major worldviews, although at the end of this book you may still find that your own beliefs and arguments do not fit neatly into any of these categories. Or you may feel that you hold a mix of views—even ones that at first seem at opposite poles, such as market liberal and social green. This does not mean that our categories are erroneous, or that you are inconsistent or hypocritical, or that you should force your views into one category. Instead, it just shows the complexity and diversity of individual views on the issues.

Our typology, moreover, does not cover all possible views, although while conscious to avoid creating dozens of labels, we do try to give a reasonable range. We include only thinkers who are *environmentalists*—that is, those who write and speak and work to maintain or improve the environment around us. This includes those highly critical of so-called environmental activists or radical greens. An economist at the World Bank is, in our view, just as much an environmentalist as a volunteer at Greenpeace, as long as the economist believes she or he is working for a better environment (however that is defined). Also, we focus principally on economic and political arguments, and tend to give less attention to philosophical and moral ones. Within the political and economic literature, we stress arguments and theories that try to *explain* global environmental change—that is, the literature that looks at an environmental problem and asks: Why is that happening? What is causing it? And what can be done?

With those introductory remarks, we now turn to our typology.

Market Liberals

The analysis of market liberals is grounded in neoclassical economics and scientific research. Market liberals believe that economic growth and high per capita incomes are essential for human welfare and the maintenance of sustainable development. Sustainable development is generally defined by these thinkers along the lines of the 1987 World Commission on Environment and Development (WCED): "development that meets the needs of the present without compromising the ability of future generations to meet their own needs."[3] In terms of improving global environmental conditions, market liberals argue that economic growth (production and consumption) creates higher incomes, which in turn generate the funds and political will to improve environmental conditions. Rapid growth may exacerbate inequalities, as some of the rich become super rich, but in the long run all will be better off. In other words, all boats will rise. Market liberal analysis along these lines is commonly found, for example, in publications of the World Bank, the World Trade Organization (WTO), and the World Business Council for Sustainable Development (WBCSD), as well as in the media in publications such as *The Economist*.

Market liberals see globalization as a positive force, because it promotes economic growth as well as global integration. They concede that as states pursue economic growth, environmental conditions—such as air and water quality—may deteriorate as governments and citizens give firms more scope to pursue short-term profits, thus stimulating further economic growth. But once a society becomes wealthy, citizens (and in turn governments and business) will raise environmental standards and expectations. *The Economist* magazine explains the global pattern: "Where most of the economic growth has occurred—the rich countries—the environment has become cleaner and healthier. It is in the poor countries, where growth has been generally meagre, that air and water pollution is an increasing hazard to health."[4] The key, market liberals argue, is good policy to ensure that economic growth improves the environment in all countries.

The main drivers of environmental degradation, according to market liberals, are a lack of economic growth, poverty, distortions and failures of the market, and bad policies. Poor people are not viewed as unconcerned or ignorant. Rather, to survive—to eat, to build homes, to earn a living—they must exploit the natural resources around them. They are, according to the World Bank, both "victims and agents of environmental damage."[5] It is unrealistic—perhaps even unjust—to ask poor people to consider the implications of their survival for future generations. The only way out of this vicious cycle is to alleviate poverty, for which economic growth is essential. Restrictive trade and investment policies and a lack of secure property rights all hamper the ability of the market to foster growth and reduce poverty. Market failures—instances where the free market results in an environmentally suboptimal outcome—are viewed as possible causes of some environmental problems, although these are seen as relatively rare in practice. More often, market liberals argue, inappropriate government policies—especially those that distort the market, such as subsidies—are the problem.

Market liberals frequently draw on more moderate estimates of environmental damage and more optimistic scenarios for the future. A few have become famous for declaring that the global environment is nowhere near a state of crisis, such as late economist Julian Simon,[6] columnist Gregg Easterbrook,[7] and political scientist Bjørn Lomborg.[8] But most recognize that many environmental problems are indeed serious, although all reject the image of the world spinning toward a catastrophic ecological crash. Instead, market liberals tend to stress our scientific achievements, our progress, and our ability to reverse and repair environmental problems with ingenuity, technology, cooperation, and adaptation. For these thinkers, population growth and resource scarcity are not major concerns when it comes to environmental quality. A glance at the historical trend of better environmental conditions for all confirms this (especially statistics from the developed world). So do the global data on human well-being, such as medical advances, longer life expectancy, and greater food production. Furthermore, most environmental problems, if not currently responding to efforts to manage them more effectively, at least have the potential to improve in the longer term.

Thinkers from the market liberal tradition place great faith in the ability of modern science and technology to help societies slip out of any environmental binds that may occur (if, e.g., unavoidable market failures occur). Human ingenuity is seen to have no limits. If resources become scarce, or if pollution becomes a problem, humans will discover substitutes and develop new, more environmentally friendly technologies. Market liberals see advances in agricultural biotechnology, for example, as a key answer to providing more food for a growing world population. Their belief in science leaves most market liberals wary of precautionary policies that restrict the use of new technology, unless there is clear scientific evidence to demonstrate that it is harmful.

Market liberals believe open and globally integrated markets promote growth, which in turn helps societies find ways to improve or repair environmental conditions. To achieve these goals, market liberals call for policy reforms to liberalize trade and investment, foster specialization, and reduce government subsidies that distort markets and waste resources. Governments, too, need to strengthen some institutions, such as institutions to secure property rights or institutions to educate and train the poor to protect the environment. Governments are encouraged to use market-

based tools—for example, environmental taxes or tradable pollution permits—to correct situations of genuine market failure. Innovative environmental markets—like a global scheme to trade carbon emissions or niche markets for environmental products such as timber from sustainable sources—and voluntary corporate measures to promote environmental stewardship are also reasonable ways to improve environmental management. But in most cases, they believe, it is best to let the market allocate resources efficiently. Market liberals, such as economist Jagdish Bhagwati[9] and business executive Stephan Schmidheiny,[10] argue that it makes economic sense for firms to improve their environmental performance, and for this reason it makes sense to let the market guide them.

Institutionalists

The ideas of institutionalists are grounded in the fields of political science and international relations. They share many of the broad assumptions and arguments of market liberals—especially the belief in the value of economic growth, globalization, trade, foreign investment, technology, and the notion of sustainable development. Indeed, moderate institutionalists sit close to moderate market liberals. It is a matter of emphasis. Market liberals stress more the benefits and dynamic solutions of free markets and technology; institutionalists emphasize the need for stronger global institutions and norms as well as sufficient state and local capacity to constrain and direct the global political economy. Institutions provide a crucial route to transfer technology and funds to the poorest parts of the planet.[11] Institutionalists also worry far more than market liberals about environmental scarcity, population growth, and the growing inequalities between and within states. But they do not see these problems as beyond hope. To address them, they stress the need for strong institutions and norms to protect the common good. Institutionalist analysis is found in publications by organizations such as the UN Environment Programme (UNEP) and by many academics who focus their analysis on international organizations and "regimes" (international environmental agreements and norms, defined more precisely in chapter 3) in the fields of political science and law.

Institutionalists see a lack of global cooperation as a key source of environmental degradation. The failure of the 2009 Copenhagen climate summit to reach a bold new international agreement to address climate change, for example, was a deep disappointment to many institutionalists, because without global cooperation, the problem only promises to worsen. Ineffective cooperation as exemplified by the climate case partly arises because of the nature of the sovereign state system, which gives a state supreme authority within its boundaries. In such a system, states tend to act in their own interest, generally leaving aside the interest of the global commons. Yet like market liberals, institutionalists *do not* reject the way we have organized political and economic life on the planet. Instead, they believe we can overcome the problem of sovereignty as the organizing principle of the international system by building and strengthening global and local institutions that promote state adherence to collective goals and norms. This can be most effectively carried out through global-level environmental agreements and organizations.

The process of globalization makes global cooperation increasingly essential (and increasingly inevitable). But institutionalists stress that unfettered globalization can add to the pressures on the global environment. The task for those worried about the state of the global environment, then, is to guide and channel globalization so that it enhances environmental cooperation and better environmental management. This point has been stressed most forcefully by key policy figures such as former Norwegian Prime Minister Gro Harlem Brundtland in her role in the 1980s as head of the WCED, Canadian diplomat Maurice Strong as an organizer of global environmental conferences, and Yvo de Boer as executive secretary of the UN Framework Convention on Climate Change. The aim of this approach is to ensure that global economic policies work to both improve the environment and raise living standards.[12] Controls at all levels of governance, from the local to the national to the global, can help to direct globalization, enhancing the benefits and limiting the drawbacks.[13]

For the global environment, institutionalists believe that institutions need to internalize the principles of sustainable development, including into the decision-making processes of state bureaucracies, corporations, and international organizations. Only then will we be able to manage economies and environments effectively—especially for common resources. For many institutionalist academics, like political scientist Oran Young, the most effective and practical means is to negotiate and

strengthen international environmental regimes.[14] Many within the policy world, such as in the United Nations Environment Programme, add the need to enhance state and local capacity in developing countries.[15] Thus, many institutionalists call for more and better "environmental aid" for the developing world.[16] It should be stressed, however, that institutionalists do not necessarily support all institutions uncritically. Some point to badly constructed institutions as a source of problems. Many point as well to the difficulty of trying to measure the implementation and effectiveness of an international agreement or institution.[17] But a defining characteristic of institutionalists is the assumption that institutions matter—that they are valuable—and that what we need to do is reform, not overthrow, them.[18]

Institutionalists also argue that strong global institutions and cooperative norms can help enhance the capacity of *all* states to manage environmental resources. What is needed, from this perspective, is to embed environmental norms into international cooperative agreements and organizations as well as state policies. Along these lines, many institutionalists support a precautionary approach, in which states agree to collective action in the face of some scientific uncertainty. Institutionalists also advocate the transfer of knowledge, finances, and technology to developing countries. Organizations like the World Bank, the United Nations Environment Programme, and the Global Environment Facility already play a role here. And many institutionalists point to the creation of and changes within these organizations as evidence of progress.

Bioenvironmentalists

Inspired by the laws of physical science, bioenvironmentalists stress the biological limits of the earth to support life. The planet is fragile, an ecosystem like any other. Some even see the earth as behaving like a living being, a self-regulating, complex, and holistic superorganism—the so-called Gaia hypothesis, as articulated by environmental scientist James Lovelock.[19] The earth can support life, but only to a certain limit, often referred to as the earth's "carrying capacity." Many bioenvironmentalists see humans as anthropocentric and selfish (or at least self-interested) animals. Some, like the academic William Rees, even see humans as having "a genetic predisposition for unsustainability."[20] All bioenvironmentalists agree that humans as a species now consume far too much of the earth's resources, such that we are near, or indeed have already overstepped, the earth's carrying capacity. Such behavior, without drastic changes, will push the planet toward a fate not much different from the ecological calamity of Easter Island of three hundred years ago—where a once-thriving people became over a few centuries "about 2000 wretched individuals . . . eking out a sparse existence from a denuded landscape and cannibalistic raids on each other's camps."[21] These scholars stress the environmental disasters around us, often citing shocking figures on such problems as overfishing, deforestation, species loss, and unstable weather patterns. Publications of the Worldwatch Institute and the WWF Network (WWF, formerly the World Wildlife Fund/World Wide Fund for Nature) are illustrative of this perspective.

For most bioenvironmentalists, population growth is a key source of stress on the earth's limits. The ideas of Thomas Malthus (1766–1834), who in "An Essay on the Principle of Population"[22] predicted that the human population would soon outstrip food supply, were revived in the late 1960s by writers such as biologist Paul Ehrlich.[23] Sometimes known as neo-Malthusians, these writers argue that global environmental problems ultimately stem from too many people on a planet with finite resources. The principle of sovereignty, which divides the world into artificial territories, aggravates the effects of too many humans because it violates the principles of ecology and creates what academic Garrett Hardin famously called a "tragedy of the commons." For him, too many people without overarching rules on how to use the commons creates a situation in which individuals, rationally seeking to maximize their own gain at the expense of others, overuse and ultimately destroy the commons.[24] This point, stressed by many bioenvironmentalists, is also made by many institutionalists, as discussed earlier.

Many bioenvironmentalists stress, too, that the neoclassical economic assumption of infinite economic growth is a key source of today's global environmental crisis. For these thinkers, a relentless drive to produce ever more in the name of economic growth is exhausting our resources and polluting the planet. Many argue that the drive to pursue ever more economic growth is what has taken the earth beyond its carrying capacity. For bioenvironmentalists, increasing human consumption is as great

a problem as population growth, and the two are seen as inextricably linked. Together, they argue, rising populations and consumption are drawing down the earth's limited resources: we must respect the biophysical limits to growth, both for people and economies.[25]

Not all bioenvironmentalists engage directly in discussions on economic globalization, but those that do tend to see globalization as a negative force for the environment. They agree with market liberals that globalization enhances economic growth. But instead of seeing this as positive for the environment, they see it as contributing to further environmental degradation. For them, more growth only means more consumption of natural resources and more stress on waste sinks. Globalization is blamed, too, for spreading Western patterns of consumption into the developing world. With much larger populations and often more fragile ecosystems (especially in the tropics), this spread of consumerism is accelerating the collapse of the global ecosystem.[26] Globalization is also seen to encourage environmentally harmful production processes in poor countries that have lower environmental standards.[27] For these reasons, these bioenvironmentalists argue that we must curtail economic globalization to save the planet.

Solutions proposed by bioenvironmentalists flow logically from their analysis of the causes of environmental damage: we need to curb economic and population growth. Those who focus on the limits to economic growth have been a core group in the field of ecological economics, pioneered by thinkers such as economist Herman Daly[28] and published in journals such as *Ecological Economics*. This group combines ideas from the physical sciences and economics to develop proposals to revamp economic models to include the notion of physical limits, which involves changing our measures of "progress" and the methods we use to promote it. Only then, these thinkers argue, can we reduce the impact of humans on the planet and prod the world toward a more sustainable global economy. Those bioenvironmentalists who focus more on overpopulation call for measures to lower population growth, like expanding family planning programs in poor countries, and for curbs on immigration to rich countries where consumption problems are the worst. At the more extreme end, some see a world government with coercive powers as the best way to control the human lust to fill all ecological space, destroying it, often inadvertently, in the process.[29]

Social Greens

Social greens, drawing primarily on radical social and economic theories, see social and environmental problems as inseparable. Inequality and domination, exacerbated by economic globalization, are seen as leading to unequal access to resources as well as unequal exposure to environmental harms. Although these views have long been important in debates over environment and development, and are themselves a mix of a variety of radical views, scholars in international political economy have only recently recognized them as a distinct perspective.[30]

Many social greens from a more activist stance focus on the destructive effects of the global spread of large-scale industrial life.[31] Accelerated by the process of globalization, large-scale industrialism is seen to encourage inequality characterized by overconsumption by the wealthy, while at the same time contributing to poverty and environmental degradation. While agreeing broadly with this analysis, other, more academic social greens draw on Marxist thought, pointing specifically to capitalism as a primary driver of social and environmental injustice in a globalized world. They argue that capitalism, and its global spread via neocolonial relations between rich and poor countries, not only leads to an unequal distribution of global income, power, and environmental problems, but is also a threat to human survival.[32] Also inspired by Marxist thought, some social greens take a neo-Gramscian or historical materialist perspective, focusing on the way those in power frame and influence ecological problems, primarily hegemonic blocs consisting of large corporations and industrial country governments.[33] Other social greens like Vandana Shiva draw heavily from feminist theory to argue that patriarchal relationships in the global economy are intricately tied to ecological destruction.[34] The key concern of all of these strands of social green thought, then, is inequality and the environmental consequences related to it. Social green analysis can be found in magazines such as *The Ecologist* and in reports of groups such as the International Forum on Globalization (IFG) and the Third World Network (TWN).

Social greens sympathize with bioenvironmentalist arguments that physical limits to economic growth exist. Overconsumption, particularly in rich industrialized countries, is seen by social greens to put a great strain on the global environment. Many, perhaps most prominently

Wolfgang Sachs[35] and Edward Goldsmith,[36] see this problem as accelerating in an era of economic globalization. The arguments of social greens on growth and consumption, and on the role of the global economy in accelerating both, are close to bioenvironmentalist arguments. But few social greens accept bioenvironmentalist arguments regarding population growth, maintaining instead that overconsumption, particularly among the rich in the First World, is a far greater problem.[37] Unlike bioenvironmentalists, most social greens see population-control policies as a threat to the self-determination of women and the poor.[38]

Whether it is viewed as spreading industrialism or capitalism (or both), social greens uniformly oppose economic globalization, arguing that it is a key factor behind much of what is wrong with the global system.[39] In addition to feeding environmentally destructive growth and consumption, globalization is seen to breed injustice in a number of ways. It exacerbates the inequality within and between countries. It reinforces the domination of the global rich and the marginalization of women, indigenous peoples, and the poor. It assists corporate exploitation of the developing world (especially labor and natural resources). It weakens local community autonomy and imposes new forms of domination that are Western and patriarchal (local customs, norms, and knowledge are lost, replaced by new forms unsuited to these new locations). Globalization is also seen to destroy local livelihoods, leaving large numbers of people disconnected from the environment in both rich and poor countries. This globalization is viewed by many social greens as a continuation of earlier waves of domination and control. In the words of the prominent antiglobalization activist Vandana Shiva, "The 'global' of today reflects a modern version of the global reach of the handful of British merchant adventurers who, as the East India Company, later, the British Empire raided and looted large areas of the world."[40]

From this analysis, it is not surprising that social greens reject the current global economy. Reactive crisis management in a globalized world, social greens believe, will not suffice to save the planet: tinkering will just momentarily stall the crash. In many instances, the environmental solutions of market liberals and institutionalists, because they assume globalization brings environmental benefits, are part of the problem. For social greens, major reforms are necessary, well beyond, for example, just strengthening institutions or internalizing environmental and social costs into the price of traded goods. Thus social greens, as the work of the International Forum on Globalization exemplifies, call for a dismantling of current global economic structures and institutions.[41] To replace this, many social greens advocate a return to local community autonomy to rejuvenate social relations and restore the natural environment. Localization activist Colin Hines has mapped out a model for how this could occur. It entails a retreat from the large-scale industrial and capitalist life and a move toward local, self-reliant, small-scale economies.[42] These thinkers stress the need to, in the words of some, "think globally, act locally." In other words, understand the global context, while at the same time acting in ways suitable to the local context. These thinkers advocate bioregional and small-scale community development because they firmly believe that a stronger sense of community will fulfill basic needs and enhance people's quality of life. Such development would help reduce inequities and levels of consumption that are out of balance with the world's natural limits.[43]

As part of their strategy for promoting community autonomy and localization, social greens also stress the need to empower voices marginalized by the process of economic globalization. They embrace indigenous knowledge systems, for example, arguing that these are equally if not more valid than the Western scientific method. The process of economic "development," these critics argue, foists the latter onto the developing world, thus threatening ecologically sound local systems. Many social greens regard local cultural diversity as essential to maintain biological diversity. The erosion of one is seen to lead to the erosion of the other. In advocating local and indigenous empowerment and input, social greens emphasize that effective solutions to environmental problems will continue to remain elusive unless the voices of women, indigenous peoples, and the poor are integrated into the global dialog on environmental and social justice, as well as into locally specific contexts.[44]

Conclusion

Table 1.1 summarizes the main assumptions and arguments of market liberals, institutionalists, bioenvironmentalists, and social greens. We have tried hard to present these views fairly and accurately based on our reading of a variety of works by policy makers, activists, academics, and business leaders across a range of perspectives. Yet we should also stress again that these are "ideal" categories, and within each there are a range of views and more subtle debates. Some authors you will read will fit neatly into one of these categories, and others are more difficult to classify. This variance in ease of classification just demonstrates the range of possible views. Moreover, there are alliances between various views on different issues, which makes the terrain difficult to map at times. For example, market liberals and institutionalists agree with one another that economic growth and globalization have positive implications for the environment, and social greens and bioenvironmentalists hold the opposite view. And institutionalists and bioenvironmentalists agree that population growth poses a problem for the world's resources, while market liberals and social greens put far less emphasis on this factor.

We do not want to leave the impression that any one of these is the "correct" view. Each, we believe, contains insights into the sources of today's environmental problems, as well as into potential solutions. Each view has its own logic, which fits with its assumptions. Understanding these views help to explain, too, the often markedly different interpretations of the condition of the global environment. One article, for example, may well declare climate change the most serious threat confronting today's governments. The next article may declare such a statement exaggerated or unnecessarily alarmist, perhaps even a ploy to raise funds or scare world leaders into action. This, we believe, does not mean that there are no facts—or causality—or analysis—or statistics. It also does not mean that some authors lie and deceive. Rather it merely shows how different interpretations and different values—that is, different worldviews—can shape which information an analyst chooses to *emphasize*.

This book does not aim to provide you with one answer as to how we can achieve a "green world." Rather, it seeks to provide you with tools to assess for yourself what the most appropriate path forward might be. As you proceed through the rest of this book, we urge you to keep an open mind regarding the debates and evidence about the consequences of the global political economy for global environmental change. This is certainly not easy. These are emotional issues. And the evidence

Table 1.1
Environmental perspectives

	Market liberals	Institutionalists
Focus	Economies	Institutions
A global environmental crisis?	No. Some inevitable problems, but overall modern science, technology, ingenuity and money are improving the global environment.	Not yet. Potential for crisis unless we act now to enhance state capacity and improve the effectiveness of regimes and global institutions.
Causes of problems	Poverty and weak economic growth. Market failures and poor government policy (i.e., market distortions such as subsidies as well as unclear property rights) are also partly to blame.	Weak institutions and inadequate global cooperation to correct environmental failures, underdevelopment, and perverse effects of state sovereignty.
Impact of globalization	Fostering economic growth, a source of progress that will improve the environment in the long run.	Enhancing opportunities for cooperation. Guided globalization enhances human welfare.
The way forward	Promote growth, alleviate poverty and enhance efficiency, best pursued with globalization. Correct market and policy failures, and use market-based incentives to encourage clean technologies. Promote voluntary corporate greening.	Harness globalization and promote strong global institutions, norms and regimes that manage the global environment and distribute technology and funds more effectively to developing countries. Build state capacity. Employ precautionary principle.

Table 1.1
(continued)

	Bioenvironmentalists	Social greens
Focus	Ecosystems	Justice
A global environmental crisis?	Yes. Near or beyond earth's carrying capacity. Ecological crisis threatens human survival.	Yes. Social injustice at both local and global levels feeds environmental crisis.
Causes of problems	Human instinct to overfill ecological space, as seen by overpopulation, excessive economic growth, and overconsumption.	Large-scale industrial life (some say global capitalism), which feeds exploitation (of labor, women, indigenous peoples, the poor, and the environment) and grossly unequal patterns of consumption.
Impact of globalization	Driving unsustainable growth, trade, investment, and debt. Accelerating depletion of natural resources and filling of sinks.	Accelerating exploitation, inequalities, and ecological injustice while concurrently eroding local community autonomy.
The way forward	Create a new global economy within limits to growth. Limit population growth and reduce consumption. Internalize the value of nonhuman life into institutions and policies. Agree to collective coercion (e.g., some advocate world government) to control greed, exploitation, and reproduction.	Reject industrialism (and/or capitalism) and reverse economic globalization. Restore local community autonomy and empower those whose voices have been marginalized. Promote ecological justice and local and indigenous knowledge systems.

and arguments are often contradictory, almost as if analysts live in different worlds. Our hope, if you do keep an open mind until the end of the book, is not to confuse you, but to leave you with a better understanding of your own assumptions and arguments. Moreover, if you then decide to reject the arguments of others, you will do so with a genuine understanding of the complexity and historical sources of those views. Only then can the debates truly move forward.

2

The Ecological Consequences of Globalization

We are now in an era of globalization. As a process that touches on most aspects of our economies, societies, and cultures, it is important to investigate how it interfaces with global environmental change. This chapter begins with a general overview of globalization—what it is and its implications. It then sketches broadly contrasting environmental pictures of today's globalizing world: first, a world of progress, of better lives that result from globalization—that is, the world as seen by market liberals and to a lesser extent institutionalists; and second, a world of failure, of crisis and looming ecological and social catastrophe unless immediate action is taken to reverse globalization—that is, the world as seen by bioenvironmentalists, and by many social greens. The first picture stresses the social and environmental benefits of macroeconomic growth, capitalism, new technologies, and international cooperation. The second picture stresses the biological strains of overpopulation and unequal consumption on a planet with too much industrial and agricultural production. Within each broad picture, however, the different perspectives emphasize different points. The chapter concludes with a sketch of how the different lenses of the market liberals, institutionalists, bioenvironmentalists, and social greens provide different insights into the process of globalization. Chapter 3 will then more specifically address globalization and the development of global environmental discourse, as well as the evolution of environmental actors, institutions, and norms at the international level.

What Is Globalization?

Some scholars question whether globalization is actually occurring. Others question whether the word has any real meaning. Still others see the definitions that swirl around the term as analytically imprecise. Without doubt, the concept of globalization is vaguely and imprecisely defined at times, and some authors seem to use the concept inconsistently. In spite of these difficulties, we believe it is still a valuable concept and helps to illuminate the relationship between global environmental change and the global political economy. In fact, we rather boldly assert that globalization is a critical force shaping global affairs.[1] Rather than getting bogged down in the debate over the various possible definitions of globalization, we will just outline our understanding of the term.

Globalization is a multidimensional process, broadly restructuring and integrating the world's economies, institutions, and civil societies. It is a dynamic, ongoing, and accelerating process that is increasing the links among actors, as well as the structures within which they operate, both within states and across borders. Trade, production, and finance are now more globally integrated than ever before, as are global organizations and social movements. These extensive linkages are making global interactions more intense and complex. Globalization is also changing the nature of space, collapsing some space—for example, the effective distance between individuals in Africa and Asia—while creating sharper spatial barriers elsewhere—for instance, between neighbors on a suburban street, customers in a megasupermarket, and commuters on a subway. It is also affecting time, speeding up the nature of these interactions as communication hardware advances. In simple terms, globalization means that the events and actions in one part of the world are affecting people in distant lands much more quickly, and with greater frequency and intensity.

Globalization is partly an extension of processes that began long ago, including modernization and colonization. But it is more than just a new term for old phenomena. It is also more than just internationalization, which focuses on the importance of greater economic and political interconnectedness among states, and which can be seen as a subprocess of globalization. It suggests instead "that human lives are increasingly played out in *the world as a single place*."[2] It points to the growing significance of a global community with global concerns, and it stresses the decreasing importance of geographic distance and increasing importance of transnational actors and forces. It further suggests that we are moving toward an effectively borderless world (especially for ideas and money), and that this process is accelerating (see box 2.1 for selected definitions of globalization).

If this is globalization, what is driving it? Separating the drivers from the consequences of globalization is difficult, because the consequences are themselves constantly reshaping the drivers. New and faster communication technologies—such as telephones, faxes, and e-mail—partly drive it. These technologies create a faster and more efficient transmission belt for people, money, and ideas, as well as for the knowledge to build other technologies—such as cars and industrial equipment. The first transborder telephone call was in 1891, between London and Paris. Since then, the number of telephone mainlines has grown steadily, from 150 million in 1965, to 546 million in 1991, and to 1,278 million in 2007. The number of cellular subscribers, meanwhile, has gone from 215 million in 1997 to 3,305 million in 2007.[3] Over the last hundred years, the cost of international phone calls fell precipitously, from US$235 for a three-minute call from New York to London in 1930 to just 35 cents in 1998 (in constant 1990 U.S. dollars).[4] With more lines and cheaper rates, the volume of international calls continues to rise steadily, from 12.7 billion minutes spent on the phone in 1982 to 183 billion in 2006.[5] The Internet is the latest revolutionary change in global communication. The World Wide Web had just 50 pages in 1993; by the end of the 1990s there were 50 million; and in 2010, Google's index alone logged over 15 billion pages, with the total number of indexed pages at more than 21 billion.[6] The number of Internet users went from 25 million in 1995 to 400 million in 2001 to well over 1.5 billion people by 2010. Perhaps most remarkable is the increase in the volume and speed of information flows on the Internet. Already in 2001, one Internet cable carried more information in a second than was sent over the whole Internet in a month in 1997.[7]

Faster and cheaper transportation facilitates communication as well. Here, too, remarkable changes have occurred over the past half century. In 1950 there were 25 million air passengers; by 1996 there were 400 million; and by 2000 there were 1.4 billion. By 2010 there were over 2.2 billion, a figure Qantas Airlines CEO Geoff Dixon has predicted will double by 2020.[8] One reason for this growth is that air travel has become much cheaper. In constant 1990 U.S. dollars, for example, the cost fell from $0.68 per passenger mile in 1930 to $0.11 in 1990.[9]

Box 2.1
Definitions of globalization

New York Times columnist Thomas Friedman writes that "globalization is the integration of everything with everything else. . . . [It] is the integration of markets, finance, and technology in a way that shrinks the world from a size medium to a size small. Globalization enables each of us, wherever we live, to reach around the world farther, faster, deeper, and cheaper than ever before and at the same time allows the world to reach into each of us farther, faster, deeper, and cheaper than ever before."[1]

Political scientist David Held and his colleagues define globalization initially as "the widening, deepening and speeding up of worldwide interconnectedness in all aspects of contemporary social life, from the cultural to the criminal, the financial to the spiritual." Later they add more precision, offering this definition: "a process (or set of processes) which embodies a transformation in the spatial organization of social relations and transactions—assessed in terms of their extensity, intensity, velocity and impact—generating transcontinental or interregional flows and networks of activity, interaction, and the exercise of power."[2]

Economist John Helliwell notes, "For protestors in Seattle, Gothenberg, and Genoa, globalization represents a state of the world wherein international organizations implement the wishes of transnational corporations, ensuring that free trade rules will combine with global market pressures to eliminate the ability of local and national governments to implement policies. . . . Within corporations and business groups, globalization usually refers to a global market reach and to an imperative that firms must 'globalize or die.' . . . Among economists, globalization refers to a situation where the so-called 'law of one price' applies on a global basis. This assumes that goods and services will be freely and costlessly traded over space and borders."[3]

Sociologist Saskia Sassen stresses the importance of what is happening inside "the national." "A good part of globalization," she writes, "consists of an enormous variety of micro-processes that begin to denationalize what had been constructed as national—whether policies, capital, political subjectivities, urban spaces, temporal frames, or any other of a variety of dynamics and domains. Sometimes these processes of denationalization allow, enable, or push the construction of new types of global scaling of dynamics and institutions; other times they continue to inhabit the realm of what is still largely national."[4]

Notes

1. Friedman 2002.
2. Held et al. 1999, 2, 16.
3. Helliwell 2002, 15–16.
4. Sassen 2006, 1.

Globalization is driven also by the increasing dominance of capitalism and Western ideologies, which stress the value of global economic integration. States themselves try to embed these values in global norms and institutions, in the hope that this will enhance their economic position in the global economy. A host of nongovernmental and intergovernmental organizations, as well as the global media and multinational corporate networks—involving a complex array of both reinforcing and contradictory pressures—collectively enhance global linkages. Global economic institutions and norms reflecting these values encourage further global integration of trade, production, and financial relationships.

Further, global environmental problems themselves—such as climate change, ozone depletion, and species extinction—also strengthen global consciousness. Because such problems do not respect borders, they are by definition "global-level" problems that require global solutions. Conferences such as the 1972 UN Conference on the Human Environment (Stockholm Conference) and the 1992 UN Conference on Environment and Development (Rio Conference), which focused attention on global environmental problems, raised the profile of these issues among the world's political elites (for details, see chapter 3).

When did globalization begin? We cannot answer this question precisely, because the process is an extension of historical, political, and socioeconomic processes that started long ago. It is reasonable, however, to point to the 1960s as the beginning of the rapid intensification of the process of contemporary globalization. Political scientist Jan Aart Scholte would agree here, having stated that "globalization did not figure continually, comprehensively, intensely, and with rapidly increasing frequency in the lives of a large proportion of humanity until around the 1960s."[10]

Economic trends since the 1960s reveal the extent of globalization. Cross-border financial flows, foreign direct investment (FDI), and international trade have grown phenomenally, as governments have actively removed barriers to these types of transactions. About US$3 trillion in foreign currency is traded per day, about twice as high as in the late 1990s, and significantly more than the average of US$10–20 billion in the 1970s.[11] The number of transnational parent firms has grown from 7,000 in 1970 to over 79,000 in 2007, with over 790,000 affiliate firms. These firms make up one-tenth of world gross domestic product (GDP) and one-third of world exports. FDI flows into developing countries have been rising as transnational corporations continue to expand: from US$22 billion in 1990 to US$380 billion in 2006. Meanwhile, the value of world merchandise exports has gone from US$58 billion in 1948 to over US$6 trillion in 2000 to over US$16 trillion in 2008.[12]

Since the 1960s, globalization has to some extent been promoting greater political, economic, cultural, and technological uniformity across the globe. Most governments now claim to be democratic[13] and many have introduced political structures similar to the ones in Western Europe and North America.[14] Most governments strive to integrate their domestic economic production into the global financial structure, introducing policies to foster foreign direct investment and trade, generally with an underlying goal of promoting higher rates of macroeconomic growth. A superficial glance at the cities around the world reveals striking similarities associated with this economic growth and with globalization: cars, roads, concrete, and steel. And consumer preferences are converging, with blue jeans, American sitcoms, and McDonald's just three of the many possible examples. The process of increasing uniformity, however, is far from inevitable. Fragmentation of whole countries and decentralization within particular countries are occurring alongside of and as part of globalization.

Some argue that globalization is an inevitable process; others argue that it is possible to resist it. Governments have played a role in fostering it. Similarly, they can break global political ties or erect economic barriers. Activist and religious organizations, using some of the same technologies driving globalization, can disseminate alternative views in an effort to oppose globalization.[15] And groups within particular societies can reassert cultural identities and gain global support for their movement to do so.

Though globalization is in many ways a unifying force, with unequal access to new technologies and knowledge, its effects are highly uneven, both across and within countries, triggering great change in some places and virtually none in others. Some of these changes are positive, bettering people's everyday lives. But other changes are contributing to greater inequality in some areas as well as to deeper levels of poverty in some areas (particularly in terms of access to sustainable livelihoods, nutrition, and community well-being). Distribution of global income is one indicator. The 1 billion people who live in the rich industrialized countries receive over half of global income; meanwhile, 3.5 billion in low-income countries receive less than one-fifth of global income. The wealthiest 20 percent of the world's population accounts for over three-quarters of total private consumption expenditures. This does not just translate into far more luxury goods for the rich, like paper and cars. The richest 20 percent also consume 45 percent of all meat and fish. Also, technological research, including medical research, tends to concentrate on the needs of the wealthy.[16]

Access to the communication and transportation technologies of globalization is unequal, too. As of 2010, over 65 percent of people in high-income countries, for example, regularly use the Internet; in low-income countries, it hovers around 5 percent of the population.[17] Similar inequalities exist for the movement of people. It is relatively easy, for example, for the educated and the rich to immigrate. But it is now harder for poor people in poor countries who want to immigrate legally.[18] Such differences in access are contributing to a new form of inequality: the connected and unconnected. Using technologies to integrate into the global economic system provides numerous opportunities. Life can be tough, however, for those left unconnected—just look at the recurring famines in North Korea and the violent conflict in the Congo over the past decade. Globalization is, in some ways, a "new Industrial Revolution" that, according to the United Nations Environment Programme (UNEP), "could result in a dangerous polarization between people and countries benefiting from the system and those that are merely passive recipients of its effects."[19]

Globalization, by its very nature, is altering the underlying processes of global environmental change. Are the effects of these changes positive or negative for the overall environmental health of the planet? Before we examine the diverse range of answers to this question, it is useful to note that critics of globalization are often referring to economic globalization rather than to other forms of globalization. Many of the technologies that have facilitated economic globalization—the computer, for example—are seen as essential for the creation of global environmental norms and a global civil society, potential counterbalances to globalization that help citizens to "think globally and act locally." E-mail and the Internet are also cheap and effective tools of antiglobalization activists, enabling groups without overarching leadership to organize global protests from Seattle to Genoa to Bangkok on short notice. Many of the critics of economic globalization, although certainly not all, are actually supporters of what might be called "social globalization."

Globalization and the Global Environment

The interpretation of the effects of globalization on the globe's natural environment evokes polar reactions. The first, reflective of how market liberals and institutionalists perceive the world, stresses our achievements, our progress over time, and our ability to promote economic well-being as well as to reverse and repair environmental problems with ingenuity, technology, cooperation, and adaptation. The second, reflective of how bioenvironmentalists and social greens see the earth, emphasizes environmental disasters and human inequality, problems aggravated and in many cases driven by economic globalization.[20] We now sketch these two broad views of globalization. It should be stressed, though, that although there is a split at the broad level over whether the process of globalization is on the whole positive or negative, each worldview emphasizes different aspects of it.

Global Positive

Market liberals and institutionalists perceive globalization as an engine of wealth creation. Both argue that the globalization of trade, investment, and finance is pushing up global GDP (for a detailed explanation of this measure of growth, see chapter 4) and global per capita incomes, which they see as essential to finance sustainable development. Over the period 1970–2000, one of heightened global economic integration, world GDP (in constant 1995 U.S. dollars) nearly tripled from US$13.4 trillion to US$34.1 trillion. The annual per capita global growth rate from 1980 to 1990 was 1.4 percent; from 1990 to 1998 it was 1.1 percent. Since then, the global economy has continued to expand. From 2001 to 2006, it grew more than in any five-year period since World War II. Over this time, rich countries' economies grew on average by over 3 percent. Growth was even faster in developing countries, averaging out at around 7 percent in 2006 (following 6.6 percent in 2005 and 7.2 percent in 2004).[21] Economic growth did slow during the global financial downturn of 2007–2009. But unprecedented bailout packages for major banks and companies, along with huge government spending packages to stimulate consumption, seemed to avoid what would have likely been an even deeper global recession. For market liberals and institutionalists, this result shows not only the critical importance of continuing "healthy" growth in the world economy, but also the value of the institutions and processes of globalization for managing recessions and preventing a global crisis like the Great Depression of the 1930s.

Market liberals and institutionalists both see globalization as an overall positive force for the environment because it generates the wealth necessary to pay for environmental improvements. Market liberals argue that such improvements will follow naturally from the functioning of open and free markets. States and institutions certainly have an important role to perform in terms of making and enforcing environmental policy, but it should be a minimal and market-friendly one. Globalization, by stressing liberalization of trade, investment, and finance, is also lowering inefficient trade barriers and state subsidies. This means fewer market distortions—such as prices that undervalue a natural resource. It also means fewer barriers to corporate investment in developing countries.

Institutionalists see a somewhat greater role for the state, and also see a need to build global-level institutions and agreements to more actively guide economic globalization (which to some extent arise naturally from the process itself). The goal is to help states, for example, advance to a higher order of development with as little environmental damage as possible. The UNEP *Global Environment Outlook* nicely summarizes the institutionalist case:

> The pursuit of individual wealth on a global economic playing field made level by universal governance mechanisms to reduce market barriers can . . . open the way to a new age of affluence for all. If developing country institutions can be adapted to benefit from the new technologies and the emerging borderless economy, and if appropriate forms of global governance can be created, the rising tide of prosperity will lift everyone to new heights of well-being.[22]

Market liberals and institutionalists also stress the need to evaluate the environmental effects of globalization within a historical perspective. For them, it is particularly important to plot and analyze global trends. Political scientist Bjørn Lomborg, an outspoken and controversial scholar who is perhaps the strongest advocate of understanding the global environment in terms of global statistics, writes, "Mankind's lot has actually improved in terms of practically every measurable indicator." He warns against relying on "stories" (examples), because this can distort the analysis of progress, creating either an overly optimistic or pessimistic assessment. Lomborg continues, "Global figures summarize *all* the good stories as well as *all* the ugly ones, allowing us to evaluate how serious the over-all situation is." He further admonishes those who accept data uncritically, citing examples where scholars have come to accept a "sweeping statement" as fact when the so-called fact has no statistical base, instead resulting from "a string of articles, each slightly inaccurately referring to its predecessor" (with a far more modest original source). Lomborg argues, too, that some groups, like Greenpeace and the World-watch Institute, have a vested interest in painting a picture of a world in crisis. It is their job; it justifies their moral and financial existence.[23]

Look, say market liberals and institutionalists, at the world at the beginning of the twentieth century. Back then, life was short and full of hardship and suffering. About a third of the global population faced possible starvation. Infectious diseases like typhoid, tuberculosis, botulism, and scarlet fever (often spread in contaminated food, milk, and water) were a leading cause of death. Global life expectancy was a mere thirty years. Even in the United States it was just forty-seven years, with infant mortality at a rate of one in ten.[24] Great progress has been made since then—not coincidentally, market liberals in particular say—as the world has become more globalized. Food production has surpassed population growth, and in 2010 around 17 percent of the world's population was undernourished, compared to 37 percent in 1969–1970.[25] It is now widely argued that famines that occur today are a result of government mismanagement, not a shortage of food.[26]

Vaccines, antibiotics, and better medical care save millions of lives. So do refrigeration, pasteurization, and safer food-handling practices. As a result, global life expectancy is far higher today—now over sixty-six years. Steady increases in life expectancy (see figure 2.1) have occurred in both low- and high-income countries, and there is every reason to expect these trends to continue. Economist Julian Simon sums up the case: "The standard of living has risen along with the size of the world's population since the beginning of recorded time. There is no convincing economic reason why these trends toward a better life should not con-

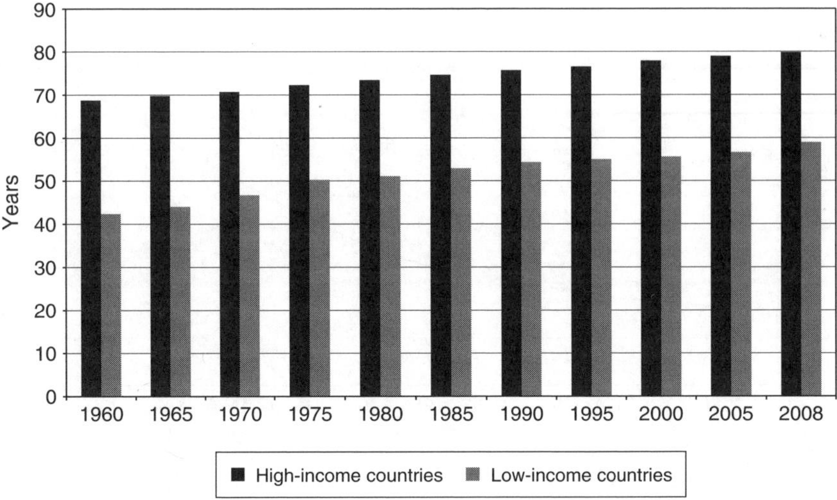

Figure 2.1
Life expectancy at birth, total (in years), 1960–2008. *Source*: World Bank World Development Indicators, <http://data.worldbank.org>.

tinue indefinitely."[27] Market liberals believe that the process of economic globalization itself will spread higher standards of living around the world because of its ability to generate economic growth. Institutionalists see the potential for globalization to raise living standards via economic growth, but argue that global institutions that promote cooperation on these fronts have been and still are necessary to ensure that this happens.

Progress on the food and health fronts has been accompanied by a much larger global population, and some see a direct causal link here. From 1 billion in 1804, world population had doubled by 1927 (123 years later), and jumped to 3 billion by 1960 (33 years later). By 1999 it had doubled again to 6 billion (just 39 years later; see figure 2.2).

The global population by early 2010 was over 6.8 billion and was making steady progress toward 7 billion. Most analysts expect that it will continue to rise in the foreseeable future, leveling off at 8–11 billion by 2050. Figure 2.3 provides three population projections for 2050: a low of 7.9 billion, a middle of 9.3 billion, and a high of 10.9 billion. (The difference is based on alternative possibilities of future birth and mortality rates.) The most likely scenario is the middle one. But for market liberals, even the highest estimate shows that the global community has managed to overcome the threat of exponential population growth.

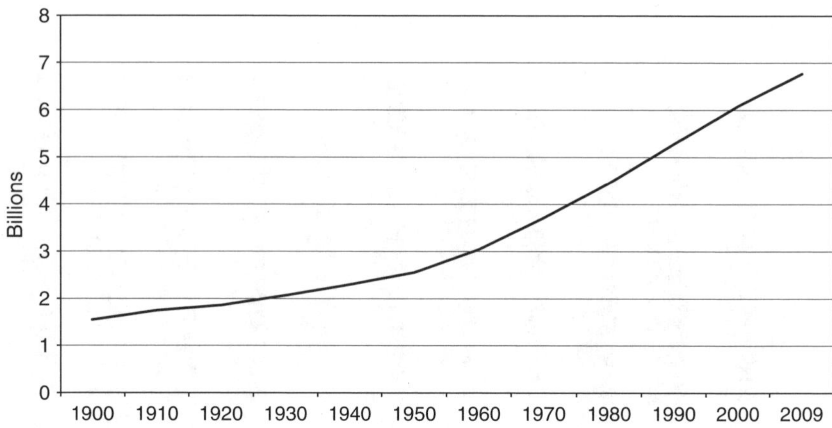

Figure 2.2
World population growth. *Source*: U.S. Census Bureau, International Data Base, <http://www.census.gov/ipc/www/idb/region.php>.

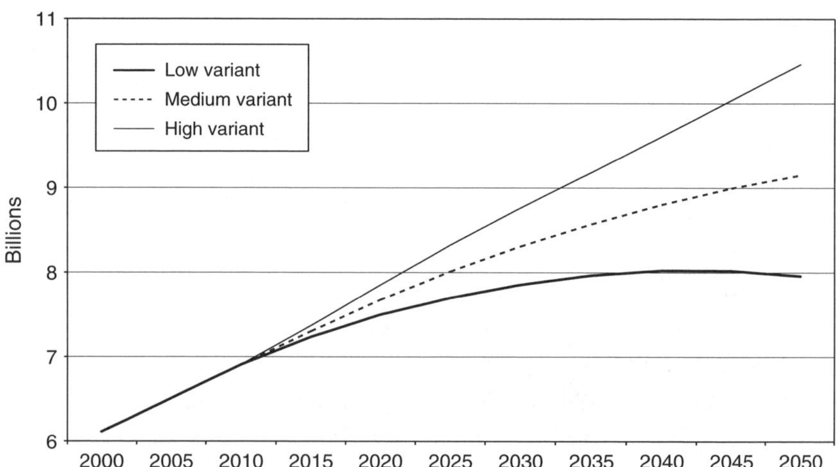

Figure 2.3
World population prospects. *Source*: United Nations Population Division, <http://esa.un.org/unpp/index.asp>.

An earth with nearly 7 billion people, market liberals point out, is not anywhere close to reaching the limits of its capacity. Abundant resources remain. And waste sinks are far from full. Market liberals assume humankind will be able to provide a decent standard of living for all well into the future, provided that a free and open global economy is encouraged. Higher incomes and more modern economies allow for higher levels of education for the general population and empower women with more choices. This naturally lowers birth rates. The history of the world's most advanced economies in Europe, North America, and Asia since the Industrial Revolution some three hundred years ago demonstrates this conclusively. Institutionalists are more cautious, recognizing that there may be some scarcities associated with population growth, but they argue that global cooperation on improved education, economic development, and family planning can help address the problem.[28]

For both institutionalists and market liberals, the global community has demonstrated that it can solve global environmental problems. Biotechnology has, for example, produced crops able to resist insects and diseases as well as to grow in dry and inhospitable areas, and both groups see this innovation as having great potential for improving global food supplies. Market liberals argue that these technologies will spread via market mechanisms and benefit farmers in rich and poor countries alike.[29] Institutionalists have a bit less faith in the global marketplace to spread these benefits equally. Although they see the potential of such crops to address future food needs, they argue for public funding for intergovernmental research that focuses on developing bioengineered crops that will benefit the poorest countries.[30]

Perhaps the most successful global cooperative effort was the one to reduce the amount of chlorofluorocarbons (CFCs) released into the atmosphere. CFCs were invented in 1928. Production and consumption rose quickly from the 1950s to the 1970s, mainly for use in aerosols, refrigerators, insulation, and solvents. In 1974 scientists discovered that CFCs were drifting into the atmosphere and depleting the ozone layer. This layer protects us from the harmful effects of ultraviolet sun rays, which can contribute to skin cancer and cataracts, decrease our immunity to diseases, and make plants less productive. In the decade after 1974, global negotiators worked on a collective response to this problem. This effort gained momentum in 1985 after a "hole" (really a thinning)

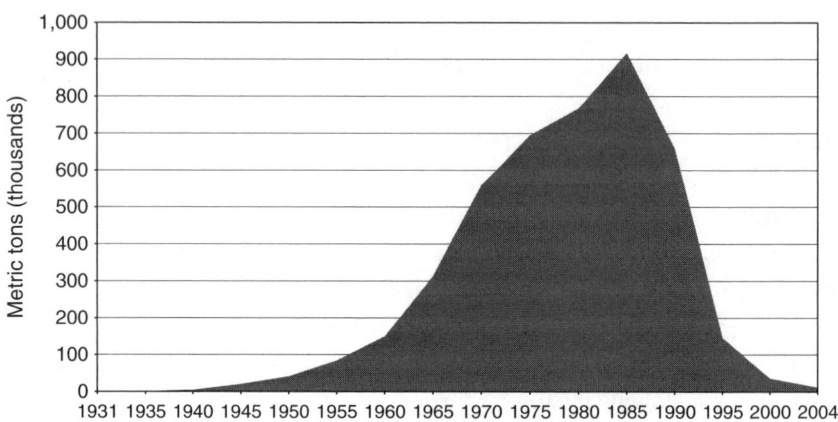

Figure 2.4
Global CFC production. *Source*: Alternative Fluorocarbons Environmental Acceptability Study, <http://www.afeas.org>. *Note*: Data no longer collected post 2004 given low levels.

in the ozone layer was discovered over the Antarctic.[31] In that year, the global community adopted the Vienna Convention for the Protection of the Ozone Layer as a framework agreement to address ozone layer depletion. The 1987 Montreal Protocol on Substances That Deplete the Ozone Layer was concluded within two years under the Vienna Convention and set mandatory targets to reduce the production of CFCs. The Montreal Protocol entered into force in 1989. Amendments to strengthen the Montreal Protocol were made in London in 1990, Copenhagen in 1992, Montreal in 1997, and Beijing in 1999. The result was a significant lowering of CFC production (see figure 2.4) and a phaseout of over 95 percent of ozone-depleting substances globally.

UNEP predicts that the ozone layer will repair itself and return to pre-1980 levels by 2050.[32] Institutionalists in particular see this as a resounding example of global cooperation to solve an environmental problem. There has been wide acceptance of the global agreements and strong compliance with them.[33] Market liberals also like this example because it demonstrates the ability of markets to respond to a global environmental problem with the development of substitutes that are less harmful.

Global Negative

Unlike market liberals and institutionalists, bioenvironmentalists and social greens see economic globalization as the cause of many of the world's environmental and social ills, rather than as its potential savior. Both bioenvironmentalists and social greens agree with market liberals and institutionalists that economic globalization is driving global macroeconomic growth, but this growth is driving the overconsumption of natural resources and the filling of waste sinks. Both also see the 2007–2009 food and financial crises as related to one another (linked to rising trade in agricultural derivatives on financial markets that itself was in response to financial instability), and as predictable symptoms of an irresponsible global order designed to spur unsustainable and unequal economic growth.[34] Moreover, economic growth is not enough to ensure well-being in a society, and government measures to stimulate consumption and spending during the crisis have merely added to growing global inequality and ecological instability. For bioenvironmentalists, economic growth also partly explains the exponential growth of the global population, which for them is of primary concern. But social greens reject the bioenvironmentalists' population argument and focus instead on the way globalization, in their view, contributes to global economic inequalities that exacerbate environmental problems.

Bioenvironmentalists and social greens are far more critical of the so-called progress of our so-called creations and advancements in the era of globalization. Edward Goldsmith and Nicholas Hildyard, former editors of *The Ecologist* magazine, put it bluntly: "There is a direct, historical link between the increasingly serious environmental problems we are experiencing today and the 'modernization' of our economic activities."[35] Humans certainly live longer. And higher incomes and better medicine and sanitation have undoubtedly made the lives of some more comfortable. Yet such data can also hide disturbing trends. Especially worrying is the steady rise in global cancer rates, even when adjusting rates for an aging population. Every year more than 7 million people die of cancer. Twelve million more are diagnosed with cancer. Cancer is now on track to overtake heart disease to soon become the world's leading cause of death. The World Health Organization's International Agency for Research on Cancer predicts that the number of people dying of cancer will increase to 17 million by 2030, and that the number of new patients will rise to 27 million a year.[36]

Scholars like philosopher Peter Wenz believe that the probable causes of higher cancer rates arise from the artificial changes to our living environments. There seems to be little cancer in traditional societies. Yet cancer rates in industrial countries like the United States, even after adjusting for longer life expectancies and excluding lung cancer, have been rising steadily since the 1950s (several studies estimate the increase at about 35 percent).[37] The American Cancer Society estimates the lifetime chance of getting cancer in the United States at almost one in two for men and just over one in three for women.[38] One of the most significant causes of cancer for scholars like Wenz is the increasing volume of pesticides in the global ecosystem. In the United States alone, farmers were using over a billion pounds of pesticides per year by the beginning of the twenty-first century, compared with 600 million per year in the 1970s and 50 million per year in the 1940s. Globally, from 1961 to 1999 the use of nitrogenous fertilizers went up over 600 percent, phosphate fertilizers over 200 percent and pesticides over 800 percent. Cancer specialists, Wenz argues, concentrate on finding the "cure" rather than the "cause" of cancer, partly because uncovering the causes would challenge "our entire way of life."[39]

While market liberals and institutionalists tend to focus on the social and political history of the last few hundred years, bioenvironmentalists in particular like to look at the impact of globalization against the background of geological time. The universe is about 15 billion years old. The earth is about 4.6 billion years old. Modern humans have existed for just over 100,000 years and civilization for about 10,000 years. Within this time frame, it is clear that humans have become a threat to the planet in a remarkably short period of geological time. Philosopher Louis Pojman provides a vivid image of this:

> If we compacted the history of the Earth into a movie lasting one year, running 146 years per second, life would not appear until March, multicellular organisms not until November, dinosaurs not until December 13 (lasting until December 26), mammals not until December 15, *Homo sapiens* (our species) not until 11 minutes before midnight of December 31, and civilization not until one minute before the movie ended. Yet in a very short time, say less than 200 years, a mere 0.000002% of Earth's life, humans have become capable of seriously altering the entire biosphere. In some respects we have already altered it more profoundly than it has changed in the past *billion* years.[40]

With this outlook on time, bioenvironmentalists argue that we will soon reach the limits of the earth's biological capacity to support human life.

For bioenvironmentalists, human population growth is often the most central part of the problem—for them linked to both economic growth and globalization. They argue that a much better reflection of population

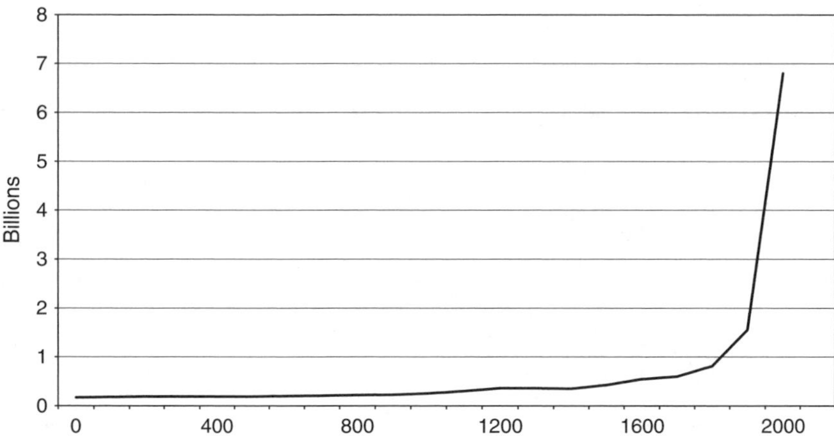

Figure 2.5
World population, 1 AD–2008 AD. *Source*: U.S. Census Bureau, International Data Base, <http://www.census.gov/ipc/www/idb/worldpop.php>.

patterns arises when we look back a few hundred years (see figure 2.5). Every month the global population is rising on average by a little over 6 million—equal to adding a large city or two. Since 1950, an era of rapid globalization, the world's population has grown more than it had in all of human history.[41] Even accepting the midrange estimate of population growth in figure 2.3, by the middle of this century it will exceed 9 billion—another 3 billion people to feed, clothe, and house. Over 85 percent of these people will live in the poor countries. Half of the population growth will occur in just six countries: India, China, Pakistan, Nigeria, Bangladesh, and Indonesia. Over this time, the populations of the forty-eight least-developed countries will triple in size—these already contain areas of acute poverty and hardship.[42]

Bioenvironmentalists stress that the current rate of population growth means the planet is increasingly awash with consumers, each demanding a portion of the earth's resources. Global consumption expenditures keep rising, as shown in figure 2.6. One indicator of rising consumption in an age of globalization is the steady increase in the number of motor vehicles. Every year automakers roll out another 60–70 million passenger cars, station wagons, and light commercial vehicles, adding another 10–20 million to the more than 850 million already on the road. Analysts now expect the number of light vehicles to pass the 2 billion mark by the middle of this century.[43] Environmental trends confirm the disastrous

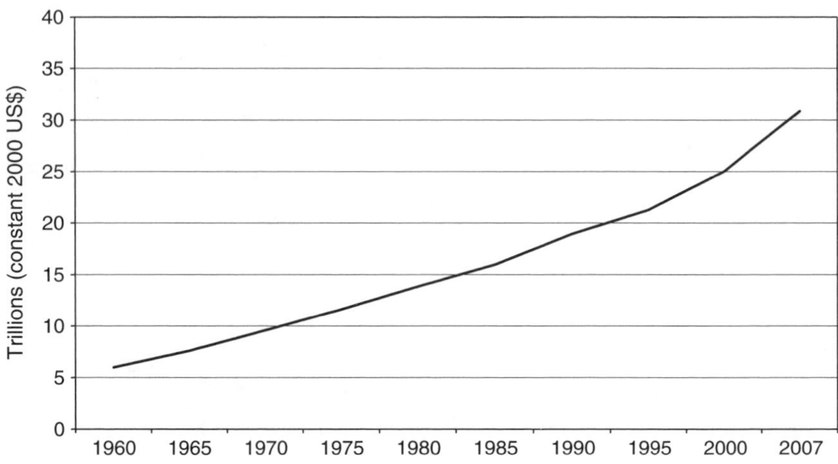

Figure 2.6
World final consumption expenditure, 1960–2007. *Source*: World Bank World Development Indicators, <http://data.worldbank.org>.

consequences of such consumption. The air in the rich industrialized countries has undoubtedly become less polluted since the 1960s, but it has become much worse in the developing world—especially in megacities like Tehran, New Delhi, Cairo, Manila, Jakarta, and Beijing. The planet has absorbed 100,000 new chemicals since 1900. Over the twentieth century, total carbon dioxide emissions grew twelvefold. In the developing world, more than 90 percent of sewage and 70 percent of industrial waste is still disposed of (untreated) into surface water.[44]

In addition to air pollution, there is a growing mountain of solid waste associated with rising consumption. A growing proportion of this waste is plastic, which takes hundreds if not thousands of years to break down, and has been implicated in the leaching of toxins into the environment. The discovery in the late 1990s of the Great Pacific Garbage Patch in the Pacific Ocean—a floating blob of plastic waste at least twice the size of the state of Texas, composed of water bottles, plastic bags, toys, and tiny plastic beads called "nurdles"—has become a symbol of an overconsumptive society.[45] Discarded consumer electronics—including cell phones, computers, and video game equipment—is another fast-growing waste stream wreaking particular havoc as these types of waste often end up in landfills or are dismantled in environmentally unsound ways after being shipped overseas for "recycling" or "reuse."[46]

Natural resources are also threatened by rising consumption. Since 1950, meat consumption has increased more than fivefold, with per capita consumption more than doubling. In 2009 people consumed over 285 million tons of meat—a figure that some analysts predict will exceed 465 million tons by 2050.[47] Around 70 percent of commercial global marine fish stocks are overfished or are at the biological limit. The once plentiful northern cod, for example, is now endangered, with stocks falling by 99 percent over the last four decades. One-fifth of the world's freshwater fish species are extinct or endangered. Meanwhile, global water consumption rose six times during the twentieth century.[48]

So much water consumption has contributed to dams, canals, and diversions disrupting about 60 percent of the world's largest rivers. Deserts threaten one-third of the world's land surface and the World Health Organization estimates that over 1 billion people still do not have access to clean water.[49] Diseases from bad water—like dysentery and cholera—kill 3 million people each year.[50] During the twentieth century, the world lost half of its wetlands. Already by 2003, 30 percent of the world's coral reefs were gone and 18 percent were at risk. Every day 68 million metric tons of topsoil washes away. Every day 50–150 species become extinct. The Worldwatch Institute sees global biodiversity in a dire state, noting that "prominent scientists consider the world to be in the midst of the biggest wave of animal extinctions since the dinosaurs disappeared 65 million years ago."[51] The future of global biodiversity under current trends looks bleak. A study by nineteen scientists published in the journal *Nature* predicts that a temperature rise from global warming of 0.8°–2.0°C will "commit" 18–35 percent of animal and plant species "to extinction" by the middle of the twenty-first century.[52] Other factors attributed to climate change—such as higher concentrations of carbon dioxide—could mean an even higher rate of extinction.[53]

The history of deforestation is typical of the destructive impact of the global political economy. Ten thousand years ago, forests covered much of the world's land outside of the polar regions. Today, around half of these forests are gone, with deforestation since 1950 roughly equal to all of the previous loss. In temperate regions, governments from China to the United States are now imposing stricter logging rules, reforesting land, and establishing more parks. Significant problems remain; nevertheless, forest cover over the last few decades has been increasing, even if much of this is plantations. At the same time, however, deforestation in many tropical and boreal regions remains as bad, or is worse. In recent years, Brazil alone has been losing more than 2.5 million hectares a year and Indonesia has been losing about 500,000 hectares a year. The net result, bioenvironmentalists argue, is an accelerating decline in the earth's capacity to sustain biodiversity and store carbon dioxide. For example, deforestation, now over 13 million hectares a year, is one of the most significant sources of climate change, accounting for as much as one-fifth of annual emissions of carbon dioxide globally.[54]

Social greens also point to the steady decline in the state of the global environment in an age of globalization. They agree strongly with bioenvironmentalists that economic growth and overconsumption are serious threats to the planet. Yet unlike bioenvironmentalists, they do not focus on population growth as a primary driver of such problems. They point out that growth in the consumption of forest products, food, and water, for example, far exceeded population growth over the past thirty years.[55] Instead, they place their main focus on global inequality and its associated environmental problems—those linked to both overconsumption among the wealthy and the dispossession of the poor from their traditional lands.

For social greens, economic globalization is the primary cause of this inequality.[56] It is seen as reinforcing neocolonial relationships between rich and poor countries, as well as changing production patterns in complex ways that have serious environmental implications.[57] The International Forum on Globalization, an international group of activists and academics opposed to globalization, argues that today's world is "in a crisis of such magnitude that it threatens the fabric of civilization and the survival of the species—a world of rapidly growing inequality, erosion of relationships of trust and caring, and failing planetary life support systems."[58] Life may be easier—and perhaps even better—in sections of the wealthier countries (although not in the slums). But in much of the developing world it is worse. Since 1972, the number of people living in extreme poverty (less than US$1 a day) has grown to 1.1 billion people. Around 40 percent of the world's population (2.6 billion) survives on less than US$2 per day.

There is not enough food in many countries. In fact, even though global food production is now higher, in many parts of the world, such as in Africa, food production has lagged behind population growth over the last two decades.[59] In India, where one-third of the population live in poverty, over half of the children are undernourished. Africa alone has 300 million people who are undernourished. Worldwide, over 1 billion people suffer from malnutrition, which contributes to 60 percent of all childhood deaths.[60] And around 9 million children die unnecessarily every year (that is, from preventable and treatable causes).[61]

At the same time, obesity rates are rising fast. About 1.6 billion adults were overweight as of 2006, with at least 400 million of them obese. The World Health Organization predicts that by 2015 about 2.3 billion adults will be overweight and more than 700 million of these people will be obese.[62] The United States has some of the highest rates of obesity, with two-thirds of adults now overweight, with half of those in this category obese. This is a significant increase from just twenty years ago when, for example, less than half of adults were overweight. Obesity and physical inactivity now account for over 300,000 premature deaths annually in the United States alone. In late 2001 the U.S. Surgeon General issued a "call to action" to address the crisis of obesity, claiming "overweight and obesity may soon cause as much preventable disease and death as cigarette smoking."[63]

Great inequality is also seen in the distribution of disease. Over 13 million people die each year from diseases like AIDS (acquired immune deficiency syndrome), malaria, tuberculosis, cholera, measles, and respiratory diseases—most of these in the developing world. These high rates of death from such diseases only contribute to further poverty. In some African countries, life expectancy has fallen, mostly because of AIDS, malaria, and war. The AIDS pandemic started over two decades ago. Despite significant advances to slow, prevent, and manage the disease, over 33 million people are infected with HIV (human immunodeficiency virus) and the disease continues to spread. In 2007 AIDS killed over 2 million people. Of these, three quarters lived in sub-Saharan Africa. Already AIDS has lowered the average life expectancy in this region from more than sixty years to less than fifty years, undermining the social fabric of much of the region. The death rates from AIDS are much higher in the developing world because of that region's lack of access to the latest drugs and treatments that are currently available in the rich industrialized countries.[64]

Social greens argue that this represents unnecessary suffering on the part of the majority of the planet's inhabitants. The inequality that has accompanied economic globalization is the driver not just of poor social conditions for many, but also of environmental problems. The poor are put in situations where they are uprooted from their lands and forced to degrade what lands they can just to make ends meet. And the rich, meanwhile, overconsume and contribute to a worsening of industrial pollution. For these thinkers, it is not surprising that pollution levels may be falling in rich countries because globalization has enabled the types of production that pollute most to migrate to developing countries, in the form of foreign direct investment. The rising levels of pollution in the rapidly industrializing developing countries of Southeast Asia and Latin America are evidence of this, according to social greens.[65] This transfer of environmental degradation from rich to poor countries has direct and negative impact on the lives of the world's poorest people.

The technological solutions offered up by Western science are not a potential savior for bioenvironmentalists and social greens. The technological transfers of globalization are in fact often a deceptive solution, the heroin of market liberals and institutionalists, temporarily allowing societies to deflect a problem into the future or into another ecosystem. They also have much less faith than institutionalists in the ability of global institutions to guide globalization to ensure that it does not cause further environmental problems. Social greens in particular worry that global organizations and agreements can become props of capitalism, as is the case with the World Bank, the International Monetary Fund (IMF), and the WTO.

Agricultural biotechnology is just one issue regarding which social greens and bioenvironmentalists are skeptical of Western technology and institutions. Critics of genetically altered crops argue that they pose enormous ecological and health risks—from genetic pollution and erosion, to potential allergic reactions.[66] These crops are also seen as reinforcing inequalities and as giving power to the transnational corporations that patent them.[67] The World Bank, for example, has continued to promote these technologies as beneficial, despite growing concern over their use. Many social greens, however, still see a need for global institutions, just not the ones we currently have. The International Forum on Globalization summarizes this position succinctly:

> There is certainly a need for international institutions to facilitate cooperative exchange and the working through of inevitable competing national interests toward solutions to global problems. These institutions must, however, be transparent and democratic and support the rights of people, communities and nations to self-determination. The World Bank, the IMF, and the World Trade Organization violate each of these conditions to such an extent that [we] recommend that they be decommissioned and new institutions built under the authority of a strengthened and reformed United Nations.[68]

Bioenvironmentalists and social greens also have less confidence than other thinkers in the value of international regimes to slow or resolve global environmental problems. Certainly most would agree that the global effort to reduce the production of CFCs was successful. But this was an exceptional case, one that tells us little about our ability to handle future global environmental crises. The causes and consequences of the depletion of the ozone layer were relatively straightforward. Skin cancer, one of the most visible consequences of less ozone, was a particular worry. In the mid-1980s, CFCs were produced by only twenty-one firms in sixteen countries, and developed countries were responsible for about 88 percent of production. Especially important, the chemical company DuPont, which accounted for one-quarter of global CFC production, alongside other firms had found an affordable substitute to CFCs by the time the international ozone regime was created. Despite initial fears about the high economic costs of phasing out CFCs, these turned out to be manageable.[69]

Most global environmental problems, say bioenvironmentalists and social greens, involve far greater complexities and uncertainties than the ozone case and will require far greater sacrifices to solve. Consider global warming. UNEP reports that the mean global surface temperature has risen by 0.3°–0.6°C in the last hundred years. This may not seem like much. But it was the largest increase of any century over the last millennium. The problem appears to be getting worse. The 1990s was the second warmest decade, January 2000 to December 2009 was the warmest decade, and 2005 was the warmest year of at least the last hundred years. The five warmest years on record are all since 1998, with the National Aeronautics and Space Administration (NASA) measuring 2009 as the second warmest on record (just a fraction below 2005).[70] For bioenvironmentalists, trends like the one in figure 2.7 indicate our failure to act.

Global warming especially alarms bioenvironmentalists and social greens because the three main greenhouse gases (carbon dioxide, methane,

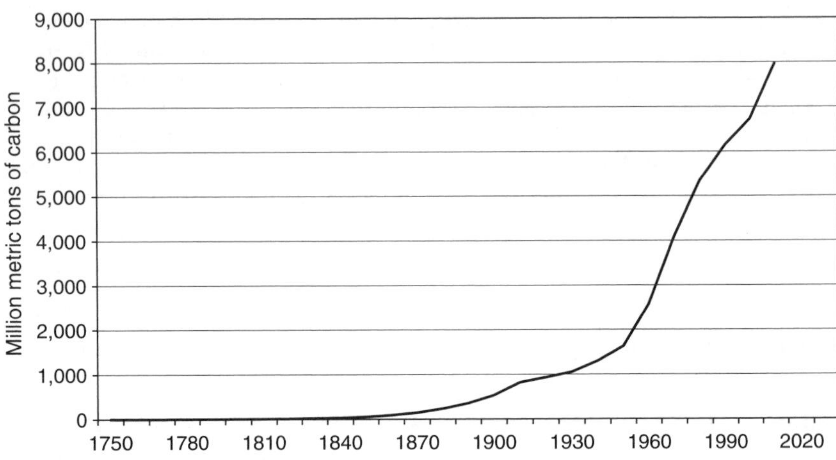

Figure 2.7
Global CO_2 emissions from fossil-fuel burning, cement manufacture, and gas flaring. *Source*: Carbon Dioxide Information Analysis Center, <http://cdiac.esd.ornl.gov/ftp/ndp030/global.1751_2006.ems>.

and nitrous oxide) arise from core economic activities (automobile use, electricity generation, factories, agriculture, and deforestation), and the main consequences (rising seas, severe storms, drought, and desertification) are beyond the lifetimes of politicians and business leaders—occurring in perhaps fifty to one hundred years. And the impacts, when they are most severe, will be mostly felt by the poor, marginalized peoples of the world. Obviously, lowering greenhouse gas emissions will involve major changes to global economic production and consumption patterns. It will also require governmental, corporate, and personal sacrifices. Substituting CFCs, say these thinkers, is just not a comparable sacrifice and it is overly naive to believe that it demonstrates that the global community has the collective capacity to act quickly, coherently, and effectively whenever a problem is deemed urgent enough. Any progress on climate change agreements, rather, is more likely linked in some way to corporate strategies to break into renewable energy markets.[71]

Conclusion

This chapter has shown the power of globalization as a force of economic and, at a slower rate, societal and political change. Each worldview brings its own lens to the process, illuminating different aspects—some more positive, others more negative. Each view thus brings different insights into the implications of globalization for the environment.

The market liberal lens focuses on the benefits of growth and the power of free markets to foster it. For market liberals, globalization is a force of good—an engine of progress. It promotes efficient production and trade of goods, diffusing appropriate technologies to areas with natural labor and resource advantages. It promotes investment that brings in environmental technologies, critical funds, and better management. This advances the economies toward less destructive activities by diversifying the sources of economic growth, which in the long run allows economies to shift away from a heavy reliance on natural resource exports. Globalization also means more macroeconomic growth, raising per capita incomes throughout the world. The higher national per capita incomes that arise from embracing globalization are creating the societal and political will to tackle national, and ultimately, global environmental problems. In the long run, economic growth extends lives, improves health care, and raises national environmental standards. Globalization allows for even more wealth. Such wealth will allow for even more economies to shift away from agricultural and industrial production and toward knowledge-based economies.

The institutionalist lens is sympathetic to the points raised by market liberals, but focuses more specifically on the international cooperation necessary to bring those benefits to fruition. Institutionalists agree that economic growth and free markets can have enormous benefits for the environment. Yet they argue that globalization should not be seen as a panacea. The process, like any dynamic force, is producing *both* positive and negative changes for the global environment. It is good in macroeconomic terms. It is raising global wealth; it is fostering innovation. But the rewards are not always equitable. At times, globalization seems to aggravate pockets of environmental degradation and inequality. From an institutionalist perspective, it is therefore best to guide globalization—to use institutions and cooperation and intelligence to maximize the economic benefits and minimize the social inequities. This requires strong national and local governments as well as powerful global organizations. The communication technologies of globalization are advancing this cause, by integrating cultures, fostering awareness, and creating global norms, standards, and sympathies that encourage cooperative efforts to address national problems like malaria or global problems like climate change. The globalization of norms, codes of conduct, environmental markets, environmental organizations, and international law is deepening global environmental governance, adding another layer of control above national government regulations.

Bioenvironmentalists look at globalization through a very different lens. Instead of seeing opportunity in globalization, they focus on the scarcity it exacerbates. For these thinkers, the future is bleak. Globalization props up unsustainable population growth in developing countries, as the rich drop food and medical supplies into poor countries with population growth rates embedded in a culture of poverty. Globalization is also herding populations into overcrowded cities or fragile ecosystems (such as onto land cleared of its rainforests), and turning rural areas into vast and specialized farms to feed the cities, often in distant lands. At the same time, the globalization of trade, investment, and financing is accelerating global economic growth. This creates more output and more consumption. It also creates more opportunities for the rich to overconsume and waste resources, deflecting the ecological impacts of this consumption overseas or into the global commons. Globalization here is little more than "eco-apartheid."[72] It is deepening the global culture of consumption for consumption's sake. People are losing the sense of "enough," as the rising rates of obesity in the developed world show. Ingenuity and technology can certainly help to mitigate particular problems. Yet technology cannot solve the ecological consequences of globalization. Solutions can begin to occur only once the human species accepts that it is now beyond its carrying capacity.

Social greens agree with much of the bioenvironmentalist analysis of the ecological consequences of globalization. Yet their lens focuses much more on the social injustice arising from globalization than on scarcity per se. Social greens see globalization as the driving force behind the spread of global inequality and large-scale capitalism and industrial life, which for them are core causes and consequences of the global ecological crisis. Inequality and the imposition of industrialism contribute to the eradication of the rights of indigenous peoples, women, and the poor, and to the destruction of culture as capitalists reconstruct societies into new markets and production nodes. These forces are eroding the autonomy of communities and creating a consumer monoculture. Globalization, for many social greens, is little more than ecoimperialism, a process to siphon off autonomy and knowledge from the local to the global. In this way, globalization is reinforcing patterns of economic, environmental, and social injustice. The globalization of production and trade, moreover, further distances an individual's ability to perceive the ecological and social impacts of these behaviors. People are increasingly unable to see (or at least are able to forget) how their everyday choices damage the environment or injure workers.

The four worldviews, in short, have markedly different interpretations of the ecological effects of globalization. Each offers its own unique insights by focusing on particular aspects of the process. The rest of this book explores the debates between these worldviews in far greater depth, drawing out subtleties of opinion both within and across the worldviews. The rather stark characterization of worldviews on globalization in this chapter will, we hope, help to guide you through the thicket of the more specific, and at times overlapping, arguments in the remaining chapters. We begin in the next chapter with what all four worldviews agree is one of the most significant effects of globalization—the global spread of environmental norms, ideas, and institutions over the past half century.

3

The Globalization of Environmentalism

What has the global community done to tackle the environmental problems discussed in the previous chapter? Although there is a rich history of formal actions on the part of states to address these problems in the international arena, we must also remember that the history of global environmental politics is inextricably tied to contests of ideas: battles of worldviews and discourses.[1] We have seen new environmental ideas and language enter into mainstream discourse as global awareness rises and as environmental conditions deteriorate. These views are also affected by wider developments in the global political economy. Until recently, most accounts of the history of global environmental politics—particularly those in the field of international relations—focused on environmental diplomacy. Such studies tended to explain the evolution of environmental discourse primarily in terms of the outcomes of environmental summits, commissions, international agreements, and organizations.[2] These political-diplomatic efforts have certainly been important. But a deeper look at this history reveals that parallel developments in the global political economy have, directly and indirectly, often set the wider agendas for these formal discussions, affecting the evolution of environmental ideas at a broader level. Tracing this history of both environmental diplomacy and the global economy, then, helps uncover when and why the various arguments of bioenvironmentalists, institutionalists, market liberals, and social greens first emerged. It also helps us to better understand the roles of various actors today in global environmental governance.

The first part of this chapter provides a brief history of the globalization of environmentalism with a focus on the past sixty years, highlighting the connections of this diplomatic, interstate history to events in the global economy and to the emergence of different perspectives on the global environment. The second part of the chapter outlines the various mechanisms of global environmental governance put into place over this period, and highlights the significance of each in today's global environmental discourse.

The Evolution of Global Discourse on Environment and Development

Homo sapiens seem to have had only a marginal environmental impact until relatively recently. From 500,000 years ago to 10,000 years ago, the global population was low and seems to have been fairly stable, and the main use of natural resources was for stone, bone, and wooden tools. The shift from hunting and gathering to settled agriculture that began about 10,000 years ago was a profound turning point in global environmental history. The global population at this time was probably a mere ten million or so. The food surpluses from settled agriculture helped populations expand into civilizations (cities). These civilizations, with more elaborate divisions of labor, in turn became great sources of technological advances. The plow (animal-drawn) and the wheel as well as writing systems and the use of numbers were invented during this time. These civilizations were also able to extract and use larger and larger amounts of natural resources—like metals and wood. Not all of them, however, could manage their natural environments in the long term. For example, the collapse more than 4,000 years ago of Mesopotamia, a land between the Tigris and Euphrates rivers, was connected to an irrigation system that poisoned the soil with salt.

Yet it was not until some three hundred years ago, with the dawn of the Industrial Revolution, that human activities began to accelerate toward a scale able to alter the global environment. The worsening impact of the environmental consequences of industrialization—such as the burning of coal—along with rising populations partly explains the emerging concern with environmental conditions over the last several centuries. The first of a series of killer fogs (air pollution from coal burning) struck London in 1873, killing more than 1,150 people. Overhunting and habitat loss brought the plains bison of North America to near extinction by the mid-1800s. The last passenger pigeon—a bird that once migrated through eastern North America in the millions—died in a zoo in 1914.[3] Poor land management was seen as a factor behind the "Dust Bowl" conditions (drought, dust storms, and agricultural collapse) in Western Canada and the United States in the 1930s. By 1950 the world population was over 2.5 billion. The effect of industrialization on the health of these growing populations was becoming an increasing concern. Smog in London killed four thousand people in 1952. A year later, two hundred died from smog in New York City. In 1956 scientists officially "discovered" the mercury poisoning of villagers in Minamata, Japan (commonly called the Minamata disease). The poison came from the local seafood, which had absorbed methyl mercury dumped into the sea by the Chisso industrial plant. Hundreds eventually died, and hundreds more were born with brain damage and other birth defects.[4]

The emergence of environmentalism from the 1700s to the mid-1900s, can thus be traced in part to environmental turmoil within the industrialized world. It was also, however, tied to the experience of the colonial powers at the time. Extensive economic and political control characterized European colonies in Latin America, Africa, and Asia. Other countries, such as the United States, exerted economic if not political control in some of the same parts of the world—what some label economic imperialism. Colonialism and imperialism had enormous environmental consequences, especially as many previously unconnected areas of the globe became primary sources of raw materials for Western industrialization. Much of this environmental damage is still evident.[5] The colonies were also the testing grounds for many Western environmental ideas, and even today the results of these "tests" continue to shape global environmental discourse. The primary environmental goal of many colonial governments was to preserve and better manage nature and wildlife to maximize its aesthetic and economic value. The conservationist ideas of this era were certainly "Western" in their origin, and often administrators from countries like Britain, France, and the United States simply imposed these views on their colonies. In other cases, environmental ideas first emerged within the colonies as colonial administrators attempted to "better" manage resource extraction and economic development. Only later were these ideas brought back to the imperial centers.[6]

Many colonists were critical of how indigenous peoples treated nature, believing them ignorant and wasteful. The British and French, for example, believing indigenous slash-and-burn agriculture (swidden farming) was primitive, put in place new bureaucracies to manage forests.[7] The purpose of these bureaucracies was to ensure the colonies managed the natural environment "efficiently" and "rationally." Colonial administrators became concerned too with hunting in Africa during the nineteenth century, as the continent was stripped of much of its wildlife. In response, legislation was enacted to create parks and game preserves as well as to regulate hunting. The latter was typically denied to Africans on the grounds that sport hunting was less destructive than that for livelihood.[8]

The development and diffusion of Western environmental ideas within the colonies and from the colonies to the imperial countries can be seen as a first step in the creation of a more "global" environmental discourse. Often, colonizers attempted to back these views with the language of science, developing models like sustainable yield management, with the purported goal of managing resources seen as fully renewable. Many scholars have since dubbed this "pseudoscience," however, arguing that these views were paternalistic, ideological, and racist.[9]

Efforts to conserve nature in both the colonies and in the metropole countries continued into the twentieth century, though the two world wars and the Great Depression of the 1930s hampered global coordination. The newly formed United Nations (UN) took on such efforts shortly after the end of the Second World War, contributing to the creation of the International Union for the Protection of Nature in 1948, which was renamed the International Union for Conservation of Nature and Natural Resources (IUCN) in 1956.[10] This organization, which in 2010 comprised over two hundred government organizations and over eight hundred nongovernmental organizations (NGOs), was designed to coordinate global efforts to conserve and preserve nature.[11]

Silent Spring and the 1960s and Early 1970s

The global economy boomed for nearly twenty years after World War II, ultimately fueling the environmental movement. Concern with the environmental abuses of rapid industrialization gained momentum in the First World in the 1960s. There were protests against nuclear weapons and chemical pollution. The WWF was founded in 1961 to work to preserve global biodiversity. Rachel Carson's bestseller *Silent Spring* (1962) was particularly influential with the public. Her message was a powerful one: the increasing use of chemicals, particularly pesticides, was killing nature and wildlife. It told of a future spring without the songs of birds. The scientific community lost no time in attacking Carson in the media, accusing her and her supporters of overstating the case against

DDT (dichlorodiphenyltrichloroethane) and other chemicals.[12] But the strength and simplicity of her message endured with the general public, and indeed research has since shown that her concern over DDT and other synthetic chemicals, many of which are now banned from production and use, was warranted.

The need to protect nature (so that, e.g., people could enjoy the "outdoors") remained a significant concern for the environmental movement in the First World. But increasing public concern with the effects of industrial production began to give the movement a new focus. Also emerging were worries among environmentalists about the cumulative impact of local problems on the health of the planet. More and more people began to see the planet as fragile and interconnected, an image reinforced as pictures of earth from space became more common. For many, the most memorable picture was the one that became available as astronauts Neil Armstrong, Michael Collins, and Edwin Aldrin journeyed to the moon in July 1969 (see box 3.1). Even today, the image of the earth from space—Spaceship Earth with no visible borders and a common atmosphere and oceans—remains a powerful symbol for global environmental consciousness.[13]

Other developments shaped the discourse in equally profound ways. The early 1960s saw rapid economic growth and greater global integration of trade and investment. Global growth rates were as high as 5–6 percent in the early to mid-1960s (though this growth was not evenly distributed among countries or regions).[14] The global economic infrastructure codified by the International Monetary Fund (IMF), the World Bank, and the General Agreement on Tariffs and Trade (GATT)—all set up at the end of World War II—was widely seen as critical to fostering this growth and integration. Advances in communication and transportation technologies also contributed to the economic growth in this period. The wave of decolonization in Asia and Africa from the late 1940s to the late 1960s put many newly independent colonies on the global stage. Most joined the global economic institutions in an effort to promote economic growth. It was a time of optimism, and many countries believed that development (often understood as rapid industrialization and economic growth) would follow automatically from participation in the global economy.

But by the late 1960s and early 1970s, many of these "developing" countries realized that replicating the industrialization in the First World

Box 3.1
The earth from space

Photograph of the earth on July 16, 1969, taken from Apollo 11 at 180,000 kilometers. It shows most of Africa and the Middle East as well as parts of Europe and western Asia.
Source: NASA photo ID AS11–36–5355.

was far from automatic. Global growth rates had been strong, but the Third World still lagged far behind the First World in terms of industrialization and industrial exports.[15] Critics from developing countries began to argue that the postwar global economic infrastructure still reflected colonial and imperial interests. Dependency theorists argued that the imperial nature of capitalist relations between rich and poor countries made the kind of industrial development seen in the capitalist North difficult, if not impossible, to replicate in the South.[16] The rich countries were seen to be dependent on the developing countries for cheap supplies of raw materials, and the elites in developing countries dependent on capitalist patrons in the rich countries. Other critics also argued that it was particularly difficult to break out of a cycle of being mere suppliers of raw materials for overseas production, mainly because of the difficulty of earning enough money from those exports to foster industrialization. This was due to a serious problem faced by developing countries, referred to as "declining terms of trade," which occurs when the prices for exports (in this case raw commodities) are generally low or falling, while prices for imports (in this case industrial goods) are generally high and rising. Under these conditions, earning enough from raw material exports to purchase the capital equipment necessary to industrialize is extremely difficult. This focus on unequal terms for rich and poor countries in the global economy has been a longstanding perspective in international political economy, and has been important in informing the views of the social greens and bioenvironmentalists.

Views critical of the inherent inequality of the global economy spread as global growth rates began to falter in the early 1970s. Rising inflation and a U.S. trade deficit spurred President Richard Nixon in 1971 to take the U.S. dollar off the gold exchange standard established at Bretton Woods in 1944. This action just further undermined confidence in the global economy.[17] This precarious global economic situation added to the growing sense of global interconnectedness, as well as mutual vulnerability. Many environmentalists in the First World began to argue that because environmental problems and economic performance were intertwined, a single global economy was surely tied to the fate of a single global environment.

Environmentalism in rich countries at this time began to reflect the economic theme of mutual (North–South) vulnerability. Books like Paul Ehrlich's *The Population Bomb* (1968) became popular bestsellers. Ehrlich argued that population growth in the developing world, spurred in part by its economic predicament, would one day threaten the globe's resource base. Inequalities would continue to grow in such a world, although both the rich and the poor would eventually suffer from mass starvation, violence, and ecological destruction. He warned, "It is obvious that we cannot exist unaffected by the fate of our fellows on the other end of the good ship Earth. If their end of the ship sinks, we shall at the very least have to put up with the spectacle of their drowning and listen to their screams."[18] The book sold three million copies and went through twenty-two printings in the three years following its release. Coming from a scientific framework and focusing on unsustainable population levels, Ehrlich was one of the key thinkers in the emerging bioenvironmentalist perspective, which began to gain popularity in industrialized countries.

Environmental concern continued to rise in the North in the late 1960s and early 1970s. On November 30, 1969, Gladwin Hill of the *New York Times* reported on the situation in the United States: "Rising concern about the environmental crisis is sweeping the nation's campuses with an intensity that may be on its way to eclipsing student discontent over the war in Vietnam."[19] Twenty million rallied in the United States at the first Earth Day on April 22, 1970, one of the biggest organized demonstrations in the history of the United States. In that year the U.S. government established the Environmental Protection Agency (EPA). Canada followed a year later with a Department of the Environment. The Canadian group Greenpeace made headlines in the same year when it sailed toward Amchitka Island, Alaska, to protest against underground nuclear tests.

Feeding into this rising interest in environmentalism in the North was another bestselling and controversial book, *The Limits to Growth*.[20] This book reported in more popular language the findings of a technical study by a research team at the Massachusetts Institute of Technology. It was part of a wider project of the Club of Rome, an international research and policy group. Its conclusions were similar to Ehrlich's. Using formal modeling and computers, it simulated future prospects for five major, interconnected trends: industrialization, human population growth, natural resource depletion, malnutrition, and pollution. It predicted that if these trends continued, the earth would reach its limits within one hundred years. If allowed, the book argued, "The most probable result will be a sudden and uncontrollable decline in both population and industrial capacity."[21] But the authors did not forecast inevitable doom. They felt it was possible to alter the growth trends to foster sustainability, both ecologically and economically. In making such arguments, the book was a key outline of the early bioenvironmentalist views. It has sold over thirty million copies and has been translated into more than thirty languages. Many people criticized both the work of Ehrlich and the Club of Rome for what they saw as exaggerations.[22] Donella Meadows, one of the original authors of *Limits to Growth*, on reflection, saw such virulent criticisms of her work as inevitable given the dominance of the economic growth paradigm. As she notes, "A book called *The Limits to Growth* could have been filled with nothing but blank pages, and some people would still have thought it was an anticapitalist plot or an anti-Communist or an anti-Keynesian or an anti–Third World one. They never saw the book for its cover."[23]

The following year saw the publication of another extraordinarily influential book in the West: E. F. Schumacher's *Small Is Beautiful: Economics as if People Mattered*. This book helped shape the development of social green thought with its focus on the need for small, local economies. Schumacher was deeply influenced by Indian anticolonial activist Mahatma Ghandi, whose nonviolent means of protest in the lead-up to Indian independence in 1947 won him international acclaim. Ghandi also outlined a vision for a free India organized around self-reliant and self-governing villages, which encouraged local household-level production, rather than large-scale industrial society.[24] Schumacher was also influenced by Buddhism, arguing that a traditional economist

> is used to measuring the "standard of living" by the amount of annual consumption, assuming all the time that a man who consumes more is "better off" than a man who consumes less. A Buddhist economist would consider this approach excessively irrational: since consumption is merely a means to human well-being, the aim should be to obtain the maximum of well-being with the minimum of consumption. . . . The less toil there is, the more time and strength is left for artistic creativity. Modern economics, on the other hand, considers consumption to be the sole end and purpose of all economic activity.[25]

In making these arguments, Schumacher also helped build the foundation of the field of ecological economics.

These views on population, consumption, and limits to growth in the 1960s and 1970s provided a foundation for the bioenvironmentalist perspective and have been a strong continuing influence in the environmental movement in the North. Though the ecological collapse predicted by many early bioenvironmentalists did not occur, thinkers in this tradition—such as Paul and Anne Ehrlich, Lester Brown (founder and president of the Earth Policy Institute), and William Rees—continue to argue that population pressures and limits to growth do exist, and that human activity is imperiling the future of the planet.[26] The bioenvironmentalist view of the ecological interdependence of national economies strengthened the support for global institutions and agreements as the only viable response to emerging global environmental problems in a highly interdependent world.

Social green thought also began to emerge at this time. Although social greens agreed with much of the bioenvironmentalists' arguments regarding consumption and growth, they were critical of the focus on population growth in the South as a cause of environmental harm. This gave their views more appeal to environmentalists in the developing world. Social greens also felt that local solutions were superior to global ones, drawing on the inequality arguments of dependency thinkers as well as on the visions for a local economy of Ghandi and later Schumacher. The next section examines global environmental cooperation and institutions since the early 1970s.

The Stockholm Conference and the 1970s

Debates over the links between the global economy, population growth, and environmental change on the one hand, and the sense of mutual interdependence and vulnerability on the other, were at the core of the United Nations Conference on the Human Environment, held in Stockholm, Sweden, in June 1972. This was the first global-level UN conference on the environment as well as the first world conference on a single issue. Delegates from 113 governments attended. Canada's Maurice Strong was chair (see box 3.2). Three parallel NGO conferences—the Environment Forum (with official UN support), the Peoples Forum, and Dai Dong—took place alongside the official proceedings.

Box 3.2
Maurice Strong

Maurice Strong played an important personal role in the globalization of environmentalism. Born in Canada in 1929, he started as an entrepreneur and businessman, working his way up to president of the Power Corporation of Canada. In 1966 he became head of Canada's External Aid Office (later, the Canadian International Development Agency). From 1970 to 1972 he was secretary-general of the UN Conference on the Human Environment and became the first executive director of the UN Environment Programme in 1973. He returned to Canada to become president of Petro-Canada (1976–1978). In the 1980s he was chair of the International Energy Development Corporation and the Canada Development Investment Corporation. He also served in the 1980s as a member of the World Commission on Environment and Development. After that he became secretary-general of the 1992 UN Conference on Environment and Development. In 1992 he again returned to Canada, this time as chair and CEO of Ontario Hydro, North America's largest utility company. In 1995 he was named senior advisor to the president of the World Bank, and served as a senior advisor to UN Secretary-General Kofi Annan on UN reforms from 2003 to 2005. He currently spends much of his time at Peking University as an advisor on the Chinese environment, energy, and technology sectors.

Source: Strong 2000; see also <http://www.mauricestrong.net> (accessed July 19, 2010).

The conference agenda and outcomes must be seen against the backdrop of the Vietnam War and the Cold War. The United States printed more and more dollars to finance the Vietnam War, contributing to a rise in global inflation and overall global economic concerns. Given the state of the global economy, many conference participants feared that their governments could not afford the requirements for global environmental protection. The Cold War also directly altered the political climate of the Stockholm Conference. Russia and the Eastern Bloc countries boycotted the conference because East Germany, not yet a member of the UN, was not allowed to participate. This was a blow for Strong, who was keen to ensure the participation of all countries.

The initial purpose of the conference, announced in 1968 by the UN General Assembly, was to discuss "problems of the human environment" and "identify those aspects of it that can only, or best be solved through international cooperation and agreement."[27] The focus was on environmental problems arising from industrialization, of particular concern to Northern governments. This was expanded, however, to include broader development concerns to gain the support of developing countries. The UN convened a meeting in 1971 in Founex, Switzerland, which brought together development experts and scientists from the developing world to discuss the links between environment and development. The *Founex Report on Development and Environment*, which is credited with encouraging wide participation at Stockholm among the developing countries, emphasized the environmental problems linked to poverty in the developing world, arguing that development, including industrialization, was necessary to overcome poverty-related environmental degradation.[28]

The focus of many of the debates at Stockholm, then, was on how to reconcile the economic development demanded by the South with the perceived need of the North to protect the global environment.[29] The phrase "the pollution of poverty" was coined at Stockholm as developing countries, reiterating some of the core conclusions of the *Founex Report*, argued that the greatest environmental threat was poverty. At the same time developing countries were skeptical of the North's environmental agenda, worrying it would deny poorer countries the benefits of economic growth and industrialization. Brazil, for example, declared that it had little interest in discussing industrial pollution, which it saw as a "rich man's" problem.[30] The representative from the Ivory Coast went as far as to state that the country would welcome more pollution if this would allow it to industrialize.[31] Developing countries further argued that exploitation by global capitalists was a core reason for their high levels of poverty in the first place. Global economic institutions were singled out for pushing them to export raw materials on declining terms of trade. Many developing countries called for global economic reforms as part of efforts to solve global environmental problems.

Official conference documents at Stockholm, however, did not reflect these calls. At most, these documents acknowledged the unique problems of developing countries, but no real remedies were offered.[32] The conference produced a Declaration on the Human Environment, which contained 26 principles, an Action Plan for the Human Environment, which included 109 recommendations, and a Resolution on Institutional and Financial Arrangements. These measures were "soft" international law—that is, they did not legally bind the signatory states—yet they did signal a growing concern among national governments over the global environment. The Stockholm Conference is also credited with the creation of the United Nations Environment Programme (UNEP). Officially launched in 1973, UNEP was headquartered in Nairobi, Kenya; Maurice Strong was named its first executive director.

The Stockholm outcomes—in particular, UNEP and the demand for international environmental cooperation—bolstered the views of institutionalists. Bioenvironmentalists saw the need for legislation to control individual behavior, and argued for states to rein in both economic and population growth. Institutionalists, on the other hand, argued for international legal controls on states to address particular environmental problems. This view gained considerable legitimacy in the years following Stockholm, reflected in an increasing number of governments entering into an increasing number of negotiations for "hard law," or environmental treaties, which are binding agreements.

The world economy spun into further turmoil in the years shortly after the Stockholm Conference. Restrictions on oil supplies by the Organization of the Petroleum Exporting Countries (OPEC) in 1973–1974, which led to a quadrupling of oil prices, spurred high inflation and low growth rates worldwide. This then filtered through to prices for all industrial goods produced with energy. Many developing countries were upset with the impact of this global economic downturn on their levels of debt and their prospects for industrial development. In 1974, the developing countries called on the UN to create a New International Economic Order (NIEO). They demanded reforms to counter declining terms of trade for raw commodity producers, increase voting power for developing countries in the IMF and the World Bank, put controls on transnational corporations, and relieve foreign debt. A NIEO was at the center of global negotiations throughout the 1970s, but concrete measures did not get far.[33] Nonetheless, these global negotiations reinforced the sense of global economic interdependence in both the North and South.

The economic turbulence of the 1970s relegated environmental issues to the back burner in many countries, especially poorer ones. Yet states still negotiated important global environmental treaties in this period, partly as a result of processes set in motion by the Stockholm Conference. These include the Convention on the Prevention of Marine Pollution by Dumping of Wastes and Other Matter (the London Convention, 1972), the Convention on International Trade in Endangered Species of Wild Fauna and Flora (CITES, 1973), and the Convention for the Prevention of Pollution by Ships (MARPOL, 1973).

Alongside these international diplomatic efforts, there were a growing number of local environmental movements to protect the environment, particularly in non-Western countries.[34] Many of these movements stressed the need to address poverty and inequality as a primary way to improve environmental conditions, seeing livelihoods and the environment as inextricably linked. This perspective gave many of these movements philosophical ties with the social greens. An example was the Chipko movement, which emerged in the Indian state of Uttar Pradesh in the Himalayas in the early 1970s. This movement was a response to deforestation linked to increased logging and industrialization, which stripped local peoples of their livelihoods, leading not only to more environmental damage, but also to higher levels of poverty. The movement, drawing inspiration from the Ghandian philosophy of promoting local and small-scale production, sought to resist government allocation of trees to large industries that denied the poor access to those resources. This grassroots movement applied direct-action methods, including tree hugging and protests, with the aim of protecting the forests for small-scale use by locals.[35] The Chipko movement was comparatively successful, gaining international recognition and spreading the methods of resistance to other parts of India.

In an attempt to ensure that states did not forgo environmental management in hard economic times, the IUCN, along with UNEP and the WWF, issued a joint report on the global environment in 1980, the *World Conservation Strategy*.[36] The report, a product of three years of collaborative work among these organizations, reflected the Northern environmental agenda of the 1960s and 1970s (drawing on elements of both bioenvironmentalist and institutionalist ideas). It focused on conservation and resource management on the one hand, and the impact of population pressures on those resources on the other. The *Strategy* outlined the goals and targets the world should aim for to preserve ecological processes, genetic diversity, and species and ecosystems. Importantly, it brought the concept of "sustainability" to the world's attention. It came close to integrating development and environment, although the emphasis was still more on the environment side. Many would later criticize the *Strategy* for failing to address the necessary changes—economic, social, and political—to achieve its ambitious goals. The concept of sustainability would, however, shape global discussions on environment and development in the 1980s.

The Brundtland Report and the 1980s

The 1980s brought more turmoil for Third World economies. A foreign debt crisis, in which the South is still mired, was first openly acknowledged in August 1982 when Mexico announced it could not repay its international debts. While other developing countries followed Mexico into a debt crisis in the early 1980s, First World economies began to recover. This relatively strong performance in the North bolstered the global spread of the values and neoliberal economic ideologies of Prime Minister Margaret Thatcher in Britain and President Ronald Reagan in the United States. Neoliberal economic prescriptions at this time called for deregulation, stressing the need for free and open markets as the best way to organize an economy. Neoliberalism advocated liberalization of trade, investment, and financial policies as a way of integrating the global economy, and—it was presumed—of stimulating economic growth. The spread of neoliberal economic thinking in the early 1980s strengthened support for the market liberal view of the environment, which saw these policies as compatible with environmental protection, as discussed later.

Over the course of the 1980s and 1990s, policies shifted worldwide to reflect the ascendancy of neoliberal economic thought. Though the global economy was widely seen in the 1970s as at least partly responsible for the South's weak economic growth, by the 1980s there was a shift toward the view that global economic integration into free markets was the best path to enhance economic growth. In other words, elites increasingly perceived the domestic failure to open economies as the reason for the economic troubles in developing countries, rather than pointing to a failure of the global economy to provide equal benefits. Institutions like the World Bank and IMF were seen as critical sources of support for the transition to globally integrated free-market economies. The Third World debt crisis was viewed as a crucial test for the neoliberal economic policies of these institutions. In return for emergency loans, the IMF and the World Bank required developing countries to implement economic reforms, commonly known as "structural adjustment." These reforms included liberalization of trade, investment, and exchange-rate policies, devaluation of currencies, government spending cutbacks, and privatization of public enterprises.[37]

The rise of neoliberalism and the realization of the severity of the Third World debt crisis shaped the work of the World Commission on Environment and Development (WCED). The UN General Assembly established this commission in 1984 to examine the relationship between the environment and global economic development. It is commonly known as the Brundtland Commission after its chair, Norway's prime minister Gro Harlem Brundtland (see box 3.3). Its report, *Our Common Future*, published in 1987, went further than any official international document to provide a new definition of development with the environment at its core. The report, too, was careful to set its analysis in the context of the global political economy.

The Brundtland Report changed the discourse of global environmentalism. The commission tried to chart a middle ground between the North and the South, and between market liberal and institutionalist views on growth on the one hand, and social green and bioenvironmentalist views on the other. It proposed a global development and environment strategy designed to be palatable to all. It did not see further economic growth and industrialization as necessarily harmful to the environment, and thus did not foresee any necessary "limits" to growth. At the same time, it

Box 3.3
Gro Harlem Brundtland

Gro Harlem Brundtland was born in Norway in 1939. Her career is replete with accomplishments. She began as a medical doctor, like her father before her. As a young doctor and mother, she won a scholarship to the Harvard School of Public Health, where she earned a Master of Public Health degree. In 1965 she returned to Norway and, while raising a family, worked on children's health issues at the Ministry of Health. But she was drawn to politics, having joined the Norwegian Labour Movement at just 7 years of age. In 1974 she was appointed Minister of the Environment. Seven years later, at 41, she became the youngest and first female prime minister of Norway (as head of the Labour Party). She would go on over a decade-and-a-half period to win three elections and hold onto the prime ministership for about ten years. Global recognition came as chair of the 1983–1987 World Commission on Environment and Development, where she was instrumental in the creation of the concept of sustainable development as well as the consensus for the 1992 Rio Summit. From 1998 to 2003 she was director-general of the World Health Organization, where she helped push for a more active and innovative global role for the WHO. In 2007, UN Secretary-General Ban Ki-moon appointed Brundtland as a UN Special Envoy on Climate Change.

Source: Summarized from the WHO web site at <http://www.who.ch> (accessed July 19, 2010). See also <http://www.un.org/climatechange/2007highlevel/envoys.shtml> (accessed July 19, 2010).

argued, very much in line with Third World sentiments at Stockholm, that poverty harmed the environment as much as industrialization. This poverty was in large part due to the place of developing economies within the global structure. The best way to move forward, the report contended, was to promote economic growth—not the kind of growth seen in the 1960s and 1970s, however, but environmentally sustainable growth. With this recommendation, the report popularized the term *sustainable development*, which it defined as development that "meets the needs of the present without compromising the ability of future generations to meet their own needs."[38] This definition of development did not pose a fundamental challenge to the neoliberal economic ideology of the time, because it did not suggest a need to slow the pursuit of economic growth or to slow the process of global economic integration.

Other recommendations in the Brundtland Report in some ways reflected the 1970s NIEO agenda. It called for the transfer of environmental technologies and economic assistance so that developing countries could afford to pursue sustainable development and economic growth. Other recommendations included calls for states to control population growth, promote education, ensure food security, conserve energy, promote urban sustainability, and develop cleaner industrial technologies. The report challenged some neoliberal economic views—for example, by calling for redistributive measures between rich and poor countries—but the dominance of the neoliberal economic agenda at the time ensured that the global community focused on the fully acceptable compromise of sustainable development.[39] The Brundtland definition of sustainable development is now part of mainstream international and national policies and rhetoric. It has not altered the fundamental dominance of neoclassical economics, though.[40] In fact, some would argue that the Brundtland Commission's real legacy was to create a concept that secures the hegemony of a coalition of the moderate market liberal and institutionalist views of environmental management within the global community.

Alongside the work of the Brundtland Commission, the world continued to negotiate global environmental treaties. These were generally in direct response to urgent environmental problems (often generating considerable media attention). These include the 1985 Vienna Convention for the Protection of the Ozone Layer and the 1987 Montreal Protocol on Substances That Deplete the Ozone Layer (see chapter 2), as well the 1989 Basel Convention on the Transboundary Movement of Hazardous Wastes and Their Disposal. The Basel Convention arose in response to the revelations in the mid-1980s of hazardous waste being shipped from rich to poor countries for disposal, raising both ecological and ethical concerns.[41]

Equally important to the diplomatic evolution of environmental discourse in the 1980s were developments in local and national-level environmental movements around the world (many of which began in the 1970s). Three examples are noted here. The West German Green Party, of which one of the founding members was Petra Kelly (see box 3.4), was instrumental in putting environmental issues on the German and broader European political agenda in the 1980s. The National Council of Rubber Tappers, a Brazilian NGO, was founded by Chico Mendes (see box 3.5) in 1985 to defend the livelihoods of rubber tappers threatened by deforestation. The Greenbelt Movement in Kenya was

Box 3.4
Petra Kelly

Petra Kelly, born in 1947, was one of the cofounders of the German Green Party in 1979. She became one of the national chairpersons of the German Green Party in the early 1980s, and gained international recognition at this time as a spokesperson for the causes of the Green Party. Kelly was elected to the German parliament in 1983, one of twenty-eight Green MPs elected that year. She authored a number of books and spoke widely on issues like ecology, peace, feminism, and human rights. Kelly received the Right Livelihood Award (known as the alternative Nobel Prize) in 1982. Tragically, she was killed in 1992, the exact circumstances of which remain unclear.

Source: Summarized from the Right Livelihood Award web site at <http://www.rightlivelihood.se/recip/kelly.htm> (accessed July 19, 2010).

established in the 1980s by Wangari Maathai (see box 3.6) as a grassroots women's organization committed to environmental conservation through the planting of trees.

The Earth Summit and the 1990s

In 1989 the UN General Assembly, following the recommendations of the Brundtland Report, passed a resolution to hold another world conference on the environment, the UN Conference on the Environment and Development (UNCED), to be held in 1992 in Rio de Janeiro, Brazil. Significant backdrops to UNCED, popularly known as the Earth Summit, included the ongoing debt crisis in the Third World, the collapse of the Eastern Bloc, and the accelerating pace of economic globalization. Politicians at this time saw green politics as a popular issue, and the axis at the global level for politics of this nature had shifted from East–West to North–South.

The Earth Summit was held on the twentieth anniversary of the Stockholm Conference. The secretary general of the meeting was Maurice Strong, who, as mentioned earlier, organized the conference in Stockholm. The Rio Earth Summit was the largest UN conference to date, with 179 countries participating and more than 110 heads of state

Box 3.5
Chico Mendes

Chico Mendes (Francisco Alves Mendes Filho) was born in 1944 in the western Brazilian Amazon. Mendes was the son of a poor rubber tapper, and joined the profession at the age of 9. His efforts to organize rubber tappers to resist nonviolently the destruction of the forests in the Amazon from excessive logging gained international support in the 1970s and 1980s. Mendes founded the National Council of Rubber Tappers in 1985 to further support this cause. The son of a cattle rancher assassinated Mendes in 1988. The subsequent global furor over his death is said to have influenced Brazil's decision to offer to host the 1992 Earth Summit.

Source: Summarized from *Events in the Life of Chico Mendes*, on the Environmental Defense Fund web site at <http://www.edf.org/article.cfm?ContentID=1605> (accessed July 19, 2010).

Box 3.6
Wangari Maathai

Wangari Maathai, born in Kenya in 1940, was trained in biology and veterinary science and became a professor at the University of Nairobi in the 1970s. Active in the National Council of Women of Kenya from 1976, Maathai went on to chair that organization from 1981 to 1987. As chair, she founded the Greenbelt Movement, an organization to promote grassroots action for the environment as a means of improving women's lives. The Greenbelt Movement focused on tree planting and by 1993 had planted over twenty million trees. The movement has been credited with raising environmental awareness in Kenya, gaining global recognition for its achievements. In 1984 Maathai won the Right Livelihood Award for her work with the Greenbelt Movement. In 2002 she was elected to the Kenya parliament and appointed as Assistant Minister for Environment, Natural Resources and Wildlife. In 2004 she won the Nobel Peace Prize.

Source: Summarized from the Right Livelihood Award web site at <http://www.rightlivelihood.se/recip/maathai.htm> (accessed July 19, 2010) and the Nobel Prize web site at <http://nobelprize.org/nobel_prizes/peace/laureates/2004/maathai-bio.html> (accessed July 19, 2010).

attending. There were 2,400 nongovernmental representatives as well as another 17,000 in a parallel NGO forum. The recommendations in the Brundtland Report dominated discussions at Rio, especially the notion of sustainable development. Governments at the conference found it politically easy to adhere to the Rio goals of promoting more growth with more environmental protection. Who would deny these goals when presented as mutually compatible?

Many developing countries, however, worried about taking on additional environmental commitments without concrete assurances of economic assistance to fund environmentally sustainable growth. Debates over financial transfers for sustainable development split along North–South lines. Developing countries felt strongly that industrialized countries should foot most of the extra costs of "green growth," and that this should be over and above current levels of assistance. Donors, on the other hand, were reluctant to take on further financial commitments.[42] Related to this theme, developing countries also wanted to ensure receipt of environmental technologies without extra cost, as well as ensure that industrialized countries would not be able to use environmental regulations to restrict developing-country exports.

What, then, were the main outcomes of the Rio conference? Perhaps most important, it put environment and development on the agendas of global leaders. It also reaffirmed the Brundtland view that more growth will bring a better environment, an idea supported by both market liberals and institutionalists. Participating governments adopted and signed the Rio Declaration on Environment and Development, a set of twenty-seven principles outlining the rights and responsibilities of states regarding the promotion of environment and development. These principles included far more of the South's concerns about the right to development than did the Declaration on the Human Environment adopted at Stockholm twenty years earlier. Agenda 21, a three-hundred-page action program to promote sustainable development, was also adopted. The majority of the text of Agenda 21 was negotiated at four separate preparatory committee meetings (PrepComs) in the two and a half years prior to the Earth Summit, and only the last 15 percent was finalized in Rio.[43] The conference also adopted a nonbinding Statement of Forest Principles to promote the sustainable management of tropical, temperate, and boreal forests. It was a compromise document, because it proved impossible to reach agreement on a legally binding forest treaty.[44] The conference also opened two legally binding conventions for signature: the UN Framework Convention on Climate Change and the Convention on Biological Diversity. Further, the conference established the UN Commission on Sustainable Development to monitor and evaluate the progress on meeting the Rio objectives. Negotiations also began at Rio on a treaty on desertification.[45] Finally, Rio was a "trigger" for the restructuring of the Global Environment Facility (GEF), set up to finance efforts in developing countries to protect the global environment (see chapter 7).[46] These global cooperative agreements were very much products of institutionalist influence at the conference.

There were some conspicuous failures at Rio, too. The estimated annual cost to implement Agenda 21 globally was US$625 billion, yet relatively little funding was pledged at Rio. Moreover, the developed countries were asked to cover just US$125 billion.[47] This amount is roughly equal to 0.7 percent of their total gross national product (GNP), a theoretical target the UN set for development assistance back in the 1970s, and one that most donors have never been close to reaching. In total, GEF had allocated only US$8.26 billion in grants by 2009 (although it has leveraged US$33.7 billion in cofinancing).[48]

Many NGO participants at Rio were skeptical of the official agenda. The focus on Brundtland-style solutions—in particular, the promotion of economic growth and industrialization as compatible with sustainability—was widely criticized.[49] Social green critics, many of whom were at Rio with more radical NGOs such as Greenpeace and the Third World Network, argued that the world community was ignoring the environmental consequences of economic globalization—inequality, industrialization, economic growth, and overconsumption—at its peril. Despite a record number of NGOs in attendance at the global conference, many felt that industry had far more influence on the official agenda, partly because of the close ties to Maurice Strong and industry's funding of the event.[50] A number of critics saw Rio as entrenching a "managerial" approach to solving environmental problems from above, while paying inadequate attention to local solutions from below.[51] These social green critics saw the nod to the role of the poor, women, and indigenous peoples in the official Rio documents as woefully inadequate.

The social green critics integrated some of the bioenvironmentalist concerns with economic growth and overconsumption into their analysis, but did not include a focus on population growth and resource availability. Rather, they stressed the need to reduce global economic inequalities and shift toward more local solutions that genuinely help those most affected by global economic change: women, the poor, and indigenous peoples.[52] It was at Rio that such thought began to fully enter the international diplomatic dialog on global environmental management.

Despite the critiques and the failure to attract significant extra financing to support the goal of implementing sustainable development, most analysts agree that UNCED did achieve some notable successes, especially in terms of raising environmental awareness among the general public in both the North and South. Five years later, a special session of the UN General Assembly, known as the Earth Summit +5, reviewed global progress with the implementation of Agenda 21. The conclusions were largely disappointing: the world had not met most of the Agenda 21 goals. Earth Summit +5 failed to inject new enthusiasm into reaching them.[53]

Johannesburg, Copenhagen, and Beyond

Economic globalization accelerated in the years following the Rio Earth Summit. Trade and investment as a proportion of the economies of both rich and poor countries grew markedly. The GATT concluded the Uruguay Round of trade negotiations in 1994, and the World Trade Organization (WTO) was established in 1995. The WTO has more power to enforce decisions in dispute settlement than the previous procedure under the GATT did, and many viewed this as a prioritization of trade goals above others in the global arena.[54] Despite rapid economic growth in some parts of the world, such as India and China, global economic inequalities remain.[55] Some contend that these inequalities have worsened as a direct result of globalization.[56] A global coalition of nongovernmental forces emerged in the late 1990s to oppose globalization. These groups began to protest at global economic meetings, such as the annual meetings of the IMF, the World Bank, and the WTO. The largest and most significant was at the Seattle ministerial meeting of the WTO in late 1999. Antiglobalization protesters rallied against the environmental and social fallout from globalization, blaming global economic institutions for aggravating inequalities and causing environmental destruction.[57] The WTO failed to begin a new round of trade negotiations at Seattle. The WTO finally launched the Doha Round of trade negotiations in the fall of 2001, at WTO meetings held in Doha, Qatar, which was dubbed the "development round" due to concerns over trade and development inequities between rich and poor countries.

The antiglobalization sentiment of activists, as well as the steadfast support for globalization among governments and international economic institutions, were continually in the background during the preparations for the 2002 World Summit on Sustainable Development (WSSD), held in Johannesburg, South Africa, ten years after the Rio Earth Summit. The goal was to evaluate the progress toward the Rio goals and set concrete targets to improve implementation. It also aimed to build on the International Conference on Financing and Development, held in March 2002 in Monterrey, Mexico, as well as to develop a strategy to implement the UN's Millennium Development Goals.[58] The WSSD was the biggest ever UN meeting to that date. Over 180 nations and 100 heads of state attended. There were over 10,000 delegates, 8,000 civil society representatives (i.e., representing nongovernmental interests), 4,000 members of the press, and countless ordinary citizens. The secretary general was Nitin Desai, an Indian diplomat with the UN who served as deputy secretary general of UNCED under Maurice Strong, and who continued to work at the UN at the undersecretary general level on policy coordination and sustainable development.[59]

Much like Rio and Stockholm, the conference adopted official documents about the need for global sustainability as well as a plan for implementation. These are the Johannesburg Declaration on Sustainable Development, which outlines both the challenges and general commitments for the global community, and the Johannesburg Plan of Implementation, an action plan for implementing these goals. Negotiations for these documents began long before the meeting, yet one-quarter of the text was still in dispute at the start of the conference. One of the most contentious issues, as was the case in Rio, was financing. But new and equally difficult items were added at Johannesburg: globalization's role in sustainable development, as well as timetables and specific targets for meeting goals.[60] In the end, the Johannesburg Declaration on Sustainable Development was somewhat different from previous global declarations on environment and development. Economic inequality was recognized as a major problem not only in terms of justice and human well-being, but also in terms of global security. And globalization was mentioned as a new challenge for achieving sustainable development.[61]

The Plan of Action, a sixty-five-page document, was much less ambitious than Rio's Agenda 21 and largely restated the areas where more action was needed to meet the Rio goals, as well as other previously agreed-upon goals such as the Millennium Development Goals and development financing goals.[62] The Plan of Action did include a few more specific goals and targets, focusing on several main priorities, such as water, agriculture, energy, health, and biodiversity. The Commission on Sustainable Development, created after Rio, was reaffirmed as the agency that would monitor progress on this implementation.

The conference also highlighted and promoted a new instrument for implementing sustainable development. These are public–private partnerships (dubbed PPPs, or "Type II" arrangements—"Type I" being government-to-government) between governments, NGOs, and business. These are voluntary agreements that involve stakeholders directly. The aim is to encourage the transfer of funds and technology to areas of critical need. Several hundred such partnerships were identified prior to and during the Summit, and many more have been initiated since.[63] At Johannesburg, for example, Greenpeace announced a partnership with the World Business Council for Sustainable Development (WBCSD) to promote action to mitigate climate change. This focus on PPPs gave the Johannesburg Summit a greater focus on the role of business in promoting sustainable development than either Stockholm or Rio had. The Plan of Action mentions the need to promote corporate accountability and responsibility, and the Johannesburg Declaration stresses the duty of business to promote sustainable development. This focus on business as a partner in sustainable development has remained prominent in the years since the WSSD. On the business side, this has contributed to a significant increase in policies and partnerships to promote corporate social responsibility (CSR), with advocates seeing this as a way toward a green economy.[64]

Was Johannesburg a success? Many business participants saw the outcome as constructive and realistic. Many nongovernmental participants were disappointed with the "weak" targets and timetables, noting that the conference added little to global efforts to enhance sustainability.[65] New targets were set on access to clean water and sanitation, as well as on halting depletion of fish stocks, but beyond these, not much new was added in terms of global commitments. The debate over the role of corporate globalization was divided, with industry and most governments on the one hand seeing it as positive, and many environmental and development NGOs on the other seeing it as negative. In the end, the official documents make bland and noncommittal statements about the importance of this issue without offering guidance on how to proceed.

Some argue that the Johannesburg Summit highlights the waning impact of environmentalism in global discourse.[66] Part of the reason that many of the debates at the WSSD did not make much progress was the clash of diverse views on the causes and consequences of global environmental change. Social greens and bioenvironmentalists were able to present their cases much more effectively than in previous global environmental meetings. This raised tough issues at the negotiating table, such as globalization, corporate accountability, exhaustible resources, and consumption. Yet the overall dominance of the views of market liberals and institutionalists in the official proceedings ensured that such "talk" did not result in concrete targets or plans of action. The UN, partly because Johannesburg disappointed so many, shifted its focus toward implementation and yearly reviews of progress in the years immediately following the WSSD.

The climate conference in Copenhagen in December 2009, officially the 15th Conference of the Parties (COP) to the UN Framework Convention on Climate Change, served the purpose of a major international environmental summit. Some 45,000 delegates attended the conference, and many more were present in Copenhagen for related events.[67] The Copenhagen summit aimed to reach a new international climate agreement on emissions reductions, given that those in the Kyoto Protocol last only until 2012. Despite the long build-up and high expectations that an agreement was imperative, delegates were unable to negotiate an agreement acceptable to all 192 parties. On the last day of the conference a closed door meeting of a handful of leaders—including those from the United States, China, India, Brazil, and South Africa—produced a three-page "Accord" that listed aspirations for a post-2012 agreement, but without any specific details or emission reduction targets. A number of developing countries objected to being excluded from the negotiations on the Accord, and some, including Bolivia and Venezuela, refused to back its adoption by the formal COP.

In the end the COP meeting "noted" the Copenhagen Accord, but did not formally adopt it. The Accord is not legally binding, nor is it an official UNFCCC document. The Accord establishes a goal of limiting temperature change to no more than 2°C above preindustrial levels. Rather than impose global or country-specific emission reduction targets to meet this goal, the Accord asked countries to set their own emission reduction targets for 2020. Despite the facts that the document does not have official status in the UNFCCC, that it is weak in terms of targets, and that many developing countries initially opposed the idea of the Accord, by the end of January 2010 nearly all countries had provided emission reduction targets to accompany the Copenhagen Accord. By April 2010, some 110 countries had signed onto the agreement. At the time of writing, it is still unclear whether these self-set individual country targets will be part of a new legally binding agreement, or whether they will remain merely aspirational numbers.[68]

Global Environmental Governance

The global discourse on environment and development shapes, and is shaped by, institutions and policies—from global to local, and public to private. The history of global environmental discourse and diplomacy, as presented in the first part of this chapter, highlights the fact that over time, there have been significant changes to the key actors. In today's world, not just states, but also international institutions, firms, markets, NGOs, foundations, communities and regions, and individuals shape global environmental governance.[69] These actors also employ an increasingly diverse range of policy tools. Thus, it is hardly surprising that we see a wide range of views on how to best pursue global environmental protection. Some even characterize this plethora of actors and arrangements as a "fragmentation" of global environmental governance, away from states as the primary holders of authority and legitimacy.[70] Here, to provide a more comprehensive view of the globalization of environmentalism, we provide a very brief discussion of states, regimes, NGOs, corporations, and subnational and multiscale actors in the context of emerging patterns of global environmental governance. We add a deeper understanding of the impact of these actors on global environmental management throughout the rest of the book.

States

States are core actors in global environmental governance. This is natural, given that state sovereignty, consisting of both rights and duties, has long been considered the guiding principle of international politics. The Treaty of Westphalia, signed in 1648, gave a state supreme authority to act within its territory. Foreign powers could not dictate what states can and cannot do within their borders. At the same time, to preserve these sovereign rights, states are obliged to respect the sovereignty of other states as well as protect their own citizens. States, as a glance at the wars of the last 350 years shows, do not always respect the sovereignty of other states. Nor do states always protect their citizens. Nevertheless, all states have guarded their sovereign rights with vigor and resolve, a fundamental feature of global environmental politics.[71]

There are obvious tensions between the sovereign-state system and the natural environment. Perhaps the most visible is that environmental problems commonly spill across the political boundaries of states. The actions needed to manage the global environment—to act in the interests of the planet—may well clash with state interests. This partly explains the global community's struggle to deal with problems like climate change. But if the sovereign-state system grants states the right to control activities within their borders, what are their duties toward other states if their activities damage the environments of other states or the global commons? Principle 21 of the Stockholm Declaration addresses this question, declaring that states do have a right to exploit their own resources or pollute their own environments, but only as long as this does not affect other states directly.[72] It is, however, difficult to prove that the actions of one state harm the environment of another state (or states) or the global commons. Can it be proven, for example, that deforestation in Costa Rica harms the United States because it decreases the global capacity to sink carbon dioxide? And who is to blame: the Costa Rican government, Costa Rican cattle ranchers, or the consumers of steaks and hamburgers?[73] Because of this difficulty of proof, in practice little beyond moral pressure tends to be applied to hold states to this principle.

A second tension between the sovereign-state system and the natural environment is the tendency for environmental problems to develop over a long period of time. The existence of some problems, such as depletion of the ozone layer by CFCs, took years not just to discover, but also to prove with enough certainty to convince states to act. Environmental problems also tend to be circular and dynamic in nature. They rarely have a single cause, and efforts to deal with one part of an environmental problem can easily create new problems in the future. This long-term and circular nature of environmental problems contrasts sharply with the short-term and results-oriented political lives of those in power in democratic nations. It is naturally difficult for politicians to gain the funding and support for long-term action with political cycles typically lasting five years or less.[74]

Many recognize the contradictions and tensions of the state system for effective global environmental management. Most assume that it is unrealistic, perhaps even counterproductive, to try to replace it (with, e.g., a federal world government). Instead, virtually every state has now created an environment department with national and global responsibilities. States have embraced economic globalization, which strengthens the power of the global market to transfer environmental funds and technology across state borders (although, as we have seen, many argue that this also transfers more environmental harms across borders). States have passed the authority for some environmental issues to global institutions. And finally, states have relinquished some sovereign rights by signing onto hundreds of global environmental agreements. States still dominate global environmental governance. But, as the next section shows, international institutions and regimes have grown in importance.

International Institutions/Regimes

Since Stockholm, global institutions have become increasingly active in promoting global environmental protection. UNEP is a catalyst for global action among states as well as other actors to mitigate specific environmental problems.[75] It does this in part by supporting research on and negotiation of new international environmental agreements.[76] Other arms of the UN also promote sustainable development. The Commission on Sustainable Development (CSD) monitors the implementation of Agenda 21 as well as other internationally declared environmental goals and targets.[77] The United Nations Development Programme (UNDP), the World Health Organization (WHO), the World Bank, and the Food and Agriculture Organization of the United Nations (FAO), just to mention a few, have also—if in a less direct fashion—shaped global environmental management. Table 3.1 outlines some of the key international organizations that influence global environment and development governance. In addition, many of the Regional Commissions of the UN have devised and implemented environmental strategies. The Organization for Economic Cooperation and Development (OECD) has adopted and applied legally binding regulations on the environment. A few of the many intergovernmental groups that consider environmental policies and strategies include the Group of 8, the African Ministerial Conference on the Environment, the Council of Arab Ministers Responsible for the Environment, the Association of Southeast Asian Nations, and the European Union Environment Ministers.[78]

International environmental regimes encompass established rules of international cooperation, either through a formal international treaty or agreement, or through a less formal set of rules and norms (established practices that states generally adhere to).[79] States have signed and ratified a remarkable number of global environmental agreements over the past forty years, with an especially large number signed between the Stockholm and Rio conferences (see table 3.2). In 1972, there were only a couple of dozen multilateral environmental treaties; today there are more than four hundred.[80]

Are these agreements and regimes working? Chapter 2 showed that some, such as the ones to solve depletion of the ozone layer, have been reasonably successful. At a minimum, the secretariats to the conventions are able to disseminate critical information as well as influence a state's level of concern about an environmental issue. Through aid and training, the secretariats also help raise the national capacity of developing countries struggling to comply with global agreements. Many observers, however, remain skeptical of the value of international agreements—arguing that many do not in fact alter state and corporate behavior.[81] States that do not sign or ratify are not bound by a treaty, creating obvious problems of free riders (a general drawback of international

Table 3.1
Intergovernmental organizations: Voting rules and revenue sources

Organization	Voting rules	Revenue source
United Nations	Each member of the General Assembly has one vote; decisions are made by a two-thirds majority of the members present and voting.[1]	Member contributions, as apportioned by the General Assembly.[2]
United Nations Environment Program	Governing Council is composed of 58 members of the UN elected by the UN General Assembly for three-year terms, taking into account the principle of equitable regional representation. Decisions are adopted by majority of members present and voting.[3]	The costs of servicing the Governing Council and providing the small secretariat are borne by the regular budget of the United Nations. The Environment Fund provides financing for environmental programs; altogether 178 countries have made at least one voluntary contribution.[4]
United Nations Development Program	The Executive Board of UNDP and UNFPA, composed of 36 members, is elected by the Economic and Social Council for three-year terms. Since 1994, decisions have always been adopted by consensus.[5]	Donor support, principally from members of the OECD DAC, plus local resources provided by host governments and reimbursement for services provided by the Program.[6]

1. Charter of the United Nations, Art. 18, Sec. 1 & 2, and Art. 18. Available at <http://www.un.org/aboutun/charter> (accessed July 19, 2010).
2. Ibid.
3. UN (n.d.), Rules of the Governing Council. Available at <http://www.unep.org/Documents/Default.asp?DocumentID=77&ArticleID=1157> (accessed July 19, 2010).
4. Financing of UNEP: Regular Budget (available at <http://www.unep.org/rms/en/Financing_of_UNEP/Regular_Budget/index.asp> [accessed July 19, 2010]) and Environment Fund (available at <http://www.unep.org/rms/en/Financing_of_UNEP/Environment_Fund/index.asp> [accessed July 19, 2010]).
5. UNDP and UNFPA 2002, Executive Board of UNDP and UNFPA. Available at <http://www.undp.org/execbrd/pdf/eb-overview.PDF> (accessed July 19, 2010).

Table 3.1
(continued)

Organization	Voting rules	Revenue source
World Bank	Voting power for members includes "membership votes" (identical for all members) plus additional votes based on shares of Bank stock held. Voting power equals a member's votes expressed as a percentage of the total number of votes held by all shareholders.[7]	Twenty percent of member subscriptions ("subscribed shares") of authorized capital stock in the Bank; guarantee commissions and interest on loans made by the Bank; buying and selling of securities in which the bank has invested (including borrowing the currency of any member).[8]
Global Environment Facility	Council consists of 32 member "constituencies"; 16 are represented by developing country members (6 each from Africa and Asia; 4 from Latin America), 14 by developed countries, and 2 by nations with transitional economies. Decisions are made by consensus; when this is not possible, voting is done by "double-majority" (a majority of participating countries plus 60 percent donor support).[9]	Funds from 32 donor nations are committed every 4 years in the GEF Replenishment process.[10]
International Monetary Fund	Voting power determined by quota, based on relative size in the world economy.[11]	Quota subscriptions from each member country, assigned based on its relative size in the world economy.[12]

6. UNDP budget estimates for the biennium 2004–2005. Available at <http://www.undp.org/execbrd/pdf/dp03-28e.pdf> (accessed July 19, 2010).
7. World Bank Board of Directors: Voting Powers. Available at <http://web.worldbank.org>.
8. IBRD Art. IV, Sections 4 and 5, and Art. II, Section 5. Available at <http://web.worldbank.org> (accessed July 19, 2010).
9. About GEF (available at <http://www.thegef.org/gef/whatisgef> [accessed July 19, 2010]), especially Council (available at <http://www.thegef.org/gef/council> [accessed July 19, 2010]).

Table 3.1
(continued)

Organization	Voting rules	Revenue source
World Trade Organization	Follows GATT (1947) consensus model: decisions are adopted if no member present formally objects. When voting is required, each member has one vote.[13] Common practice is the Green Room meeting, in which the chairperson of a negotiating group consults privately with delegations, either individually or in groups, in an attempt to forge a compromise.[14]	Member contributions, established according to a formula based on their share of international trade; management of trust funds contributed by members; rental fees and sales of WTO print and electronic publications.[15]

10. GEF Replenishment. Available at <http://www.thegef.org/gef/replenishment> (accessed July 19, 2010).
11. IMF Quotas: A Factsheet. Available at <http://www.imf.org/external/np/exr/facts/quotas.htm>.
12. Ibid.
13. Understanding the WTO: The Organization. Whose WTO Is It Anyway? Available at <http://www.wto.org/english/thewto_e/whatis_e/tif_e/org1_e.htm> (accessed July 19, 2010).
14. Ibid.
15. WTO 2003, Secretariat Budget for 2003. Available at <http://www.wto.org/english/thewto_e/secre_e/budget03_e.htm> (accessed July 19, 2010).

law). It is exceedingly difficult to design effective agreements and enforce compliance.[82] It is also becoming increasingly tricky to coordinate the plethora of sometimes overlapping institutional arrangements for global environmental management.[83] It is, moreover, often unclear that full compliance with a treaty will genuinely improve environmental conditions (a point commonly made about the Kyoto Protocol).

Nongovernmental Organizations and Activist Groups

We define NGOs as organizations that are autonomous from governments, and are generally nonprofit (thus excluding corporations), though the lines are not always clear. International NGOs can include a wide variety of types of groups, such as large transnational organizations (like Greenpeace or Friends of the Earth International), regional groups (like

Table 3.2
Chronology of international environmental cooperation (summary of major initiatives)

1954	International Convention for the Prevention of Pollution of the Sea by Oil is adopted. 71 parties; entered into force in 1958. Replaced my MARPOL Annex I.
1972	United Nations Conference on the Human Environment is convened in Stockholm. Stockholm Declaration is adopted.
1972	Convention on the Prevention of Marine Pollution by Dumping of Waste and Other Matter (London Dumping Convention) is adopted. 86 parties; entered into force in 1975.
1973	International Convention for the Prevention of Pollution from Ships (MARPOL) is adopted. 150 parties; entered into force in 1983.
1973	Convention on International Trade in Endangered Species of Wild Fauna and Flora (CITES) is adopted. 175 parties; entered into force in 1975.
1973	Creation of the United Nations Environment Programme (UNEP).
1979	Convention on Long Range Transboundary Air Pollution is adopted. 51 parties; entered into force in 1983.
1979	Convention on the Conservation of Migratory Species of Wild Animals is adopted. 112 parties; entered into force in 1983.
1982	United Nations Convention on the Law of the Sea is adopted. 159 parties; entered into force in 1994.
1985	Vienna Convention for the Protection of the Ozone Layer is adopted. 196 parties; entered into force in 1988.
1987	Montreal Protocol on Substances that Deplete the Ozone Layer is adopted. 196 parties; entered into force in 1989.
1987	The Brundtland Report is published.
1989	The Basel Convention on the Control of Transboundary Movements of Hazardous Wastes and their Disposal is adopted. 172 parties; entered into force in 1992.
1991	Establishment of the Global Environment Facility (GEF). 178 member countries.
1992	United Nations Conference on Environment and Development is convened in Rio de Janeiro. Agenda 21 and Rio Declaration are adopted.
1992	United Nations Framework Convention on Climate Change is adopted. 196 parties; entered into force in 1994.
1992	United Nations Convention on Biological Diversity is adopted. 191 parties; entered into force in 1993.

Table 3.2
(continued)

1992	Creation of the United Nations Commission on Sustainable Development (CSD). 53 members.
1994	International Convention to Combat Desertification is adopted. 193 parties; entered into force in 1996.
1997	Kyoto Protocol on Climate Change is adopted. 188 parties; entered into force in 2005.
1998	The Rotterdam Convention on the Prior Informed Consent procedure for Certain Hazardous Chemicals and Pesticides in International Trade is adopted. 130 parties; entered into force in 2004.
2000	Cartagena Protocol on Biosafety is adopted. 156 parties; entered into force in 2003.
2001	Stockholm Convention for the Elimination of the Persistent Organic Pollutants (POPs) is adopted. 167 parties; entered into force in 2004.
2002	World Summit on Sustainable Development held in Johannesburg. Johannesburg Declaration and Plan of Implementation are adopted.
2009	COP 15 UN Climate Change Conference in Copenhagen is held. The nonbinding Copenhagen Accord is noted by the Conference of Parties.

Note: Party counts updated as of October 2009.

the Third World Network or PANNA, the Pesticide Action Network North America), and local and grassroots groups and international networking groups that facilitate coordination of such groups (like the International Rivers Network and the many local NGOs around the world networked with it), as well as research institutes (like the International Institute for Sustainable Development, or the Worldwatch Institute) and private foundations (such as the Bill and Melinda Gates Foundation). Some of these groups have significant budgets. In 2010, the endowment assets for charitable activities of the Bill and Melinda Gates Foundation was more than US$35 billion. The budget of Greenpeace (including national offices) was close to US$150 million in 2001. The budget of the WWF (including national organizations) in 2000–2001 was around US$350 million. This compared favorably to UNEP's budget in 2001 of about US$120 million.[84]

As with international environmental agreements, the number of international NGOs has increased significantly over the past few decades. In 1990, there were 6,126 NGOs with some international characteristics. This number jumped to 23,135 in 1996 and to 51,509 in 2005.[85] To some extent, this explosion of participation of civil society—on issues ranging from protection of the environment, to the rights of minorities and indigenous peoples, to the fair treatment of women—is evidence of a growing skepticism toward the efficacy of states in the current global political economy. Environmental NGOs increasingly participate in the negotiation processes for international environmental treaties. Some have gained UN accreditation through the Economic and Social Council, but others are let in as observers (just about anyone who asks can usually attend). This participation of NGOs in the UN treaty process goes back to the formation of the UN—NGOs were recognized in 1948 as important actors that states could officially consult. NGOs cannot vote in formal UN settings or in environmental treaty negotiations/COPs. But sometimes they are included on national delegations. And sometimes they are able to address the plenary sessions, as well as most of the smaller negotiating meetings and workshops, where state officials generally hammer out details.

International environmental NGOs can have particular advantages that help them influence global agreements.[86] They can develop specialized knowledge and expertise on specific issues, which they can then use to challenge mainstream scientific assumptions. Many state delegates do not have particular expertise, and in such cases NGO interpretations can be very important in shaping states' positions. Accreditation also allows some international environmental NGOs to gain direct access to global decision-making processes. In addition, these NGOs sometimes form alliances with governments (especially those of developing countries), and help them with scientific data and negotiations. For example, in the hazardous-waste-trade negotiations, Greenpeace formed a strong alliance with Group of 77 (G-77) countries to push for a waste-trade ban.[87] In fulfilling these roles, environmental NGOs enhance the transparency of dominant actors—that is, the states, international organizations, and corporations—although it is important to point out that NGOs can also suffer from bias and opaque procedures. In negotiations, wealthier northern NGOs may also dominate NGOs from the South. Other actors can co-opt NGOs at times, and their presence at negotiations may simply "legitimate" a state-based treaty process.

Another important role of environmental NGOs is to educate and raise global environmental consciousness, even transforming global culture.[88] Daring stunts can capture global media attention and help make complex scientific issues accessible to ordinary people. Greenpeace was able to change the prevailing image of whaling from that of a heroic battle to a perception of the slaughter of the innocent.[89] NGOs can influence other nonstate actors, such as corporations. They can empower and help alleviate poverty in local communities in the South. And they can help translate local concerns into the global language of treaties and financial assistance.[90]

Nongovernmental participation in shaping global environmental politics also encompasses efforts by groups and movements that may not focus exclusively on the environment, but that do have interest in particular environmental issues in their broader agenda. These include many indigenous peoples' organizations, which actively participate in global negotiations and dialog on environmental issues that affect them. The Inuit Circumpolar Conference, for example, was instrumental in efforts to bring about a global treaty on POPs (the Stockholm Convention on Persistent Organic Pollutants).[91]

The emergence of Green parties has also influenced environmental governance nationally as well as globally. The West German Green Party, as mentioned earlier, was a national political party that aimed to further goals of a number of social movements, including obviously the environment, but also civil rights, peace, and women's rights. The success of this party in the 1980s inspired the development of national Green parties first in other European countries and then throughout the world. Though they are national political parties, they are linked through an international network with a Global Green Charter.[92]

Corporations and Private Governance Mechanisms

Firms have also increased their role and influence in the global environmental arena, in part as a response to the enhanced role of environmental NGOs. Both business advocacy groups, such as the International Chamber of Commerce (ICC) and the World Business Council for Sustainable Development, and individual transnational corporations (TNCs), like DuPont and Monsanto, have been actively engaged in global discussions on the environment and development.

Like environmental NGOs, corporate actors also play a diplomatic role in an effort to enhance interstate cooperation on environmental issues. Lobbying before state delegations head to international environmental negotiations has enabled firms to influence governmental positions from behind the scenes.[93] Business actors have also increasingly begun to lobby at the international level as well, through industry associations and industry representatives with observer status at global environmental meetings. The presence of these actors at global negotiations is now routine—for example, during negotiations on the waste trade, climate change, ozone depletion, biosafety, and persistent organic pollutants. Business actors make fewer public interventions than environmental NGOs do (e.g., in plenary sessions and smaller meetings), but they are active in the corridors, lobbying and shaping the positions of states. Business representatives also end up on national delegations. This trend can be seen in both positive and negative lights. Market liberals and institutionalists argue that business groups should take part in negotiating treaties that will affect them directly, because this will ensure greater compliance later. Critics, in particular social greens, argue that the increasing influence of corporate environmental lobbyists means that international environmental law favors economic goals over environmental ones.

Corporate actors also influence global environmental governance through other international forums as well. Firms have helped shape private forms of global environmental governance, such as industry-set voluntary codes of environmental conduct as well as market-based governance mechanisms such as carbon trading at both the national and international levels. TNCs also exert influence in more diffuse ways, such as through their role in the global discourse over defining key terms, such as sustainable development, and through their economic weight, which influences state decisions in an indirect way.[94] We expand on these key themes in more detail in chapters 6 and 7.

Multilevel and Multiscale Actors

Until recently, most of the scholarship on global environmental governance focused on NGOs, business actors, states, and institutions working internationally. There is now growing recognition of the important role played by subnational actors, particularly cities and regional governments, as well as individuals, who are increasingly networking across borders at multiple scales to implement more *local-level* legislation and initiatives that work toward meeting *global* environmental goals.[95] Actors at the subnational level are not merely responding passively to international environmental norms and agreements. Increasingly, they are the leaders and innovators in environmental governance. Such activity at the subnational level has been especially apparent on issues such as climate change, where networks of cities, partly in response to weak efforts among national governments, have emerged as leaders in promoting reductions of carbon emissions.[96] Similarly, efforts to address problems with a global reach but without an international agreement, such as growing volumes of plastic waste in the world's oceans, have seen local governments around the world banning and legislating to reduce the use of plastic shopping bags.[97]

In addition to cities and regions, informal networks of individuals are also becoming increasingly influential on the global stage. In some cases, these networks form as like-minded individuals lobby for change (such as fewer automobiles), different or new consumer products (such as organic food), or new lifestyles (such as voluntary simplicity).[98] The rapid spread of the Internet over the last decade, which provides a cheap, quick, easy, and decentralized tool for people from different backgrounds and places to communicate for collective action, has allowed a vast array of direct-action and social networks to deepen and spread globally. The Internet is also enabling larger numbers of individuals to communicate about—and recognize—the global environmental consequences of daily choices and activities.

Sometimes these informal networks can combine with more formal processes—such as people wanting to consume more responsibly with ecolabeling or Fair Trade programs—to increase global reach and influence.[99] Other times these networks of individuals remain spontaneous and organic. One example, among many, is the Critical Mass global cycling movement. Emerging in San Francisco in the early 1990s, this movement is leaderless, decentralized, and memberless. Its only formal structure is to arrange for a time (often the last Friday of every month) and a place for cyclists to gather for a noncompetitive and festive protest ride through a city. Generally, these rides occur worldwide without any coordination, although sometimes they correspond with another environmental event, such as Earth Day on April 22; often, although not always, the group follows a spontaneous route. The idea is to bring together enough riders for a "critical mass" able to disrupt traffic and highlight the dangers of cycling in cities designed for cars and trucks. Direct-action movements like Critical Mass also weave into a tapestry of other movements. One example is Reclaim the Streets, a nonviolent grassroots protest movement to oppose corporate capitalism (especially automobiles) that started in London in 1991 and has now spread worldwide, from the United States to Australia, Canada, Finland, and Mexico.[100] Another example is the "guerrilla gardening" movement, in which individuals, sometimes as part of Reclaim the Streets, illegally plant seeds or seedlings in urban spaces, such as on traffic medians, back alleys, vacant lots, and commercial land. Although difficult to track and not a "global actor" in a traditional sense, these multilevel and multiscale networks are becoming increasingly important to the fabric of environmental governance, influencing policy choices, NGO strategies, consumption patterns, and corporate decision making.

Conclusion

The globalization of environmentalism should be seen as a complex and fluid process with no neat beginning, middle, or end point. The emergence of ideas regarding the protection of the "global" environment date back hundreds of years. In the past fifty years, groundbreaking publications like *Silent Spring*, *The Population Bomb*, *The Limits to Growth*, *Small Is Beautiful*, the *Founex Report*, and the Brundtland Report were undeniably important in shaping modern conceptions of global environmentalism. So were the watershed global conferences at Stockholm, Rio, and Johannesburg. So were the negotiations leading up to and occurring during issue-specific meetings and conferences on hundreds of international environmental issues, from ozone depletion to hazardous waste to desertification to climate change (e.g., the UN Conferences on Climate Change, most recently in Copenhagen in December 2009). And so were the emergence of local and global activist groups around the world. These developments all took place in a broader political and economic context. To fully understand the rise of sustainable development and the current structures and norms (i.e., established practices and understandings) of environmental governance, we have argued that it is equally critical to examine the broader trends in the global political economy.

Global economic shifts—in the generation and distribution of wealth—have been key in influencing the evolution of perspectives not just on the economy, but also on the relationship between the global economy and the environment.

Globalization and the global interaction of states, institutions, firms, markets, communities, and individuals will continue to shape and reshape global environmental governance for decades to come. As in the past, the arguments of market liberals, institutionalists, bioenvironmentalists, and social greens will continue to swirl around each other. The result of the globalization of environmentalism has to some extent split the current international debate over how to best manage the global environment into two broad coalitions: on the one side, market liberals and institutionalists calling for more effective implementation of sustainable development; on the other side, bioenvironmentalists and social greens calling for a new global political economy built on the principle of ecological and social sustainability. For now, however, the dominance of the "Brundtland compromise" of "sustainable development" gives institutionalists and market liberals the upper hand in the global community. Many NGO and activist groups that draw more from social green and bioenvironmentalist thought will no doubt continue to be highly critical of this compromise, especially the assumption that economic growth can (and should) continue indefinitely. It is important, however, to avoid oversimplifying the cases for and against sustainable development, because there are distinct differences in assumptions, arguments, and solutions across the four worldviews. The next chapter explores in detail the different assumptions and arguments about the global ecological consequences of economic growth.

4

Economic Growth in a World of Wealth and Poverty

Does wealth lead to better or worse environmental conditions? Does poverty lead to ecological neglect as people struggle to survive, or do poor people place less of a burden on the environment than the rich? There are no easy answers to these questions. They are intensely debated, and the conclusions one reaches depend very much on one's worldview. The question of economic growth is at the core of most of the debates over the global economy and the environment because it is intricately tied to debates over wealth and poverty and their relationship to the quality of the natural environment. This chapter examines these debates in detail, thus providing a foundation for understanding the debates about the other global economic issues we discuss later in this book, such as trade, investment, and aid.

Debates over growth at first glance appear to be polarized into two camps, much like the debate on globalization: those who see growth as a positive force for the environment, primarily market liberals and institutionalists, and those who view growth as largely negative, primarily bioenvironmentalists and social greens. But when we look carefully at how each group examines the questions of poverty, wealth, and population in relation to one another, and in relation to economic growth, we see that the debate is more complex than a simple two-camp split.

Wealth and Poverty for Market Liberals and Institutionalists

The rate of economic growth is a core measure of a country's economic performance. Typically it is measured by the rate of growth of the GNP or GDP (see box 4.1), which, with UN support, became standard measures for "development" in national income accounting after World War

Box 4.1
Measures of economic development

Gross National Product (GNP), Gross Domestic Product (GDP), and Gross National Income (GNI)

GNP and GDP are the most widely used indicators of economic growth. GNP measures the total monetary value of all goods and services produced by a country's resources, regardless of where production takes place. GDP measures the total monetary value of goods and services produced within a country's territory. So what is the difference between these measures?

To be more precise, GNP measures the value of goods and services produced by a country's citizens or nationals, regardless of whether they live in that country. The example of the accounting of a Japanese auto plant in the United States will help illustrate the differences between GDP and GNP. Presuming that most of the workers are American, U.S. GNP only includes the part of the value of production that accrued to American workers—that is, the employees' wages are part of GNP but the profits are not (they are counted in Japan's GNP). For GDP a transnational corporation's income is counted in the country where the income was generated. In this example, all of the value of the plant's production is included in the U.S. GDP and none is included in the Japanese GDP.

Both indicators give a rough idea of a country's economic activity. Nevertheless, over the past decade or so most countries have switched from GNP to GDP, partly because the UN has tried to standardize economic accounting across countries (the UN's System of National Accounts uses GDP). GDP per capita is now the most common measure of "development" or "progress," as this captures the amount of wealth available per person in a country.[1] The switch to GDP as a measure of national income did result in some major shifts in that TNC income is now wholly counted in the country in which the production takes place, raising the GDP of some countries that are host to a large number of TNCs.

Some organizations, such as the World Bank, have replaced the use of the term GNP with gross national income (GNI). GNI is the income of the sectors that produced the goods and services measured in GNP, and so by definition it is identical to GNP in value, though technically it is a measure of income rather than production.

Human Development Index

HDI is a measure of development that ranks countries according to their performance on indicators linked to longevity, knowledge, and income. Life expectancy at birth is the indicator used for measuring longevity. Educational attainment is the knowledge indicator, and includes a two-

Box 4.1
(continued)

thirds weight to the adult literacy ratio and a one-third weight to primary, secondary, and tertiary enrollment ratios. Income is measured in real GDP per capita. This is determined by counting income up to US$5,000 at full value, and after $5,000 its value diminishes in the index. In this way, the GDP per capita is adjusted for its real value (i.e., what it can purchase) rather than being assessed at official exchange rates. HDI then measures the human condition in a broader sense than does income alone. It is widely accepted among development thinkers and practitioners as a valid measure of development, even though GNP/GDP measures are still predominant.

Note

1. We are grateful to Torben Drewes for his comments on this section.

II. A rise in GNP or GDP implies growth of industrial production and, in turn, growth of consumption. This in effect means that more consumption equals development, as more needs and wants are met, which presumably betters the lives of ordinary people. The loose correlation between high economic growth and indicators such as life expectancy, health, and literacy adds statistical weight to this view. Global economic institutions such as the World Bank and International Monetary Fund, both of which tend to hold market liberal outlooks on environmental issues, rely heavily on GNP and GDP figures to rank countries' economic performance.

Assessed with these measures, the global economy has performed well over the past 50 years. Global GDP has jumped from US$7.2 trillion in 1960 to US$40 trillion in 2008 (in constant 2000 dollars; see figure 4.1). Global GDP per capita has grown from approximately US$2,400 to US$6000 in that same period (see figure 4.2). These global figures, however, hide some noteworthy variations across regions. GDP in East Asia and the Pacific grew from US$128 billion in 1960 to US$3.1 trillion in 2007 (with average annual growth rates well over 5 percent), while Latin America's GDP grew from US$440 billion to US$2.6 trillion over that same period. Meanwhile, sub-Saharan Africa went from only US$97

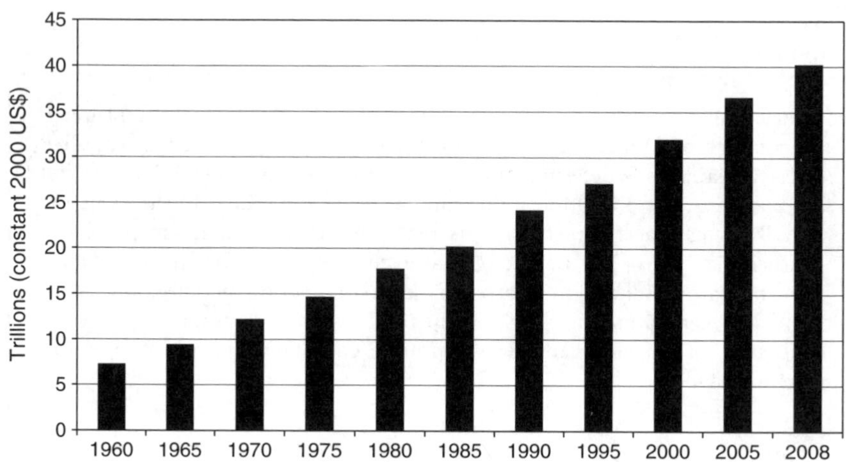

Figure 4.1
Global GDP, 1960–2008. *Source*: World Bank World Development Indicators, <http://data.worldbank.org>.

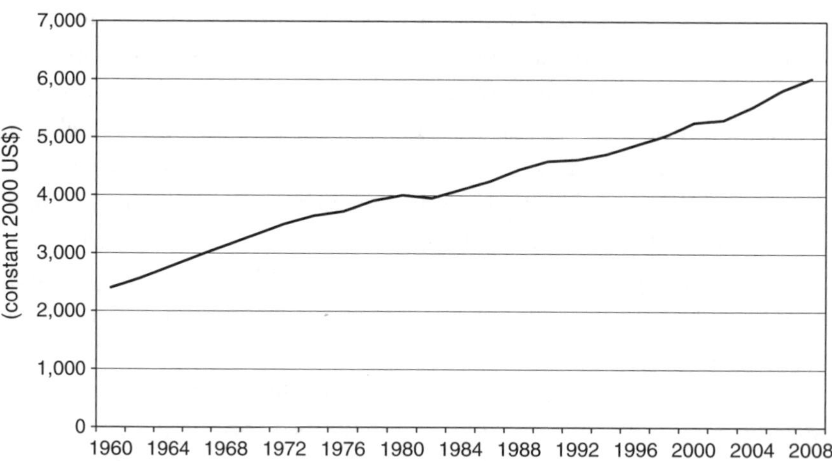

Figure 4.2
Global GDP per capita, 1960–2008. *Source*: World Bank World Development Indicators, <http://data.worldbank.org>.

billion to US$480 billion, and South Asia from US$105 billion to US$985 billion (in constant 2000 U.S. dollars).[1]

These growth figures by region mask some of the large gains in individual countries over this period. Rapidly industrializing countries—notably India, China, and Brazil—have posted much higher rates of economic growth in recent years. Some institutionalists have tried to push the global community to recognize the limitations of relying on total global GDP and global GDP per capita as the sole measures of development. Since 1990 the United Nations Development Programme's Human Development Index (HDI) has challenged the simple GDP measures, trying to create a more holistic assessment of economic performance and development. The *Human Development Report* explains the idea of human development:

> Human development is about much more than the rise or fall of national incomes. It is about creating an environment in which people can develop their full potential and lead productive, creative lives in accord with their needs and interests. People are the real wealth of nations. Development is thus about expanding the choices people have to lead lives that they value. And it is thus about much more than economic growth, which is only a means—if a very important one—of enlarging people's choices.[2]

The HDI ranks countries based on their performance on a range of indicators: life expectancy at birth, educational attainment (including literacy rate), and GDP per capita (see box 4.1). The HDI rankings are different from simple GDP per capita rankings (see figures 4.3 and 4.4), though the differences are not dramatic. The media reports these rankings widely and countries such as Iceland, Norway, Canada, and Australia have prided themselves on consistently high rankings. The HDI rankings have helped expand the mainstream analysis of economic performance somewhat, and more than 600 national, regional, and subnational human development reports have now been produced in more than 140 countries.[3] Nevertheless, GNP and GDP figures still tend to dominate national and global economic policy making.

Mild differences exist, then, between market liberals and institutionalists on how best to measure development. But both draw on neoclassical economics to address questions of the implications of economic growth. For neoclassical economists, growth within an economy occurs through the functioning of the market. The economy is seen as a circular flow system. That is, there is a closed loop between firms and households in the wider macroeconomy. Firms produce goods and services, and house-

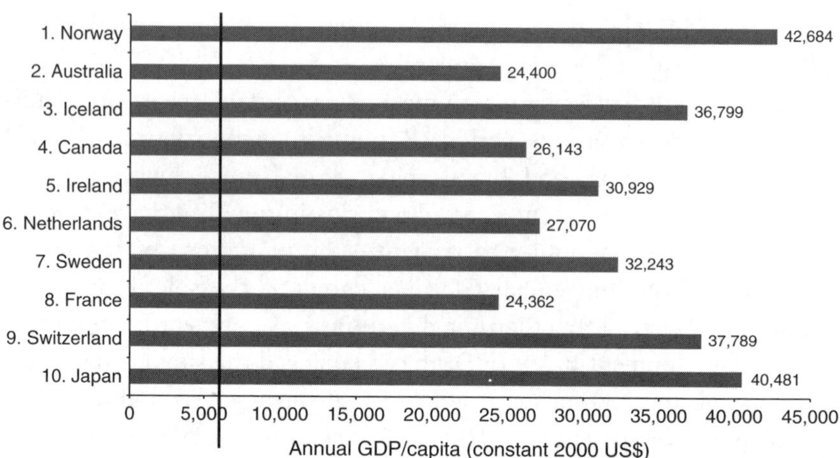

Figure 4.3
HDI and GDP, 2008, top ten HDI-ranked nations. *Source*: 2009 UNDP Statistical Update, <http://hdr.undp.org/en/statistics/>, and World Bank World Development Indicators, <http://data.worldbank.org>.

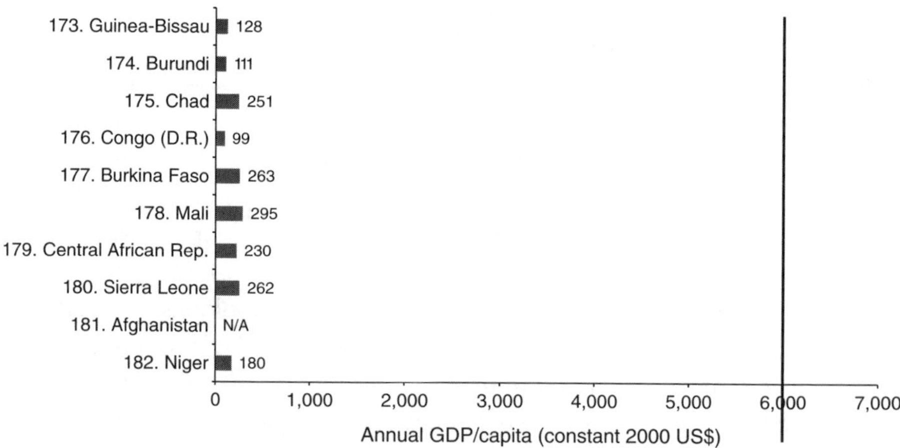

Figure 4.4
HDI and GDP, 2008, bottom ten HDI-ranked nations. *Source*: 2009 UNDP Statistical Update: <http://hdr.undp.org/en/statistics/>, and World Bank World Development Indicators, <http://data.worldbank.org>.

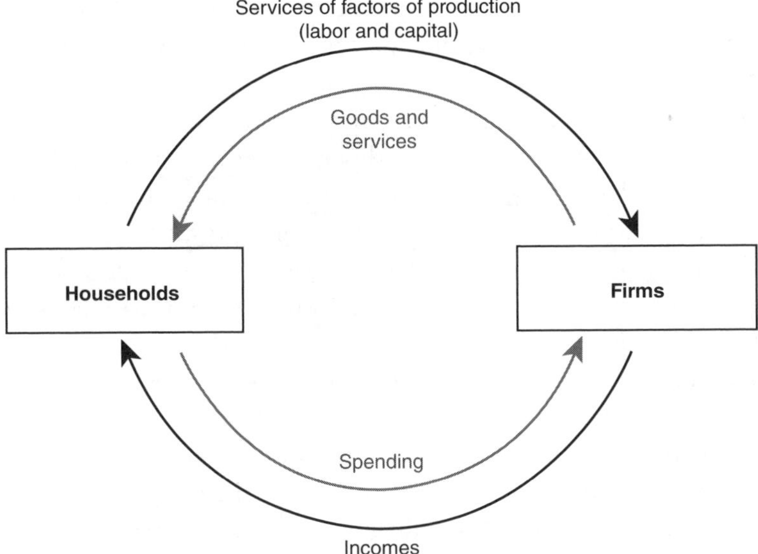

Figure 4.5
The economy as a circular flow system. *Source*: Adapted from Dornbusch, Fischer, and Sparks 1993, 28.

holds (consisting of consumers) consume them. These same households in turn make the means of production (labor in the form of workers and capital in the form of investment) available to the firms again. The firms continue the flow through the circle by producing more goods and services, and the flow moves forward indefinitely. The macroeconomy is thus seen as self-sustaining. Economic growth occurs as more goods and services are produced. In this way, the overall economy is thought to be capable of growing indefinitely (see figure 4.5).

How does the natural environment fit into this circular flow system that generates growth? The macroeconomy is seen as an all-encompassing system, within which everything else is a part, including the ecosystem. This model does not factor in the natural environment as part of the circular flow between firms and households. Rather it assumes that the natural environment is static and infinitely available. Resources are thus not a crucial factor of production, because growth is seen to occur primarily as a result of inputs of capital and labor.

Neoclassical economics primarily views environmental problems as arising from a type of market failure known as *negative externalities*.

"Market failure" broadly refers to specific instances in which markets on their own do not allocate resources efficiently. A negative externality occurs when the market does not account for the impact of an economic activity on those not directly engaged in that activity. Pollution is widely regarded as a negative externality. Although cleaner production processes may be available, firms may choose the less clean methods because they are cheaper to adopt. When this occurs, it represents a failure of the market to operate efficiently, because all of the costs of production (i.e., the cost of production plus the costs of pollution) are not fully accounted for. In this case, the health and cleanup costs of the pollution are "externalized" because they are not borne by the firm, but by governments and the general public.

Often negative externalities occur because environmental benefits and costs do not receive a sufficient monetary value in the marketplace. The ecological service from trees left standing in a forest is not typically given a market price, and thus when trees are cut down, the losses to the environment are not counted as a "cost" in the transaction. Similarly, no price is automatically assigned to a ton of CO_2 emitted into the atmosphere, making it relatively cost-free for firms to continue to emit carbon. In a free-market system, most neoclassical economists see negative externalities as relatively rare, and in cases where they do occur, they are assumed to be reversible with proper regulation, taxes, or other economic tools. This view of the natural environment in neoclassical economics is not surprising, given that the field emerged in the nineteenth century—a time of frontier economies, in which natural resources and sinks for pollution seemed limitless.

Availability of Natural Resources and Pollution Sinks

Today, market liberals and institutionalists still largely accept the foundational principles of neoclassical economics. Yet over the past forty years, many have modified the parameters, seeing pollution as a serious problem, and accepting that certain natural resources are finite. This challenges the neoclassical assumption that environmental harm is a rare market failure. Economic policies must therefore account for externalities to ensure effective environmental management, particularly for problems like air pollution, climate change, and deforestation. These thinkers note, for example, that firms commonly pollute the atmosphere and emit

CO_2 without penalty or full cost (individuals, governments, and the global commons often absorb the cost). They point out, too, that firms regularly obtain logging licenses with only a token charge for the environmental or social value of the original forest ecosystem (which explains why tropical wood is so inexpensive, despite the threat deforestation poses to climate change and much of the globe's biodiversity).

The subfield of environmental economics, which emerged in the 1970s, specifically seeks to address market failures and negative externalities related to natural resources and pollution. It works to price natural resources and sinks more accurately to avoid market failures, and to bring them more effectively into the circular flow system.[4] Environmental economists seek to correct environmental externalities through proper tax levels as well as other economic incentives, such as user fees, tradable permits, and payments for ecological services. Taxes, for example, could be levied against polluters, such as a tax per ton of carbon emitted into the atmosphere, as a disincentive to pollute. User fees, such as charges for garbage collection or for access to protected areas, can provide the state with revenues to address environmental problems. Payments could also be made to farmers who manage their lands in ways that provide ecological services, such as sinking carbon and filtering water. And a limited number of pollution permits could be issued that firms could then trade among themselves. Environmental economics assumes that market failures are a key source of unsustainable development, yet importantly these failures are correctable with such tools.

The Stern Review on the economics of climate change is a good example of environmental economics at work in the policy world. Published in 2006, the seven-hundred-page report, written by London School of Economics economist Nicholas Stern and a team of researchers, maps out the impact of climate change on the global economy and analyzes various economic policy measures that could be used to reduce carbon emissions. The key finding of the report is that climate change is an enormous market failure that will have massive economic impact, particularly for the world's poorest and most vulnerable countries. The report argues that an investment of approximately 1 percent of current GDP per year could avoid the worst impacts of climate change, and that the cost of not doing this now risks a future drop in GDP of up to 20 percent. In particular, it recommends a predictable and meaningful price on carbon as a way to encourage investment in carbon reduction.[5] The report was both celebrated and criticized upon its release.[6] These mixed reviews of the report demonstrate the controversial nature not only of climate change, but also the use of economic methods to evaluate likely costs of action and inaction.

Although environmental economic theory has contributed to more elaborate models to account for environmental change and scarcity, many of the original tenets of neoclassical economics remain. It is still widely assumed, for example, that as resources and sinks become genuinely scarce, prices for them will rise, which in turn will foster the ingenuity to find alternatives. This could involve new resources to replace the scarce ones, producing goods with less resources or waste, or discovering new ways to manage pollution, including the use of taxes and other market mechanisms. This view still places great faith in the ability of humans to use existing resources more efficiently (by improving technology), to manage existing resources effectively through environmental policy, or to discover "new" resources to drive economic growth. The late economist Julian Simon, challenging the bioenvironmentalist view of resource scarcity, argued that "the supply of natural resources is not finite in any economic sense, which is one reason why their cost can continue to fall indefinitely." For him, with current and future knowledge "we and our descendants can manipulate the elements in such fashion that we can have all the raw materials that we desire at prices ever smaller relative to other goods and to our total incomes."[7]

The depletion of natural resources and the filling of waste sinks may thus pose some environmental and economic problems for market liberals and institutionalists. Yet from their standpoint, with the aid of ingenuity and appropriate policies and institutions, these should not be acute constraints on economic growth and the accumulation of wealth. Market liberals place faith in individuals and firms to come up with solutions to scarcity and pollution. Institutionalists stress the need to foster such ingenuity through government and institutional efforts. The implication is the same for both, however: there is no need whatsoever to abandon the growth-based model of development. Growth can— and must—go on indefinitely for the sake of humanity as well as for the sake of the global environment. Growth just needs to be made more efficient and less polluting. Pricing resources and sinks properly can ensure that this occurs.

Growth Brings a Cleaner Environment: The Environmental Kuznets Curve

How, then, does steady economic growth ensure a cleaner future? The answer, say market liberals and institutionalists, is best illustrated with the *environmental Kuznets curve* (or EKC, also known as the inverted-U curve).[8] The EKC shows that economic growth will sometimes harm the environment in the short run. This occurs because of inefficiencies, inappropriate policies, insufficient funds, weak state capacity, and low political and societal will. Yet in the long run this growth will improve environmental conditions (once per capita income reaches a high enough level).

Research on the relationship between pollution intensity and per capita income by economists Gene Grossman and Alan Krueger and the World Bank in the early 1990s sparked a great deal of interest in the EKC.[9] This research shows that as a country develops and industrializes, emissions of certain pollutants—such as sulfur oxides, suspended particulates (smog), and lead—increase along with GNP growth. In other words, growth and pollution are *coupled*. Pollution rises mainly because of an emphasis on industrialization, growth, and income generation, at the expense of pollution controls. Yet when per capita income gets high enough (in the past, between US$5,000 and US$8,000), the link *decouples* as economic changes, societal pressure, and state capacity converge to raise environmental standards. At that point, emissions of these pollutants begin to fall (see figure 4.6 for an example of a hypothetical EKC).

A classic case of this relationship is Japan. After World War II, economic growth soared in Japan, resulting in an economic growth rate of 10.9 percent from 1950 to 1955, 8.7 percent from 1955 to 1960, 9.7 percent from 1960 to 1965, and 12.2 percent from 1965 to 1970.[10] Such rapid growth left Japan severely polluted by the 1960s. The notorious mercury poisoning case in the city of Minamata illustrates these appalling conditions. In the 1950s and 1960s, many people in and around Minamata contracted a mysterious neurological disease. The cause, scientists eventually discovered, was exposure to methylmercury compound, which the Chisso industrial plant had discharged in factory effluent upriver from Minamata. The mercury had worked its way up the food chain until humans ingested it in seafood. Public outcry over the Minamata disease contributed to rising citizen protests in Japan in the 1960s and 1970s over the environmental consequences of rapid

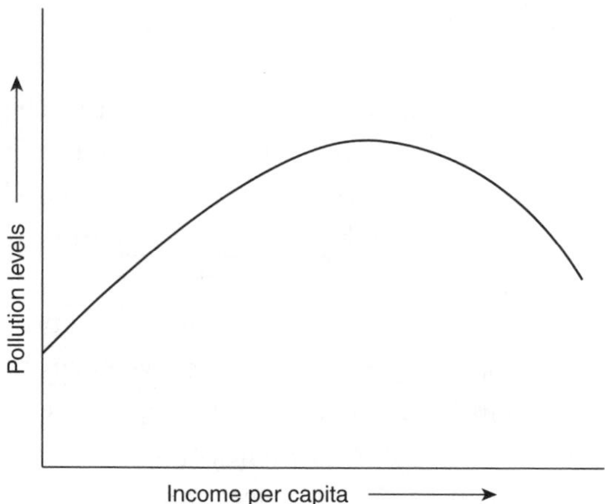

Figure 4.6
The environmental Kuznets curve

industrial growth. In a relatively short time, government and business cooperated to improve environmental policy and practice. By the end of the 1970s, Japan's environmental regulations had become among the strictest in the world, and the environment was already much cleaner.[11] Today, given Japan's population density and small land mass, it is arguably one of the world's best-managed environments.

Most of the research on the long-term correlation between economic growth and improvements to environmental quality has concentrated on pollutants. Some studies have found, in addition, that the EKC holds for certain natural resources—such as a study of deforestation and national income in Asia, Latin America, and Africa.[12] The World Bank has also determined that the proportion of the urban population without adequate sanitation and safe water supplies drops as income rises in a cross-section of countries.[13] The World Bank qualifies these findings by adding that with respect to some indicators—such as municipal waste and carbon dioxide (CO_2) emissions—problems rise along with income. Carbon emissions per capita, for example, are much higher in high-income countries than in low-income countries, as shown in figure 4.7. These caveats, however, have in no way undermined the belief among market liberals and many institutionalists that growth is good for overall environmental conditions.

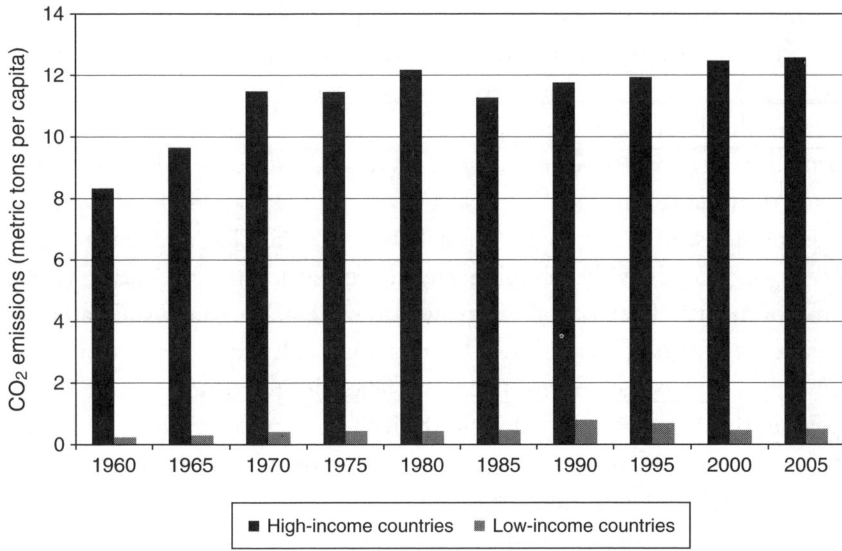

Figure 4.7
CO_2 emissions per capita (metric tons). *Source*: World Bank World Development Indicators, <http://data.worldbank.org>.

Why do pollution levels and other types of environmental problems begin to decline as income rises? Intuitively one might think that because production and consumption rise along with income, pollution levels should rise too. Yet significant counterforces account for the inverted-U relationship between economic growth and environmental degradation.[14] As an economy grows and people become richer, structural shifts can occur in the economy, such as a shift toward service- and information-based industries, which are less resource-intensive and less polluting. At the same time, societal concern over environmental issues (especially health problems) tends to rise as wealth brings education, communication technologies, and higher expectations. People begin to demand stronger government regulations to protect the environment, and are willing to pay the taxes to see these enforced. Markets and governments push firms to develop pollution-control technologies, thus reducing emissions. Consumer demand for "green" products also increases as disposable income rises.

Most market liberals and institutionalists who support the EKC hypothesis do not argue, however, that countries should blindly pursue economic growth without environmental policies and clean up only after such policies become affordable. The EKC simply points to the value of economic growth for long-term environmental management. Growth itself—although it can create problems at certain points along a country's growth trajectory—does not always have to do so, especially if environmental measures are put into place early on.[15] Policy makers aware of this relationship can strive to "tunnel" through the EKC and reach higher levels of growth without doing any more harm than necessary to the environment.[16] Then economic growth in later stages can sustain an even cleaner environment. Market liberals also advocate market-oriented measures to promote environmental efficiencies to reduce the size of the peak in the curve, in particular clarification of property rights and reduction of market distortions (discussed in more detail later in this chapter). Institutionalists focus much more on the need for international cooperation to help developing countries obtain necessary technologies to enable them to reduce the environmental impact of their economic growth. With such measures, growth can be turned into a positive force for the environment. Both market liberals and institutionalists argue, however, that slow growth or lack of growth can result in considerable environmental damage—possibly even more than rapid growth can.

Environmental Consequences of Poverty and Weak Growth

Slow growth, according to market liberals and institutionalists, does not allow for a future with better environmental conditions—just more of the same, or in some instances, cycles of increasingly weak economic growth and environmental collapses. In rich countries, it limits government and corporate investments in cleaner technologies. It is an even greater problem for developing countries because it keeps them on the wrong side of the critical point on the EKC where environmental improvements begin to occur naturally (and therefore efforts to reduce poverty tend to override environmental spending).

Slow growth reinforces the vicious cycle between poverty and environmental degradation, often the most serious barrier to achieving sustainability in the developing world.[17] Market liberals in particular focus on the environmental impacts of poverty. Institutionalists put more stress on how population growth interacts with poverty and environmental degradation (and call for better institutions to support family planning). The nexus between poverty, population growth, and environmental degradation produces a self-reinforcing downward spiral from which it is difficult to escape. In both versions of this vicious cycle, poor people are seen as agents as well as victims of environmental degradation. They often have no choice but to overuse the land and overharvest local resources for food and shelter. They must act to satisfy short-term needs even if doing so works against their long-term interests. Poor people, then, are forced into decisions with short time horizons. How can we expect them to preserve natural resources for future generations? Degradation of the land and other natural resources, regardless of the cause, also harms poor people's ability to escape from poverty. The Millennium Ecosystem Assessment puts it bluntly: "The degradation of ecosystem services is harming many of the world's poorest people and is sometimes the principal factor causing poverty."[18]

It is common for market liberals and institutionalists to link poverty with environmental degradation when analyzing rural and agricultural settlements in developing countries. The UNDP calculates that "at least 500 million of the world's poorest people live in ecologically marginal areas," and many poor people in the developing world farm ecologically fragile land.[19] Often, poor people have few options except to cultivate unsuitable land, which leads to soil erosion, land degradation, and deforestation. Growing populations in these areas reduce fallow periods as well, quickening the pace of soil degradation and deforestation.[20] The children of farmers then struggle to get increasing yields on ever-smaller patches of land. Sometimes the situation is made worse because powerful elites (perhaps descendants of colonizers) control the most productive land. Sometimes it is made worse as settlers leave overpopulated areas to occupy "empty" land (usually small numbers of indigenous peoples actually reside in these areas). Some governments have even promoted such disastrous internal migration. Indonesia under former President Suharto (1965/1967–1998), for example, implemented a "transmigration program" to move poor residents from the overcrowded island of Java to the outer islands (such as Kalimantan). Few of these people had the knowledge to farm these ecologically fragile outer islands. Many agricultural sites failed to produce adequate yields. The sites also upset the natural balance of local forest ecosystems, contributing to soil erosion as well as exposing and drying out large areas of peat moss. Appalling fires have swept through these areas (fueled in part by forests logged and degraded by firms linked to Suharto's family and cronies). The worst so far were in 1997–1998, when fires lit primarily by plantation firms to clear land cheaply swept out of control and scorched 5–10 million hectares of land (including 1 million hectares of peat). The peat fires alone released 150 million tons of carbon into the atmosphere.[21] Klaus Töpfer, then executive director of UNEP, described these fires "as one of the worst environmental disasters of the last decade of the century."[22]

Market liberals and institutionalists thus point to poor people as critical forces contributing to environmental problems. They also recognize, however, that without the means of avoiding the impacts, poor people suffer the most from environmental degradation. They are more vulnerable to bad water and poor sanitation. They often live closest to the sources of pollution in urban centers. And they rely on types of fuel, such as wood and charcoal, which produce poor air quality in their own homes. This all feeds back to make poor people even poorer as productivity declines—a situation that in turn further accelerates the vicious cycle between poverty and environmental degradation.[23] In this way, poverty and environmental degradation are mutually reinforcing, as depicted in figure 4.8. A World Bank report on economic adjustment in Africa illustrates this poverty-population-environment analysis:

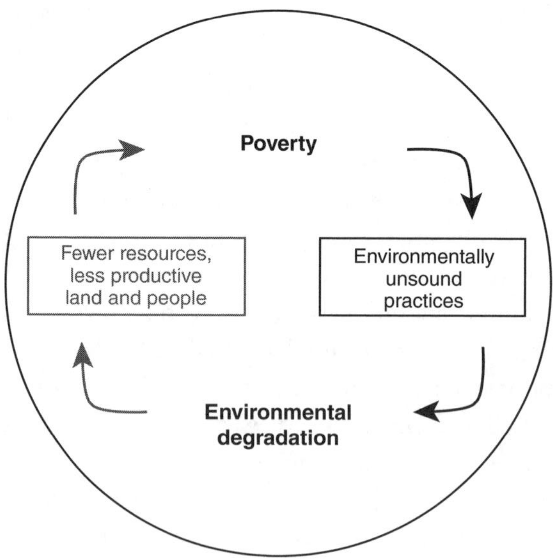

Figure 4.8
The vicious cycle of poverty and environmental degradation

> The poor are both victims and the agents of damage to the environment. Because the poor—especially poor women—tend to have access only to the more environmentally fragile resources, they often suffer high productivity declines because of soil degradation or the loss of tree cover. And because they are poor, they may have little recourse but to extract what they can from the resources available to them. The high fertility rates of poor households further strain the natural resource base.[24]

The poverty-environment spiral, market liberals and institutionalists argue, occurs for many reasons, and is highly complex and often context-specific. Market liberals focus on inadequate economic growth and insecure property rights as the key causes of poverty, while institutionalists tend to stress the role of weak government services. Weak economic growth in part arises because of insufficient domestic integration into global markets. Insecure property rights, especially in cases where arrangements for commonly-held property (common property regimes, or CPRs) break down into chaos and open access, create strong incentives for poor people to overexploit local resources. Insecure property rights also leave poor people without land title and therefore less able to obtain credit in times of need. Women in particular are often shut out of credit systems because of weak property rights in many poor countries. Inadequate government services, such as education, are also unevenly distributed among poor people, with women often being denied equal access—conditions that perpetuate unsustainable resource use.[25] Multilateral and bilateral aid programs to promote "women in development" are now common. An underlying purpose of these programs is generally to promote sustainable development.

Both market liberals and institutionalists emphasize the need to break the cycle between poverty and environmental degradation. For market liberals, this requires the integration of poor countries into the global marketplace—that is, countries need to liberalize trade, deregulate "overmanaged" industries, and improve the investment climate for transnational firms. To carry out these measures, governments must accept global standards (private property, customs, banking, taxation, environmental regulations, and so on) as well as eliminate corruption and ineptitude. Only then can these countries reduce poverty levels and move toward a future with better environmental conditions.[26] Institutionalists agree to some extent with this prescription, although they place more emphasis on the need to build the capacity of local and national governments to manage the integration into the global economy, as well as the need to incorporate gender measures, such as improving access to credit for women and making more effective family planning available. A major UN initiative on poverty and the environment, jointly run through UNDP and UNEP, promotes the mainstreaming of policies that seek to both reduce poverty and improve environmental conditions.[27]

Market Distortions, Inefficiencies, and Weak Institutions

Market liberals in particular see market distortions as a serious problem for the natural environment. Market distortions are perversions of free markets causing a social or environmental outcome that is suboptimal. These distortions create inefficient use of resources and sinks, with negative outcomes for both society and the environment.

Government subsidies to promote economic growth in certain sectors or to diversify the economy are one of the most common examples of market distortions. Governments may provide financial support (in the form of payments, tax breaks, and so on) to certain sectors as a way to encourage investment and improve their international competitiveness.[28] Many industrialized countries, for example, subsidize the mining and logging sectors by charging what many consider to be below-market prices for the rights to extract these resources. The United States and European Union (EU) countries also use a variety of support instruments to subsidize their agricultural sectors, such as direct payments, price supports, and cheap credit, to encourage food production and exports. Developing countries often subsidize certain sectors as well, most commonly as a strategy to move away from an economy with a heavy reliance on a few unprocessed natural resource exports (like bananas, gold, or logs) and toward one where domestic firms process these natural resources before export (which adds economic value and creates jobs). Energy subsidies are common in the developing world, too, as a means of promoting this kind of industrialization.

Market liberals argue that even though the ultimate goal of such subsidies may seem sound and sensible, they ultimately contribute to wasteful and inefficient domestic industries that overconsume local resources and pollute local ecosystems. Artificially low prices of timber, minerals, oil, or agricultural chemicals encourage overuse and waste, leading to environmental degradation and inefficient markets. Subsidies can also undercut incentives to upgrade environmental technologies. This is common, for example, with fuel subsidies that weaken incentives to use more fuel-efficient technologies. They can also lead to overproduction or diversion of resources from other uses, as happened as a result of the subsidization of corn-based ethanol in the United States. Increasing production of corn-based ethanol has been seen as a major factor in diverting corn from food markets, leading not only to dramatic increases to food prices in late 2007 and early 2008, but also to negative environmental impacts because of its inefficiency as an energy source and its intensive use of chemicals and land.[29] Although market liberals most often make the point that subsidies can lead to environmental harm, it should be noted that bioenvironmentalists, social greens, and institutionalists all tend to agree that governments should eliminate subsidies with negative environmental implications. Market liberals, though, are against most forms of subsidies on the grounds that they are inefficient, whereas the other groups are more open to them as tools—for example, to promote cleaner production or to pursue other goals, where environmental impacts can be controlled.

Both market liberals and institutionalists have begun to promote the notion of sustainable consumption to encapsulate the environmental drawbacks of inefficient production and consumption. This concept is often formulated so as to incorporate both the dematerialization of production and the optimization of consumption. Some refer to this as "ecoefficiency"[30] and "cleaner production," concepts that involve decoupling higher production and consumption levels from their negative impact on the natural environment. The idea is that such practices are not only good for the environment, but also more efficient and thus improve the profitability of firms as well. Business, here, is seen as an environmental leader, as the pursuit of profits becomes the pursuit of more efficient use of the environment. The notion of environmentally sustainable production and consumption seems to be gaining strength. The OECD, for example, now has a research program on Consumption, Production, and the Environment.[31] UNEP and the UN Department of Economic and Social Affairs are also leading efforts to negotiate a draft framework on sustainable consumption and production for the nineteenth session of the UN Commission on Sustainable Development in 2011. Known as the Marrakech Process, this effort was called for by the

2002 World Summit on Sustainable Development's Johannesburg Plan of Action. It is aiming to produce a 10-Year Framework of Programs with three primary goals: to support more sustainable business practices, to assist initiatives to green national economies, and to encourage consumers to act more sustainably.[32]

How, then, does the global community achieve more sustainable consumption? Market liberals argue for market-based measures that encourage cleaner technologies, such as sensible taxes, transparent and equitable user fees, and tradable permits. Also, governments need to eliminate subsidies that encourage individuals and firms to use outdated and unclean technologies. If such changes take place, market liberals argue, the free market is the best way to develop and disseminate green technologies, via global corporations.

Institutionalists argue that greater global cooperation—for example, global standardization of reasonable environmental laws, environmental certification programs, government support, and global environmental regimes—is necessary to assist with the spread of cleaner production and sustainable consumption. The global community needs to raise awareness so more individuals begin to demand greener products. International institutions such as UNEP can help promote more environmentally sound consumption. So can international agreements that help reduce the use and impact of certain industrial and natural resource inputs. Governments can provide financial support for research, development, and dissemination of green technologies. Finally, green aid can help transfer cleaner technologies to developing countries.[33]

Critiques: Bioenvironmentalists and Social Greens

Bioenvironmentalists and social greens are highly critical of mainstream neoclassical economics, particularly the emphasis on "economic growth" and GNP/GDP as measures of human well-being. Growth, in fact, is not a source of better environmental conditions, but a source of environmental degradation at all income levels. Academics known as *ecological economists* have developed some of the most compelling critiques. Many of these arguments began to emerge in the late 1960s and early 1970s, and were part of a broader questioning of the environmental implications of the global political economy (see chapter 3). Ecological economists,

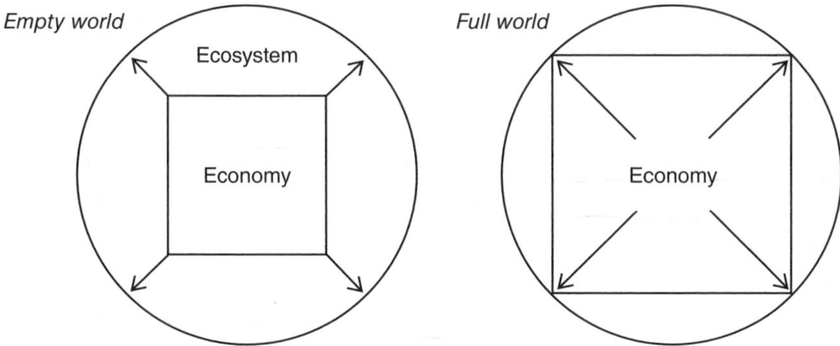

Figure 4.9
The economy as a subsystem of the ecosystem. *Source*: Adapted from Daly 1996, 29.

who tend to fall into the categories of either bioenvironmentalists or social greens, reject many of the assumptions and principles of the neoclassical economic model and propose an alternative model. Based on different assumptions, ecological economics is very different from the "environmental economics" discussed earlier, because the latter incorporates environmental concerns into what is more or less a neoclassical economic framework. Perhaps the most significant assumption of ecological economics is that the environment is not a subsystem of the macroeconomy, but rather that the economy is a subsystem of the natural environment or ecosystem (see figure 4.9).

Framing the economy-environment relationship in this way has profound implications for how ecological economists analyze the causes of, and solutions for, global environmental change. The model highlights the biological "limits" to economic growth—that is, the view that the earth is a finite resource. How soon will we reach these limits? For some ecological economists, we are already beyond them. For others, we are plummeting closer by the day. We now examine these arguments more closely.

Problems with GDP/GNP as a Measure of Development

Social greens and bioenvironmentalists both argue that GNP and GDP are generally poor measures of development and of the human condition. GNP and GDP count only things that go through the market—that is, goods or services that involve an exchange of money. A great deal of work that enhances the quality of life does not go through the market, however, and thus is not counted in GNP or GDP figures. This includes, for example, work that is done in the home, such as housekeeping and child care, and volunteer community work. Social greens in particular stress that much of this work is done by women, and that GNP and GDP figures seriously undervalue women's work.[34] The way national wealth is measured also undervalues subsistence farming, even though in some parts of the world this is the primary source of food. Linked to this critique is the fact that markets often ignore or underprice environmental goods and services, like the service of trees in absorbing carbon, fresh sources of water, or clean air. These either do not enter the calculation of GNP/GDP at accurate prices, or do not enter it at all.[35] Some argue that such services should be—indeed, must be—accurately priced to encourage the protection of these ecological services.[36]

At the same time, GNP and GDP do count things that we may consider "negatives" for the human condition, such as industrial pollution and resource extraction. Some argue that GNP and GDP should subtract such costs.[37] Similarly, these measures count spending on services to deal with human catastrophes,[38] like cleaning up after an environmental disaster (such as the Deepwater Horizon oil spill in 2010 or the toxic waste dumped illegally in Côte d'Ivoire in 2006) or terrorist attack (such as the World Trade Center attack on September 11, 2001), simply because these efforts do go through the market. GNP also tells us little about things that are not easy to quantify, such as the state of the natural environment or human rights. Nor does it tell us much about the distribution of wealth or income. Although most mainstream economists do admit that GNP is a simplified and far-from-perfect measure of human well-being, they nonetheless continue to use it.

UNDP's HDI tries to address some of these critiques by incorporating education and health indicators, but because the main HDI index does not include environmental quality and still includes economic growth as a major component, many social greens and bioenvironmentalists see it as an inadequate measure of development. They would like to see governments react to changes in ecological measures with the same vigor as with economic and social ones. Yet environmental policy making does not have a standard measure as widely recognized as GDP and GNP. Many scholars and research institutes have proposed ecological indicators in the last decade or so, but so far they have not gained significant ground among policymakers. (Chapter 8 discusses these indicators in more detail.)

Limited Natural Resources and Sinks

We noted previously that neoclassical economists do not see economies as part of ecosystems, but rather view ecosystems as part of the economy. For this reason, they do not see resources as being necessarily finite in an economic sense—a position reinforced by the steady decline in the prices of natural resources in real terms (box 4.2 summarizes the now-notorious bet between Julian Simon and Paul Ehrlich over the scarcity and prices of the globe's natural resources). These assumptions have been challenged from many angles, including by ecological economists who have drawn on the laws of thermodynamics (in particular, the "entropy law") to illuminate the link between economies and the ecosystems (see box 4.3 and figure 4.10).

These thinkers, many of whom closely adhere to the bioenvironmentalist perspective, use the laws of thermodynamics to show that the notion of perpetual growth, promoted by market liberals, is erroneous because unlimited economic growth is physically impossible. By linking economics to the physical sciences, these thinkers have opened an entirely new way of viewing economic processes. For them, the economy is bound by the energy and matter physically available to fuel the production of consumer goods, and by the Earth's ability to absorb the waste products associated with those goods. In other words, there is a physical basis for scarcity, not just an economic one.[39] They argue that unless we recognize the very tangible and real physical limits of nature and learn to live within them, we will overstep the ecological bounds of the planet. Some fear we are already beyond those bounds.[40]

Because the global political economy fails to account for these limits, it fails to send signals when resources are being used in an "unsustainable" manner. Thus, in direct opposition to Simon's argument discussed in the first part of this chapter, ecological economists do not consider prices to be a good determinant of the availability of resources.[41] For them, the discovery of new resources will not solve this inherent defect in the global political economy. All it does is to substitute the use of one limited resource for another. Terrestrial stocks of natural resources are

Box 4.2
The Simon-Ehrlich wager

In 1980 the economist Julian Simon and the ecologist Paul Ehrlich made a ten-year bet on the direction of natural resource prices. Ehrlich chose five metals—chromium, copper, nickel, tin, and tungsten—worth US$1,000 (US$200 of each metal). The wager was simple: If, after adjusting for inflation, the market price of these metals in 1990 was more than US$1,000, Ehrlich would win (because this would demonstrate growing scarcity). If the total was less than US$1,000, Simon would win, because this would confirm the economic infinity of the resources, as new substitutes and technologies kept prices low. The loser had to pay the winner the total difference in the 1990 inflation-adjusted prices. Ehrlich seemed confident at the beginning: "Simon is wrong about the economics of mineral resources. . . . The trough-like pattern long predicted for mineral resource prices has now shown up. . . . I and my colleagues, John P. Holdren . . . and John Harte [Berkeley physicists] jointly accept Simon's astonishing offer before other greedy people jump in."[1]

Simon turned out to be the more astute gambler, however, and by 1990 the market prices of all five metals were below the 1980 levels (in real terms). The price of the nickel and copper were only down to US$193 and US$163 respectively, but chrome had fallen to US$120, tungsten to US$86, and tin to US$56. Ehrlich and his colleagues paid Simon a sum of US$576.07. Reporter John Tierney asked Ehrlich, "Is there a lesson here for the future?"

"Absolutely not," Ehrlich answered. "I still think the price of those metals will go up eventually, but that's a minor point. The resource that worries me most is the declining capacity of our planet to buffer itself against human impacts. Look at the new problems that have come up, the ozone hole, acid rain, global warming. . . . If we get climate change and let the ecological systems keep running downhill, we could have a gigantic population crash."

When Tierney told Simon of Ehrlich's response, he was quick to retort: "So Ehrlich is talking about a population crash. That sounds like an even better way to make money. I'll give him heavy odds on that one."[2]

Notes

1. Ehrlich 1981, 46.
2. Tierney 1990. Also see Simon 1996; Ehrlich and Ehrlich 1996.

Box 4.3
Entropy and ecological economics

Pioneer ecological economist Nicholas Georgescu-Roegen first outlined the relevance of entropy to the field of economics.[1] Entropy is the amount of "used-up-ness" of energy or matter in an isolated system. High entropy means the energy or matter is not very useful, whereas low-entropy energy or matter is useful and can perform work. The second law of thermodynamics, also known as the law of entropy, states that the amount of usable energy and matter decreases in an isolated system as work is performed. In other words, when energy or matter is transformed to make a product, its entropy increases, making it less useful.

The earth provides us with some forms of low-entropy energy—that is, energy that is easily usable by humans to perform work. Ecological economists view the earth as a closed system (i.e., one that energy flows through while matter cycles within), and argue that the functioning of the current global economy neglects the implications of the second law of thermodynamics.[2] In this view of economics, nature provides the earth with a flow of low-entropy energy (in the form of solar energy) and a set stock of low-entropy nonrenewable energy (e.g., from terrestrial sources, such as fossil fuels). Modern economies tend to exhaust the terrestrial stock of low-entropy energy. When we use this natural resource–based energy to make products or burn it as fuel, referred to as "throughput," we end up with less useful, high-entropy energy, such as carbon dioxide, that is disbursed into the atmosphere.

Ecological economists like Herman Daly and Joshua Farley argue, following Georgescu-Roegen, that the more we use up low-entropy energy to support the functioning of our modern economy, the more high-entropy waste energy we are left with. The sun still provides a limited flow of low-entropy energy and will do so for the foreseeable future. Humans have more control over the flow rates of the use of low-entropy terrestrial stocks of energy. Daly argues that we are using these terrestrial stocks of low-entropy energy far too quickly, and that once these run out it will be extremely difficult to reorganize the economy to run on the available flow of solar energy. He uses a similar line of reasoning to argue that we are using our terrestrial stock of low-entropy matter (nonrenewable resources, such as minerals) and our flow of low-entropy matter (renewable resources, such as timber, that rely on the sun for replenishment) at a rate that is out of line with the earth's capacity to replace them.[3] In other words, there is too much physical "throughput" in the economy. The importance of entropy is illustrated in the entropy hourglass, illustrated in figure 4.10.

Notes

1. Georgescu-Roegen 1971.
2. Georgescu-Roegen 1993; Costanza et al. 1997, 56–59.
3. Daly 1996, 29–30, 193–98; Daly and Farley 2003.

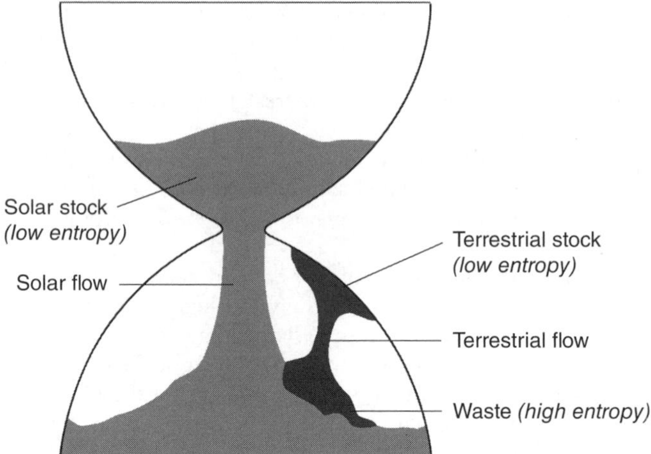

Figure 4.10
Entropy hourglass. *Source*: Georgescu-Roegen, as depicted in Daly 1996, 29.

still finite, and the flow of solar energy and renewable resources is limited at any given time as well. Such limits imply that market failures are a much greater problem than most economists acknowledge—that is, negative externalities that result in pollution and resource exhaustion are inherent in economic growth, and are not a rare occurrence that a few market-oriented mechanisms can fix.[42]

Market liberals and institutionalists have not spent a great deal of time directly engaging with these ideas. Ecological economist Herman Daly, whose work is devoted to promoting the idea of the physical limits of economic growth, worked as a senior economist in the World Bank's Environment Department from the late 1980s to the mid-1990s. Though the World Bank focused much more on environmental issues during his tenure (see chapter 7), Daly eventually left that institution, many speculating that in part it was because the World Bank was not taking his ideas seriously enough.[43]

Critique of the Environmental Kuznets Curve

Bioenvironmentalists and social greens are both highly critical of the notion that growth eventually improves environmental conditions, as illustrated by the EKC. Many of these thinkers have raised concerns with respect to the methodology used to determine the EKC.[44] Many argue, too, that the connection made between growth and pollution levels as a general indicator of environmental quality is too simplistic, and therefore that relying on the EKC to guide policy is inappropriate.

There are many specific criticisms. First, the research on environmental Kuznets curves applies to only a limited range of pollutants and resources. It does not apply, for example, to CO_2 emissions, which continue to rise with economic growth.[45] Moreover, it is insufficient to simply plot a decline in the emissions of specific pollutants as income rises. It is entirely possible that higher emissions of other pollutants in the same economy may simply be accounting for this decline. For instance, sulfur dioxide emissions may decline with a switch away from coal-fired power plants, but other types of emissions may rise, such as nuclear waste from nuclear-fueled plants. If the current population does not bear the environmental costs of such emissions, as is the case with nuclear waste, it is less likely that a rise in income will result in lower emissions of those pollutants.[46]

Similarly, a decline of particular pollutants in one country may occur because these have been shifted to other, often poorer countries. Japan, for example, partly in an effort to improve domestic environmental conditions, has transferred dirty industries, such as heavy machinery and chemicals, to developing countries in Asia and the Pacific.[47] Much of the initial data on the EKC was from OECD countries. It is not clear what would happen to the relationship between income and environmental quality if we took a global perspective. We do know, however, that as certain types of environmentally harmful production decline in rich countries, they have tended to increase in poor countries. Economic globalization, critics of the EKC argue, makes pollution displacement to poorer countries even more likely. Accounting for displacement is critical. Global pollution levels may continue to rise, even as levels fall in some wealthier countries.[48]

Critics of the EKC note further that it ignores the feedback from environmental degradation. In other words, it assumes that environmental degradation does not harm future growth.[49] Yet some forms of pollution are cumulative (CO_2, CFCs, and nuclear waste) and some degradation is permanent (biodiversity and species loss). This sort of environmental harm can and does create negative feedback, holding back future growth. If this is the case, it may be possible for a country to get "stuck" along the curve, never experiencing the decline in pollution that the EKC predicts.[50]

Poverty as a Scapegoat for Environmental Degradation

Social greens and bioenvironmentalists do not deny that poverty and environmental degradation are linked. However, they see that linkage as being far more complex than market liberals and institutionalists portray. They argue that if we look more deeply and differentiate between various types of poverty, it becomes clear that this relationship is far from straightforward and varies with the type of poverty a particular community is experiencing.[51] Although many bioenvironmentalists acknowledge this important point, it is the social greens who have focused most on debunking the idea that poor people are responsible for environmental degradation. Rather than asking simply why poor people tend to degrade the environment, social greens in particular ask why they are poor in the first place.

For social greens, it is not poverty itself that causes environmental degradation. Instead, it is the inequalities of the global economy that marginalize people and put them into precarious situations that sometimes, but not always, result in environmentally damaging outcomes.[52] For these thinkers, when significant environmental damage occurs in areas with high rates of poverty, it is a result of development itself, rather than the lack of it.[53] For social greens, then, it is not poor farmers who are responsible for the loss of biodiversity in the developing world; it is the spread of industrial agriculture and inequity in landholdings. Small traditional farmers plant many varieties of seeds in their fields, whereas modern farming practices tend to rely on monocultures, creating much more vulnerability, risk, and soil degradation.[54]

It is impossible, then, to fully understand the poverty-environment linkage until we first put poverty into its broader context, and ask how national and global policies and economic relationships affect poverty at the local level.[55] Social greens in particular argue that we need to look beneath the surface in each specific case of environmental degradation amid poverty. In other words, rather than blaming poor people, we must examine the forces that pushed them onto marginal lands in the first place, and ask whether these are the forces responsible for environmental degradation.[56]

Often, social greens argue, one of the main forces that push people onto marginal lands is the process of "enclosure"—or privatization of lands once held in common property regimes. CPRs can be quite sustainable property arrangements, drawing on local community knowledge and longstanding mechanisms of control.[57] These thinkers frequently point out that the so-called tragedy of the commons described by the bioenvironmentalist Garrett Hardin is in fact a tragedy of open-access regimes, which are quite different from commons regimes.[58] CPRs are often mistaken for open-access systems, and unnecessarily attacked as "unenvironmental."

Social greens see enclosure as part of a long historical process whereby more powerful actors, including states and corporations, have taken over control of the commons. The powerful actors in the global economy in particular have tended to confiscate the property rights of poor people (especially women and indigenous peoples), pushing them onto marginal lands with poor soils and into a fragile existence. Minorities whose lands have been seized have then often been the target of state violence to repress their resistance.[59]

Many social greens view enclosure as an outgrowth of the global spread of capitalism. This process has enabled transnational corporations or local elites to occupy the best lands for timber or export-crop production, which in turn creates pockets of poverty with severe environmental problems (like soil erosion and falling agricultural yields). Indigenous and women's knowledge is the latest enclosure of the commons. In this case, powerful actors are claiming peoples' knowledge of their local environments—for example, the medicinal value of seeds and plants—as their private property in the form of patents.[60] Activist-scholar Vandana Shiva, for instance, argues forcefully that commoditizing and privatizing such knowledge leaves many of the world's poor even poorer. Few, if any, local people benefit from this process; in some cases, locals can even lose the right to "use" their own knowledge.[61]

The agricultural practices of poor farmers, many social greens also note, are (or once were) fully sustainable. One example is shifting cultivation, in which farmers move from one plot to another after several years of cultivation, often burning the new plots to clear them. Market liberals, mirroring colonial attitudes in developing countries, conclude that this type of farming creates excessive deforestation and soil degradation in poor countries. Institutionalists and some bioenvironmentalists add that population pressures often exacerbate this problem, or at least make this once-sustainable practice unsustainable. Social greens stress, however, that this type of cultivation is still generally more ecological than industrial monoculture agriculture, resulting in richer soil for the crops with fewer weeds, and acting to prevent more serious and devastating fires.[62] Blaming poor people and population pressures in these cases might be easy, because their global voice is weak, but for social greens it is ethically *and* ecologically wrong.

A further point made, in particular in the social green literature on the poverty-and-environment linkage, is that poor people in many cases are environmental activists and environmental protectors.[63] They are often women and indigenous peoples who have been forced off their ancestral lands by state and global political and economic processes. They may lack material wealth, but they have a rich wealth in local knowledge—much of which is centered on practices that preserve the environment in the long run. Their marginalization, however, often deprives them of the opportunities to follow these practices.[64]

There has in fact been a rise in ecological resistance movements, particularly in the developing world, mainly made up of poor people who are passionate about the environment. In many cases, these activists are women whose access to resources is most threatened by industrial development. The movements in India against the Narmada Dam (discussed in chapter 7) and the Chipko movement, the Greenbelt Movement in Kenya, and the rubber tappers' movement in Brazil (discussed in chapter 3) are some important examples of this type of grassroots resistance from the 1980s and 1990s. More recently, indigenous groups in many developing countries have networked with organizations around the world to fight the extraction of oil and minerals by TNCs. In Ecuador, for example, indigenous groups have organized resistance to the activities of transnational oil companies.[65] Such activists see themselves as fighting against outside forces that contribute to environmental degradation. Political scientist Robin Broad argues that when poor people rely on the environment to survive, when there is a sense of permanence on their land, and when civil society is highly organized, their environmental activism is likely to be strong. Those without land rights, or who have been forced off their own ancestral land and onto other land, and where political regimes are repressive, are less likely to become environmental activists.[66]

Policies that flow from this analysis center on the notion of regaining community control of the commons. For this reason, most social greens are highly skeptical of the market liberal view that we need to have clearer, legally enforced property rights, and in particular private property and land title in poor countries. Social greens call for a reordering of the global economy toward a more local one. As stressed by *The Ecologist*, this involves "pushing for an erosion of enclosers' power, so that capital flows around the world can be reduced, local control increased, consumption cut, and markets limited."[67]

Wealth and Overconsumption

Bioenvironmentalists and social greens see environmental problems associated with extremes of wealth and poverty as closely related to one another. They argue that the globalization of the world economy is one of the main drivers leading to ever more skewed income and consumption inequalities. They are critical of the market liberal and institutionalist view that more globalization will raise all boats, because to them the finiteness of resources dictates a zero-sum equation. Wolfgang Sachs, a leading social green scholar and activist, puts it like this: "In a closed space with finite resources the underconsumption of one party is the necessary condition for the overconsumption of the other party. . . . The rising tide, before lifting all the boats, is likely to burst through the banks."[68]

Bioenvironmentalists and social greens both argue that growing consumption, itself tightly linked to economic growth, is devastating the planet. Consumption doubled in real terms from 1973 to 1998 (global consumption expenditures were US$23 trillion in 1998) and grew to US$31 trillion by 2007 (in constant 2000 US$).[69] Manufactured products make up a growing proportion of this rising consumption. Even though technological advances may result in efficiency gains in terms of reducing natural resource inputs per unit of production, total global consumption of resources is still rising. This phenomenon is referred to by ecological economists as the "rebound effect," whereby efficiency gains for using a resource lead to increases in the overall use of that resource. William Stanley Jevons first identified this paradox in the 1860s when he showed that efficiency gains in coal use actually led to an increase in the use of coal.[70] The rebound effect is a key reason why social greens and bioenvironmentalists are skeptical of "eco-efficiency" and "sustainable consumption" initiatives that focus on reducing material inputs into the production process. For these critics, the idea of growing our way out of environmental problems through improved efficiency of production is simply not viable because of the rebound effect. Instead, a reduction in overall consumption is needed.

Most of this reduction in consumption will have to take place in the rich countries. Rich countries have much higher rates of per capita private consumption (around US$17,500 in 2006) than developing countries do (only US$280 in 2006 for the much of the developing world). For example, North America and Europe, with less than 12 percent of the global population, account for more than 60 percent of total private consumption expenditures.[71] This disparity remains even though consumption is rising faster in developing countries—particularly India and China—than in industrialized countries.

This skewed distribution of consumption also has unequal impacts on the use of natural resources—both nonrenewable and renewable. The rich primarily benefit from this use of resources, yet the environmental costs are spread widely. There are, for example, severe problems with oil drilling to sustain the global industrial system. As reserves dwindle, the net is cast ever wider, and more and more investment is making its way into the developing world (see chapter 6 for further details). The global trade in timber and minerals also primarily affects the developing world, while most of the resources end up in industrialized countries.

One way of demonstrating this inequality in consumption and its impact on resources and sinks is ecological-footprint analysis. William Rees and Mathis Wackernagel developed this approach as an accounting tool to measure the sustainability of average lifestyles. It translates human consumption of food, housing, transportation, and consumer goods and services into hectares of productive land. This allows for individuals and states to compare their ecological footprint on the planet —a vivid way for a person to imagine his or her ecological impact on the sustainability of life on earth.[72] Ecological footprints vary enormously by region. For example, the ecological footprint of an average person in Africa is 1.4, in the Asia-Pacific region 1.6, in Latin America and the Caribbean 2.4, in the European Union 4.7, and in North America

9.2.[73] China and the United States each consumed 21 percent of the world's biocapacity in 2005, but with a population four times that of the United States, China's per person consumption was much lower. It would require four earths for everyone in the world to reach present U.S. levels of consumption.[74]

The centrality of the concept of inequality to social green and bioenvironmentalist thought goes back to the environmental movement in the late 1960s and early 1970s. As discussed in chapter 3, at Stockholm the environmental movement and developing countries pushed the theme of inequality. More recently, social greens in particular have pinpointed globalization as a key driver of inequality, which in turn exacerbates the process of environmental degradation. The antiglobalization movement, which includes many social green thinkers, has also taken up this line of argumentation. This critique of globalization has been bolstered by academic studies that pinpoint globalization as a major contributor to global economic inequality.[75]

Bioenvironmentalists tend to emphasize both overconsumption in the rich world and population growth in the developing world, because both contribute to growing global consumption.[76] They are particularly concerned with consumption rates *per person*, and on this front the industrialized world tends to take the most criticism. According to leading environmental analyst Norman Myers, "The main responsibility for grand scale consumption lies with the rich; not only do they consume a disproportionate quantity of resources, but they have more opportunity to modify their lifestyles than people struggling to raise themselves out of poverty."[77]

This line of analysis is inspired by the IPAT formula outlined by Paul Ehrlich and John Holdren in the early 1970s. This formula, $I = P \times A \times T$, postulates that environmental *impact* is a product of *population* multiplied by *affluence* (consumption per person) and *technology* (which affects the amount of resources and wastes required to produce each unit of consumption). Though the formula is simple, Ehrlich and Holdren felt that it got to the root of the environmental problem, and showed how inextricable each factor was from the others. Although in the early 1970s most bioenvironmentalists focused on population growth in the developing world as a core stress on the global ecosystem, today many worry more that increasing global trade will harmonize consumption rates upward, with developing countries eventually consuming at the rate of the developed countries (all within the context of an increasing global population).[78] The WWF points out that all of humanity's ecological footprint exceeds the planet's regenerative capacity by some 30 percent, and that 75 percent of the world's population lives in countries that consume more than their national biocapacity. Global trade is facilitating this trend of consuming beyond their means. In 2005, for example, China imported the equivalent of the biocapacity of Germany.[79]

Most social greens sharply criticize the inclusion of population in the discussion of consumption, in favor of an analysis more squarely focused on the rich world's consumption patterns. They stress the political use of population in much of the mainstream analysis. As *The Ecologist* notes: "The 'population issue' is being used politically to control groups of people who are considered a threat by those who have benefited from enclosure. Moreover, 'population control' is being used to enclose women further: it takes away their power to determine how many children they have and to choose a method of regulating births which they control and which does not threaten their health."[80] For them, then, it is dangerous to focus on the question of population, particularly in the developing world, because it disempowers women, who make up the majority of the world's poor people. Instead, they focus squarely on the rich world's consumption patterns, and also actively defend the right of all women to have control over their own bodies.[81]

On the question of consumption, both social greens and bioenvironmentalists point to what they call the problem of "overconsumption." As the rich get richer and consume more, they inflict more damage on the natural environment than is necessary. This is because much of their consumption is based on wants, not needs. Luxury goods that are far beyond what is necessary are seen to be proliferating. For many in the rich industrialized world, shopping has become a pastime rather than a necessity, creating or exacerbating a culture of consumption.[82] The growing presence of low fuel-efficiency sport utility vehicles (SUVs), crossovers, and light trucks in urban centers in North America over the past decade, as incomes were rising, is a case in point.[83]

This overconsumption by the rich is exhausting energy supplies and contributing to "overpollution" and "overgeneration of wastes." This is true of both corporate and individual consumption.[84] People now buy more things and throw more things away than in the past. This arises partly because products like computers become obsolete in just a few years, partly because engineers intentionally design some products to last for only a short time, and partly because an increasing number of people now "enjoy" replacing their possessions (a trend reinforced by advertising). Increased consumption has also come with a growing amount of packaging waste, much of it plastic. Recent reports of the growth of the Great Pacific Garbage Patch, already at least twice the size of Texas, vividly illustrate the global impact of overconsumption.[85]

Consumption in rich countries has continued to rise at an increasing pace over the last 200 years. So far, contend many social greens and bioenvironmentalists, these countries have managed to skirt the natural limits of resources by acquiring resources from other parts of the world.[86] In the past, this was done mainly through colonialism and migration to new lands. Today it is done through international trade, foreign direct investment, and foreign aid. A country's "ecological shadow" refers to its aggregate environmental impact in other parts of the world to feed domestic consumption and avoid domestic environmental problems. The globalization of corporate chains to produce and replace consumer goods is increasing the reach and size of ecological shadows, shifting the environmental burden from consumption far beyond the borders of rich countries: from the slums of India to the rainforests of Brazil to future generations of Inuit. It is also allowing people in wealthier places to consume ever more without suffering a proportionate share of the consequences of hitting ecological limits.[87] The "bigness" of the global economy supports the lengthening of supply and disposal chains for consumer products. This "distances" consumption from extraction and production and from its final resting place as waste, making the environmental implications of consumption harder for consumers to comprehend and act on.[88] One example is computers. The production and disposal of computers involve toxic metals and plastics. The hazardous part of computer production commonly occurs in developing countries. Developed countries also commonly export computer waste to developing countries for recycling.[89]

Why does this trend occur? Social greens argue in particular that the privileged position the rich countries hold in the current international political economy has enabled them to continue their high per capita consumption of natural resources by buying up the developing world's resources at cheap prices, which contributes to both poverty and environmental degradation. The ideology of growth and industrialization, part and parcel of capitalism, is blamed for contributing to the dominance of the industrialized countries and transnational corporations in an increasingly global economy in ways that weaken the power of the poor countries to resist the environmental degradation inflicted on them.[90] This situation is seen as part of a longer historical trend. In the past, rich countries extracted resources from colonies; today they do so through the process of economic globalization. Both eras are similar, in that they involve uneven development and uneven distribution of environmental degradation, with poor people suffering the most.[91]

The wealthy, argue social greens and bioenvironmentalists, must curtail consumption. No other option is reasonable and fair. Bioenvironmentalists stress the need to lower total global consumption. This will require rich countries to lower consumption, but equally important for many bioenvironmentalists is the need for poorer countries to control population growth. In making this recommendation, they are not that far from the suggestions of institutionalists, who stress sustainable consumption and population management, as discussed in the first part of this chapter, even though they arrived at this point along a different path. Social greens emphasize the need to reduce consumption inequities—that is, lower consumption among the wealthy, so more are getting what they need (sufficiency), rather than a few getting what they want (gluttony). It is critical, too, that local communities incorporating the views of women and indigenous peoples maintain control of local resources and do not relinquish this control to transnational corporations. This will require not just more individual responsibility for environmental management, but also genuine sacrifices by well-off populations as well as broader political and economic changes to reward less consumptive, more local economies.[92]

Conclusion

This chapter demonstrates the complexity of the ecological debates over the impacts of economic growth, poverty, and wealth. There is no doubt a major fault line on the question of economic growth, with a coalition of social greens and bioenvironmentalists on the critical side, and a coalition of market liberals and institutionalists on the other. A deeper look has revealed, however, that on other issues different coalitions have emerged. On a few points there is even common ground among all of the views.

On the main question of economic growth and whether it is positive or negative for the environment, market liberals and institutionalists both advocate a pro-growth strategy with the aim of reducing poverty and raising revenues to pay for environmental improvements. Market liberals stress that promotion of a free market is the best way to foster growth, because price signals will encourage innovation and greater efficiency in resource use. Alternatives to oil, for example, will develop when prices dictate, a process already underway with alternatives like solar and wind power. Market tools—such as ecotaxes, user fees, and tradable resource rights—can help address distortions and market failures. This does not mean that all environmental regulations are bad, just that whenever possible, market approaches tend to create a better long-term environmental management structure. Institutionalists are much more cautious about leaving everything to the free market. They want to see policies and structures that help economies tunnel through the EKC—in other words, more active policies to better weather the inevitable pollution en route to the cleaner world of wealth and prosperity. One such policy is rooted in the idea of sustainable consumption.

A coalition of bioenvironmentalists and social greens, on the other hand, argue that economic growth directly harms the environment. It is not going to lift poor people out of poverty, but rather, given the current structure of the global economy, will merely exacerbate inequalities. Bioenvironmentalists and social greens call for a total rethink of consumption and growth and nature. They both agree that this requires that the physical "throughput" in the economy does not outpace the supply of renewable and nonrenewable natural resources. It is still possible to provide for a "good life" for present and future generations. But doing so will require slowing the physical aspects of "growth" based on resource use to sustainable levels. This does not necessarily mean "zero economic growth," but it would mean a radical cutback. Social greens take the point further by calling for a redistribution of global economic resources and a refocus on local community development instead of emphasizing economic growth per se. For them, there are many ways to work toward this broader concept of development, including local economic trading systems, reinstatement of common property regimes in the developing world, and tight restrictions on transnational corporations and trade.

On other issues, such as the relationship of population to poverty and environmental degradation, we see a different convergence of arguments among the worldviews. Bioenvironmentalists remain particularly concerned with population growth, viewing it as intricately tied to consumption and economic growth. Institutionalists share many of these concerns, seeing population growth as tightly tied to the vicious cycle of poverty and environmental degradation. Thus, both feel that policy initiatives to promote sustainability should incorporate some sort of population-control measures. On the other hand, social greens view the inclusion of population in the discussion of consumption and growth as dangerous, because often it becomes a political tool to control women and poor people. For different reasons, market liberals also do not focus much on the links between population growth and the environment. For them, resource scarcity is not a pressing environmental issue. Therefore, neither is the number of humans.

This examination of the debates over growth, poverty, and wealth shows, too, a convergence of all views on some points. This includes most prominently the idea of reducing environmentally harmful subsidies, the promotion of green taxes, and payments for environmental services. Though environmental economists (many of whom adhere to market liberal or institutionalist views) have proposed many of these ideas, ecological economists (many of whom adhere to bioenvironmentalist or social green views) embrace them, too. The Green Party in many countries, for example, has called for a "tax shift" toward environmentally sound taxes.[93] The different groups we have been surveying have arrived at this policy prescription using different assumptions and arguments. Social greens and bioenvironmentalists, for example, are skeptical of growth in general and see the tax shift as a way to slow down growth—especially ecologically destructive growth. Market liberals see growth as environmentally positive, and efficient growth, fostered by a reduction in subsidies and green taxes, as even more positive. Institutionalists see these tools as a way for institutions such as the state to enhance their capacity to promote environmentally sound growth.

This chapter has shown that when it comes to these particular debates, each worldview has its own unique take and the coalitions are not the same on all issues. The next chapter on trade further demonstrates the value of comparing the particular arguments and assumptions of each worldview.

5

Global Trade and the Environment

Is global trade saving the planet? Or destroying it? There are cogent arguments for both positions, as well as some in between.[1] Market liberals see a free flow of goods around the world as a positive environmental force. From this perspective, global trade has many benefits. Free trade, through the logic of comparative advantage, creates global wealth and prosperity. It is efficient, avoiding waste and duplication. It also provides less industrialized regions of the world with advanced technologies and goods in exchange for relatively simple products (like minerals or timber). Global trade allows a country with a comparative ecological and labor advantage in, say, forest management, to develop a viable economy around timber exports. The ensuing societal wealth then enables citizens to purchase imports (like cars and computers), many of which would be well beyond the capacity of the country to manufacture (or, at best, terribly inefficient to do so). Poorer countries, moreover, have little choice in an era of globalization, because restricting trade leads only to hardship and greater inequalities. Instead, states, indeed the entire global community, need to liberalize trade as much as possible. Free-trade institutions like the WTO and regional trade agreements like the North American Free Trade Agreement (NAFTA) are therefore critical to the health of the planet, because these are helping remove the inefficiencies and market distortions that lie behind many environmental problems.

Bioenvironmentalists and social greens challenge the logic and evidence behind these market liberal arguments. Bioenvironmentalists see global trade (along with free-trade agreements) as a root cause of the global environmental crisis. The apparent wealth that seems to arise from free international trade in fact arises from the shortsighted exploitation of nature. Free trade externalizes environmental and social costs—that is, the prices of traded goods do not reflect their full environmental and social value (e.g., the price of a hardwood board does not reflect the value of a tree as a home for birds or a sink for carbon). Often, the price of products ignores, too, the environmental cost of the production process—for example, water and air pollution from a factory. In addition, *global* trade *distances* the point of manufacturing from the point of consumption, which leaves consumers less able or willing to accept the environmental consequences of their personal consumption habits. Like bioenvironmentalists, social greens also see trade as a root cause of global environmental problems. They agree with the points raised by bioenvironmentalists, but also stress the ways that trade exploits peoples and locales in unfair and unequal ways, resulting in uneven distribution of environmental and social problems. In particular, social greens argue that the global trading system contributes to the exploitation of developing countries (with respect to both workers and local environments).

Institutionalists take somewhat of a middle ground between these two broad positions on trade. On the one hand, they agree with the broad market liberal view that trade can enhance prosperity, generate efficiencies, and result in higher overall capacity to manage environmental affairs. Yet institutionalists add that it is critical to manage trade judiciously. They argue that there are some cases in which trade is directly tied to undesirable environmental outcomes—for example, trade in endangered species, hazardous wastes, and dangerous chemicals—where governments can and should be able to control trade in order to protect the environment. For institutionalists, then, free-trade agreements need environmental clauses, and in some cases global environmental regimes need to override trade agreements. The key for institutionalists is for states to coordinate and cooperate on these issues. Following a brief overview of some basic data on globalization and trade, this chapter examines these three schools of thought on trade and the environment. To explore these arguments further in the context of real-world trade issues, the chapter then considers specific trade agreements, including the WTO as a global example, as well as several regional trade agreements.

Globalization and Trade

The process of economic globalization over the last fifty years, as we explained in chapter 2, has contributed to a sharp rise in the value and volume of global exports of goods and services (see figure 5.1). World trade as a percentage of GDP has also steadily grown from 24 percent in 1960 to 58 percent in 2007 (see figure 5.2). This high proportion of global GDP that is accounted for by trade indicates the growing importance of trade in the world economy. Indeed, from 1950 to 2008 the volume of global trade increased by a factor of thirty-two, while the value of exports of goods and services increased by a factor of 260.[2] This growth has occurred even with the onset of a global economic downturn in 2007.[3]

The value of world merchandise trade was over US$16 trillion in 2008, dwarfing the value in 1948 (US$58 billion; see figure 5.3).[4] Much of this trade is between developed states, although rapidly industrializing countries such as China and India have gained a growing share of global trade in recent years.[5]

Although trade has grown remarkably on a global scale, there are still wide differences between and within regions (see figure 5.4). In 2007, Europe (42 percent), Asia (28 percent), and North America (14 percent) together accounted for 84 percent of regional trade flows in world mer-

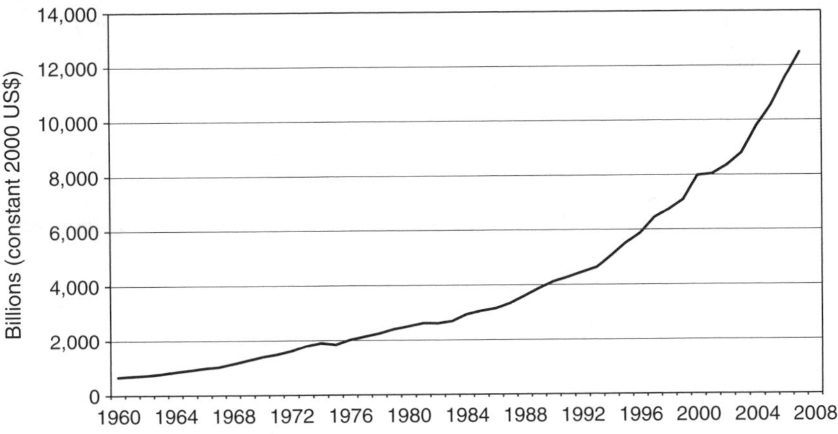

Figure 5.1
World exports of goods and services, 1960–2008. *Source*: World Bank World Development Indicators, <http://data.worldbank.org>.

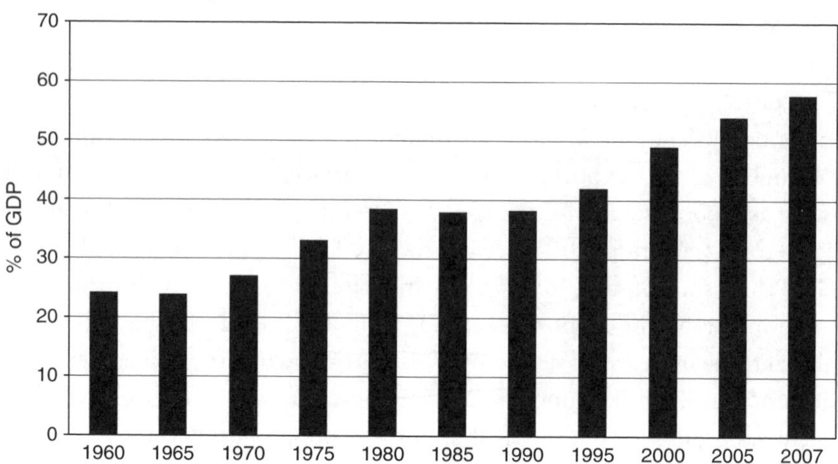

Figure 5.2
World trade as percentage of global GDP, 1960–2007. *Source*: World Bank World Development Indicators, <http://data.worldbank.org>.

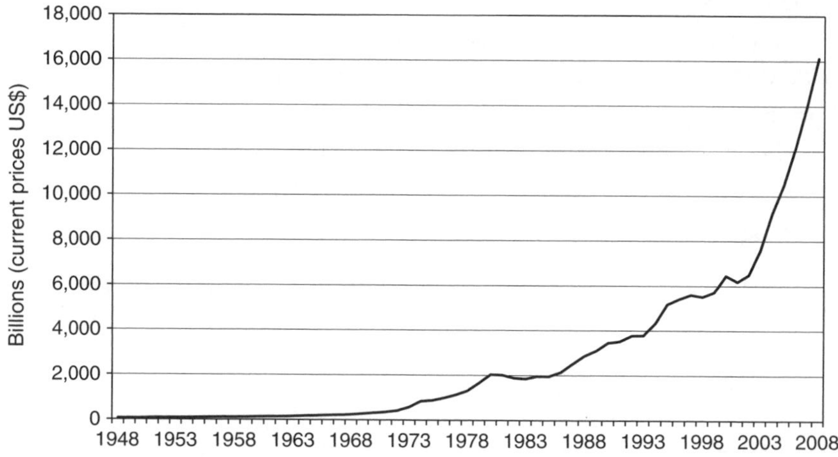

Figure 5.3
Global total merchandise trade, 1948–2008. *Source*: World Bank World Development Indicators, <http://data.worldbank.org>.

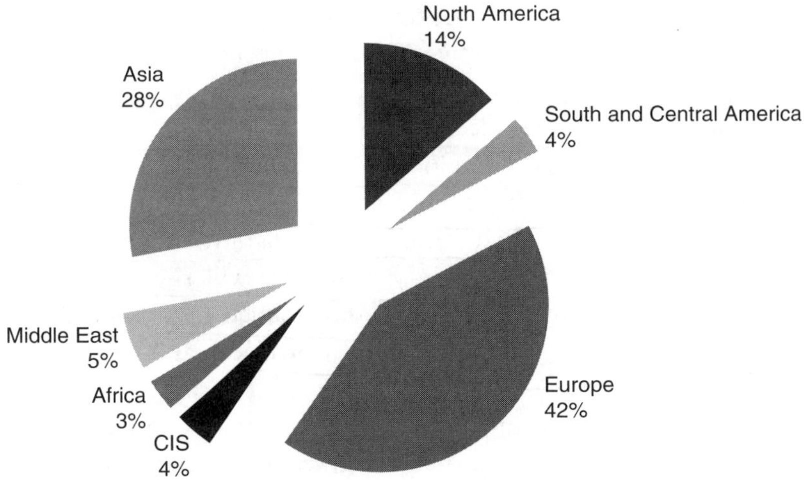

Figure 5.4
Shares of regional trade flows in world merchandise exports, 2007. *Source*: WTO, <http://www.wto.org/english/res_e/statis_e/its2008_e/its08_toc_e.htm>.

chandise exports.[6] The Middle East accounted for 5 percent of these flows, South and Central America and the Commonwealth of Independent States (CIS) states for 4 percent each, and Africa accounted for just 3 percent. There are equally wide differences over the value of trade for national economies within regions. In Africa in 2007, for instance, Burundi's trade in goods and services accounted for 7 percent of its GDP, whereas in Equatorial Guinea trade accounted for 94 percent of GDP.[7] As the example of Equatorial Guinea shows, although Africa's proportion of total global trade is small, some economies rely heavily on trade and are thus quite vulnerable to market fluctuations.

Trade's Impact on the Environment: Three Schools of Thought

Trade Is Ultimately Good for the Environment

Most mainstream economists assume that in the long run the globalization of trade will enhance, rather than hinder, global environmental management. A few, like Columbia University professor Jagdish Bhagwati, have actively challenged the environmental critics of global trade.[8] These supporters, at a general level, accept the neoliberal economic assumption that trade promotes economic growth, and that economic growth in turn is crucial for preserving and managing the natural environment (as discussed in chapter 4). The promotion of trade for market liberals is closely tied to this growth argument. But there are also more specific arguments as to why trade and environmental goals are mutually supportive. Trade, according to market liberals, enhances efficiency and diffuses cleaner technologies and standards. Each of these, in turn, promotes better environmental conditions. The efficiency from trade is based on the economic theory of comparative advantage (see box 5.1). This theory states that global welfare will rise (because of efficiency gains) if countries specialize in the products that they are relatively better at producing compared to other products, and then engage in trade with one another.

Though the theory of comparative advantage was not originally tied to global environmental management, market liberals have noted the environmental benefits of the efficiency gains from specialization. More efficient production optimizes the use of resources in the production process, meaning the world gets more goods using fewer resources.[9] More efficiency means more production per unit of inputs, which in theory should improve economic growth and raise per capita incomes. Higher incomes, market liberals argue, lead citizens to demand a cleaner environment (see chapter 4). More resources mean that more can be spent to preserve the environment as well as to enforce environmental regulations. Market liberals do not deny that rising production may initially harm the environment. The specialization needed to pursue comparative advantage may in fact result in some countries specializing in dirty industries. And if production rises faster than efficiency gains, pollution may also rise,[10] although often the increase in societal demand for a cleaner environment from higher incomes should in time counteract this. The key issue for market liberals is how the money is spent. As the WTO notes in a report on the linkages between trade and the environment, "The income gain associated with trade could in principle pay for the necessary abatement costs and still leave an economic surplus."[11]

More global trade, contend market liberals, generates other efficiencies beneficial to the environment as well. Trade policies that distort markets (such as tariffs, quotas, and export subsidies) can create situations in which the prices of natural resources do not reflect real scarcity, which in turn drives overuse and overconsumption. Trade liberalization aims to remove these distortions and is thus needed to correct the under-

Box 5.1
Absolute and comparative advantage

The neoclassical economic assumption of the logical benefits of trade is grounded in two theories of advantage: absolute and comparative.

Absolute advantage suggests that if country A is able to produce a good (e.g., wheat) more cheaply than country B can produce it (perhaps for ecological, labor, technological, or societal reasons), and if country B is able to produce a different good (e.g., cloth) at a lower cost than country A can produce it, it is in the financial and material interests of *both* countries to specialize in the good they have the absolute advantage in producing and then trade these goods with each other.[1] In this case, then, both countries will have more of both goods (wheat and cloth) if they specialize their production in the product they have the absolute advantage in and engage in trade.

The theory of *comparative advantage*, first formalized in numerical terms by David Ricardo in 1817,[2] is consistent with absolute advantage, but applies another layer of logic that expands the instances when trade is beneficial. Imagine that country A has no absolute production advantage over country B in any product. Country B then produces both cloth and wheat more cheaply than country A can produce those goods. Under the theory of absolute advantage, it does not seem logical for the countries to trade. Yet, as Ricardo shows, by thinking in terms of *comparative* advantage, trade is still logical, and indeed enhances the welfare of both countries. Ricardo (and many others since) demonstrated that if countries specialize in the goods that they themselves are *comparatively* best at producing (i.e., the products that they are relatively more efficient at producing compared to other goods), there will still be gains from trade.

To continue our example, if country A specializes in the good that it is relatively more efficient at producing (e.g., it is relatively better at producing wheat than it is at producing cloth), and then country A trades with country B for the product that country A is relatively less efficient at producing (in this case cloth), *total* productive output will still rise. This holds even though country B is absolutely more efficient than country A at making both cloth and wheat. This specialization and trade will create efficiencies, which in turn will increase the output of both products. And the competition among traders with similar or identical comparative advantages may foster further innovation and efficiencies.

Notes

1. Adam Smith (1776) outlines some of the most influential logic of trade.

2. Ricardo 1817. John Stuart Mill's 1848 work *Principles of Political Economy* ensured that comparative advantage would become central to the field of international political economy.

pricing of resources. Liberalization also diffuses more environmentally friendly production processes, because firms that produce goods behind trade barriers that protect them from competition have less incentive to adopt cleaner technologies.[12] The World Bank summed up this position well early on in the debate: "Liberalized trade fosters greater efficiency and higher productivity and may actually reduce pollution by encouraging the growth of less-polluting industries and the adoption and diffusion of cleaner technologies."[13] A joint report of the WTO and UNEP on climate and trade similarly notes the importance of free trade for the diffusion of climate-friendly production technologies.[14]

Market liberals argue further that countries with more liberal trade policies are more likely to adopt higher environmental standards so as to better compete in markets with higher standards. This is referred to by Berkeley professor David Vogel as the "California effect," whereby trade acts as a mechanism to encourage exporting countries to raise their environmental standards to be the same as importing countries.[15] The U.S. Government Accountability Office found, for example, that U.S. standards for pesticide regulations are often taken into account by countries seeking to export agricultural products to the United States.[16] Another example is Mexico, where environmental regulations have become more stringent since NAFTA came into force in 1994.[17]

For these reasons, market liberals see the use of trade tools—such as sanctions (restrictions on trade to pursue other objectives) or taxes on imports of goods based on the exporting country's carbon emission levels—to advance environmental goals as counterproductive. The fear is that such tactics could encourage countries to erect trade barriers under the guise of environmental protection. Bhagwati gives the example of the Ontario–United States "beer war" over a 10-cent-a-can tax on beer, which Ontario claimed was intended to discourage littering. The United States argued that the tax was discriminatory, because it was not levied on juice and soup cans and was borne disproportionately by the United States, as U.S. exporters supplied most of the canned beer for Ontario. The result was not a better environment in Ontario, but straightforward trade protection for Canadian producers of bottled beer.[18]

Partly to avoid discriminatory protectionism, market liberals are skeptical of the use of a precautionary principle that would enable countries to restrict trade in the absence of full scientific information on the grounds that an activity *may* harm the environment. Market liberals, for example, opposed the European Union's 1998–2004 de facto moratorium on the import of genetically modified organisms (GMOs), seeing the use of the precautionary principle in this case as little more than governments protecting domestic markets. Interfering with trade, for market liberals, is rarely—if ever—the best tool to address environmental problems. "Trade sanctions are at best crude weapons," trade specialists Michael Weinstein and Steve Charnovitz argue, "and environmentalists should reconsider their enthusiasm for them."[19] For market liberals, what is needed is sound environmental policy that zeros in on the source of the problem: indirect approaches through trade are inevitably inefficient and can even worsen environmental problems. Government bans on raw log exports (to force exporters to process logs at home), for example, have made forest management worse in many countries. In Indonesia such a ban in the 1980s and 1990s allowed firms to export huge quantities of sawn wood and plywood, which wasteful and inefficient domestic firms processed with little regard for sustainable management.[20]

Certain market liberals have gone so far as to say that some countries have a higher capacity to absorb pollution than others, and that this is part of their natural comparative advantage.[21] The WTO notes that the existence of differing ecological conditions means that different national standards are bound to develop. It further notes that "even if ecological conditions were identical, international variations in standards may be desirable in order to reflect differences in income and ability to pay for environmental quality."[22] For market liberals, the use of trade sanctions to force a single, harmonized set of global environmental regulations would therefore destroy some of the natural advantages in some of the world's poorest regions. This would lower global welfare and ultimately hurt the natural environment.

In short, market liberals do not see free trade as a source of environmental problems. Instead, they argue that free trade brings efficiencies and growth. It reduces wasteful use of resources. It provides firms and governments with the funds to adopt or require environmental technologies. And it raises environmental standards in developing countries. Governments must address environmental problems with proper policies and incentives, and restricting trade does not create these policies and incentives. Critics of market liberalism, as discussed in the next section, have challenged all of these arguments, in many cases reaching the opposite conclusions.

Trade Is Bad for the Environment

Bioenvironmentalists and social greens are both skeptical of global free trade. In a nutshell, they associate many environmental problems with global trade: the large scale and rate of economic growth under a free-trade regime; the unequal and unfair effects of the export-import synergies between the rich industrialized and poor developing countries; the hidden environmental and social costs of trade; and the "dumbing down" of environmental standards.

Like market liberals, many bioenvironmentalists and social greens see global free trade as a potential catalyst for further global economic growth. But for bioenvironmentalists and social greens, as explained in chapters 2 through 4, the world cannot sustain such growth. A global-free-trade regime may well make production more "efficient," but such rapid and ever-upward growth will inevitably outstrip any gains in efficiency, so that the net impact on the natural environment is negative. This is referred to as the *scale effect*, whereby efficiency gains themselves can result in the use of even more resources. This can occur because more efficient production can lower consumer prices, which in turn can raise demand. The consumer in such cases may become thoughtless and wasteful (in effect thinking of the product as having little real value). The classic example is paper—supplies are plentiful and cheap. How many of us print stacks of papers we never read? How many of us would do that if each sheet of paper were worth a bottle of wine? Demand for a product may also simply outpace any potential efficiency gains. Cars, for example, are far more fuel-efficient now than they were three decades ago, yet there are so many now that the total amount of resource use is much higher. The expansive nature of international trade is so closely tied to the growth of consumption that the ecological impact of the sheer increase in volumes far exceeds any potential efficiency gains that derive from more trade.[23]

Social greens in particular add that global "free" trade does not benefit all trading partners equally. The original theories of the benefits of trade under conditions of absolute and comparative advantage assume that goods, but not capital or labor, are mobile. This may have made sense when the theories were first postulated several hundred years ago. But this premise, social greens argue, is no longer valid. Today's highly mobile capital fundamentally alters the relationship between trade and welfare, as well as trade and the environment.[24] Today, specialization for trade concentrates pollution in production zones—and usually it is developing countries that absorb most of the environmental costs of free trade by producing the most polluting items, while rich countries enjoy the benefits of those products. Moreover, global trade merely delays the inevitable limits to growth by distributing environmental costs worldwide, allowing some countries to consume far beyond the carrying capacity of their natural environment, at the same time using up the carrying capacity of other countries. "Free trade," ecological economist Herman Daly argues, "allows the ecological burden to be spread more evenly across the globe, thereby buying time before facing up to limits, but at the cost of eventually having to face the problem simultaneously and globally rather than sequentially and nationally."[25]

Production for export in the developing world in many cases relies heavily either on the unsustainable use of local natural resources or on dirty and unsafe factories (relying on cheap labor). Coffee, cotton, banana, sugar, tea, and palm oil (for cooking oils and margarine) plantations in Latin America, Africa, and the Asia-Pacific region show the environmental harm of export-oriented crops.[26] Forestry, fishing, and mining in these regions highlight the ecological damage of large-scale natural resource extraction for export.[27] Across the developing world, textile and electronics factories pollute the environment with mass production of cheap goods for consumers in wealthy countries.[28] Trade, in other words, allows countries to import or export carrying capacity, so rich people may live beyond their carrying capacity, while poor people live well under it. Some, as the previous chapter mentions, refer to this as part of a country's "ecological shadow";[29] others call this "distancing."[30] Such trade in effect casts ecological harm from consumers onto countries that are net exporters of natural resources or pollution-intensive goods (or both). When the shadow cast is long and the distance is great (either in geographic or mental terms), there is little ecological feedback to the consuming countries.

There are other innate drawbacks of too much trade. The prices of goods and services in a free-trade regime often do not reflect the true environmental and social costs of production. This problem generally begins with the extraction of natural resources. The market value of an old-growth tree, for example, primarily reflects the cost of access to the forests (perhaps a tax or flat fee), along with the cost of harvesting and transportation to the mill. The tree (now a log) is sold to a mill for a "profit" when the buyer's price is higher than the total monetary costs for the logger. Yet little, if any, effort is made by governments to assign a monetary value to the hundreds of years it took for the tree to grow. Or to the aesthetic or ecological value of the tree in a forest. Or to the global environmental value of the tree as a store and absorber of carbon dioxide. Instead, the tree is worth what the government of the day sees as its financial value to the people of the nation-state. In many cases it is worth almost nothing. In fact, getting it out of the forest and into "production" for the international market creates jobs as well as income and spurs economic growth, and most governments merely assign a token fee for access. The stumpage value—or the total state fees for a tree in the forest—was, for example, typically less than 1 percent of the consumer price in the Asia-Pacific region in the 1990s.[31] Accounting for the "true value" of timber would create a very different trading structure: one in which few, if any, old-growth trees would ever enter into world trade.[32]

The externalization of the environmental costs of traded goods does not end here, but continues throughout the production process (from extraction to the exporter to the importer to the factory floor to the retail outlet). The price that a consumer pays, for example, generally does not include the full costs of transportation. This occurs partly because governments tend to subsidize energy inputs (in a bid to spur growth and industrialization), and partly because energy sources themselves (like oil and coal) generally do not reflect the full environmental costs (e.g., the climatic effects of the release of carbon dioxide into the atmosphere). The underpricing of the transportation of goods over long distances—via ships, trucks, or airplanes—encourages more trade, and the end result is more fuel use, greenhouse gas emissions, and smog. Moreover, even if properly priced, the transportation effects of trade will cause significant environmental disruptions.[33] A study of the impact of NAFTA on transportation, for example, found that greenhouse gas emissions linked to trade were expected to rise substantially by 2020, and that much of that increase would be concentrated in certain transportation corridors linked to trade routes.[34]

Critics further argue that trade puts downward pressure on environmental standards. This occurs because—in a bid to reduce costs to become more competitive in international markets—countries sometimes lower or at least fail to strengthen environmental regulations as well as enforcement. Some refer to this as the "race to the bottom."[35] Others point out that many countries appear "stuck at the bottom."[36] Critics argue that weak regulations and enforcement in order to gain a competitive edge simply encourage firms to externalize environmental costs. Many critics further argue that the global environmental standards that trade agreements rely on may be less strict than the domestic standards of the original signatories, thus pushing down environmental regulations in advanced economies. For example, Codex Alimentarius—a UN commission that sets global food-safety standards for trade—adopted standards for pesticides in the 1990s that were 42 percent lower than the standards of the EPA and the U.S. Food and Drug Administration (FDA). The Codex standards allowed farmers to spray fifty times more DDT on peaches and bananas than the U.S. standards allowed.[37] Similar issues have arisen within the European Union, with EU-level food safety standards less stringent than those in some member countries.[38]

Social greens in particular argue further that most global trade agreements, even ones with environmental clauses, prioritize economic concerns over ecological ones, partly because corporate interests—sometimes through firms and sometimes through state agencies—wield so much power over the agendas and negotiations.[39] Corporate executives often sit on advisory panels in trade negotiations, and in that capacity can have significant influence over trade negotiations. A study by the NGO ActionAid found that in the United States 93 percent of the 743 approved advisers on U.S. Trade Advisory Committees were representing corporations or business lobby groups.[40]

In view of the numerous problems associated with trade, many social greens and bioenvironmentalists argue that it is perfectly legitimate—indeed, it is often necessary—for states and global institutions to restrict trade to achieve environmental goals. Trade sanctions (or the threat of sanctions), for instance, can be an effective (and nonviolent) way to punish states that duck global environmental standards to be "free riders" on the global benefits of higher environmental standards in other countries. Some social greens warn, however, that certain measures can be grossly unfair, such as carbon taxes levied against high-emitting countries like China based on total rather than per capita emissions. Such measures end up privileging exports from rich countries over poor ones, which in turn increases poverty. Further, they are in effect a way for the rich countries to evade their commitments under the Kyoto protocol to pay for climate actions taken in developing countries.[41] A way to get around these problems and to work toward a more sustainable global economy is to limit international free trade and its associated growth.[42]

Moreover, both bioenvironmentalists and social greens support the use of the precautionary principle in trade agreements to prevent future environmental harm.[43] Waiting for full scientific consensus before taking action, they argue, will often, if not always, mean the action is too late to prevent serious (and perhaps irreparable) ecological damage. Scientists, for example, fully understood the harms caused by CFCs and persistent organic pollutants such as DDT and polychlorinated biphenyls (PCBs) only long after their invention. Newer products may pose similar risks. Genetically modified grains are one example.[44] Another is the family of chemicals called *perfluorochemicals*, which are used to make nonstick pots and pans and to make rugs and raincoats grease-resistant, stain-resistant, and waterproof. Everything from pizza boxes to microwave popcorn bags to shaving creams contains these chemicals. Their potential to harm human health is hotly debated. Some scientists see little of concern given the low doses now in the environment and in people's blood. Recent tests on animals, however, have linked low levels to damage to thyroid glands, breast cancer, and liver cancer.[45]

Managed Trade Can Be Good for the Environment

Arguments over whether trade is good or bad for the environment dominated the initial debate over trade and the environment throughout most of the early 1990s. A compromise position, along the lines of institutionalist thought, garnered increasing support after the Rio Earth Summit in 1992. The summit's declarations went to great pains to emphasize that free trade and environmental protection were not in conflict. Principle 12 of the Rio Declaration, for example, says that "states should cooperate to promote a supportive and open international economic system" and that "trade policy measures for environmental purposes should not constitute a means of arbitrary or unjustifiable discrimination or disguised restriction on international trade." Further, Agenda 21 states that "the international economy should provide a supportive international climate for achieving environment and development goals by . . . promoting sustainable development through trade liberalization."[46] Today, these views dominate the discourse on trade and the environment.

Advocates of "managed trade" accept the broad claims of market liberals with respect to the environment.[47] But they qualify those claims. They argue that international trade can in some cases be harmful, but ultimately need not be any more harmful to the environment than regional or local trade. Each kind of trade involves a mixture of benefits and problems. For example, although greater import ties between developing countries and more pollution-efficient countries does have spillover effects in terms of improvements in pollution levels in the former, greater export ties between those same countries do not have the same effect.[48] States are no doubt tempted to weaken environmental standards in order to gain a competitive advantage in trade, yet global agreements can largely resolve this drawback of free trade.[49] Global cooperation, in other words, is an effective mediating force that can produce positive outcomes for both trade and the environment.[50] Such cooperation is in fact essential because efficient trade requires healthy environments, and healthy environments (at least in an era of globalization) will degrade and collapse without trade (a glance at the environment of North Korea confirms this).

Sometimes, institutionalists argue, global agreements warrant trade measures, and on occasion even trade sanctions. Obvious cases include the need to control the global trade in hazardous wastes and dangerous chemicals (such as persistent organic pollutants, ozone-depleting substances, and banned chemical pesticides). The idea of using trade measures to address climate change is gaining more proponents, too.[51] Yet the global community should rely on such measures only in clear instances of environmental harm resulting from trade. At the same time, institutionalists argue that it is important to liberalize trade to improve global environmental management. The elimination of subsidies on pesticides and fuels is an example of how liberalization of trade can help promote cleaner environments.[52]

Furthermore, global cooperation can create green markets—that is, markets in which the prices throughout a trade chain better reflect the full environmental and social costs. This can lower consumption and increase the amount of funds available for sustainable management. Global ecolabels for products from "sustainable sources" or with "sustainable production processes," for example, can encourage consumers (as well as perhaps wholesalers and retailers) to voluntarily pay more for a product that does less damage to the global environment while also promoting more equitable social development. The market growth of timber certified as sustainable and coffee, sugar, and bananas labeled as organic or fair trade suggest the potential here for win-win trading arrangements. The nonprofit Forest Stewardship Council (FSC), founded in 1993 to promote better forest management, is an example of an organization designed to construct a green market. Its members include environmental NGOs, forest industries, indigenous and community groups, and timber-certification bodies. The FSC accredits and monitors national certifying agencies. Its logo on a wood product is meant to signal to the consumer that "the product comes from a well-managed forest."[53] Market demand is seen to primarily drive the internalization of environmental costs for fair trade and eco-certified products, with producers and consumers opting in voluntarily, rather than as a government requirement.[54] Because they are voluntary and do not violate trade agreements, these programs can flourish both domestically and via international trade.

Institutionalists who back managed trade also tend to advocate a limited use of the precautionary principle. Although they do see the potential for those states wishing to protect domestic industries to abuse this principle, they also believe there are some instances where precaution is warranted.[55] They also support the "polluter-pays principle." This principle states that the polluter should bear the full cost of pollution prevention and control, rather than that cost being pushed onto society or the government. This may require governments to consider restricting (such as with tariffs) or banning the import of polluting goods, or on goods produced with considerable environmental harm.[56] A number of countries are now considering a border carbon adjustment on imports—such as a tax, tariff, or surcharge—to force cost internalization on foreign firms that operate under jurisdictions with high carbon emissions. As more countries commit to reducing carbon emissions through carbon taxes or trading schemes, domestic producers in those countries do not want to be disadvantaged by higher costs associated with those policies, and governments do not want firms to relocate abroad in order to take advantage of weaker emission standards (which would result in what is called carbon leakage). A border carbon adjustment on imports has been proposed as a means by which to address this dilemma.[57] With such a policy, firms and governments could take the profits from such adjustments and invest them in sustainable production methods—or perhaps, more radically, use these funds to compensate countries for past environmental degradation.[58]

Proponents of managed trade see the greening of international trade agreements as one of the most constructive and practical ways to move toward both a richer and a cleaner world. And they argue that such agreements should coordinate with international environmental agreements in ways that produce win-win situations.

The WTO and the Environment

Theoretical differences over the environmental merits and demerits of trade continue to influence international debates and negotiations. At the same time, however, much of this debate now takes place in the context of real-world trading arrangements, many of which now explicitly address environmental management. This section explores these debates through a detailed case study of the WTO as well as discussion of several regional trade agreements, including NAFTA and the European Union.

The WTO: Principles of Trade

The World Trade Organization, founded in 1995, evolved out of the General Agreement on Tariffs and Trade (GATT).[59] The GATT was established in 1947 as a global trade agreement that aimed to achieve freer trade. It was founded on the neoclassical economic assumption that free trade will improve global welfare, following on the ideas of David Ricardo (see box 5.1). The principles of the GATT are still in force today under the rubric of the WTO. One of the key principles is the most-favored-nation (MFN) principle, which requires each GATT signatory to accord to all other GATT signatories the same trading conditions as it accords to its "most favored nation." Signatories must also provide so-called national treatment to other GATT signatories. This requires them to treat "like products" the same regardless of the country of origin or the process of production. The GATT also calls for a general reduction in the level of tariffs and an elimination of quantitative restrictions on trade. The GATT secretariat was an interim body to oversee the trade agreement.

The GATT was periodically renegotiated through eight trade "rounds." The 1986 Uruguay Round, completed in late 1994, created the WTO as a permanent body to oversee global trade rules. The WTO adopted the rules negotiated under the GATT (which deals with trade in goods), as well as a number of other agreements, including the Trade-Related Intellectual Property Rights Agreement (TRIPS), the General Agreement on Trade in Services (GATS), the Agreement on Technical Barriers to Trade (commonly referred to as the TBT Agreement), the Agreement on the Application of Sanitary and Phytosanitary Measures (commonly referred to as the SPS Agreement), and the Agreement on Agriculture (AoA), which were all negotiated as part of the Uruguay Round. The WTO also created a Dispute Settlement Body and adopted a dispute-resolution process that is compulsory and binding, properties that the GATT dispute-resolution process lacked.[60] A permanent Appellate Body to hear appeals to WTO dispute-panel decisions was also created. Countries found in violation of the WTO rules do not have to remove the offensive trade restriction if they are willing to compensate complainants by allowing them to apply retaliatory trade sanctions.

Following the failed attempt to launch a new round of trade negotiations under the WTO in Seattle in late 1999 (see chapter 3), a new round was finally launched in 2001 in Doha, Qatar (referred to as the Doha Round).[61] Talks are ongoing, but have been on and off again since they began, due to disagreements both among industrial countries and between industrial and developing countries over a number of issues, including most prominently revisions to the AoA.[62] Today 153 countries are members of the WTO, and GATT/WTO trade rules now cover over 97 percent of world trade.[63] WTO members pledge to obey the provisions of the WTO agreements. If one country believes another country has violated these provisions (resulting in a negative impact on the first country), it may turn to the dispute panel for a ruling.

The GATT was created well before major concerns arose over the state of the global environment. It therefore contains few direct references to the environment. The preamble to the WTO does mention the need to pursue "sustainable development," making it somewhat more progressive than the original GATT agreement. But the GATT is still the main trade agreement for goods under the WTO. The main article in the GATT that appears to potentially accommodate environmental concerns is Article XX (see box 5.2). This article sets out cases eligible for exceptions to the GATT rules. It is not terribly clear, however, on whether exceptions are allowed for environmental reasons.

Article XX allows exceptions to the GATT rules to protect human, animal, or plant life or health, or to ensure conservation of natural resources. Yet even these exceptions are qualified. If a country uses trade restrictions to protect the health of humans, animals, or plants under XX(b), such measures must be considered "necessary." That is, if other measures that do not restrict trade are available, then trade restrictions are not seen as necessary. If a country restricts trade to conserve exhaustible natural resources under XX(g), such measures must "relate strictly to natural resource depletion, and domestic restrictions must also be instituted at the same time." That is, if a country restricts trade, it *must* simultaneously impose domestic restrictions.[64] Moreover, any trade

Box 5.2
GATT article XX (the general exceptions clause of the GATT)

Subject to the requirement that such measures are not applied in a manner which would constitute a means of arbitrary or unjustifiable discrimination between countries where the same conditions prevail, or a disguised restriction on international trade, nothing in this Agreement shall be construed to prevent the adoption of enforcement by any contracting part of measures: . . .

b) necessary to protect human, animal or plant life or health; . . .

g) relating to the conservation of exhaustible natural resources if such measures are made effective in conjunction with restrictions on domestic production or consumption.

measures under this article "cannot be arbitrary or unjustifiable," as stated in the introductory paragraph (referred to as the *chapeau*).

More broadly, Article XX does not cover global environmental concerns particularly well. Many environmental problems do not fit into its narrow terms, such as global warming or disposal at sea, and thus do not qualify for exceptions to GATT rules. Measures seeking to protect natural resources outside of the restricting country's border also are not allowed—that is, the trade measures cannot be extrajurisdictional or unilateral, and can apply only to natural resource depletion or animal, plant, or human health in the country imposing the measure. That said, more recent dispute panel decisions have broadened the scope of WTO exceptions somewhat (as discussed later in this chapter).

Article XX and Environment Disputes at the GATT/WTO

Given the potential ambiguity of GATT/WTO rules for environmental protection, and given the stringent conditions that must be met, it is not surprising that there have been a number of environment-related trade disputes at the GATT/WTO.[65] Some countries have cited Article XX as a rationale to justify trade restrictions for environmental reasons, although in early cases these were nearly all ruled as GATT-illegal. The Tuna-Dolphin cases of 1991 and 1994 are two of the best known of the trade-environment disputes at the GATT/WTO. In the 1991 dispute, Mexico complained that the U.S. Marine Mammal Protection Act (MMPA) called for an embargo against imported tuna from countries that could not prove that their fishing methods did not harm dolphins. The GATT panel decided against the United States, and in favor of Mexico. The United States tried to justify its action based on GATT Article XX(b), by arguing that its actions were "necessary" to protect human or animal life; it also relied on Article XX(g), by claiming that it undertook the measures to conserve natural resources. Specifically, the United States argued that the regulation of non-product-related production and processing methods (PPMs)—in this case, tuna fishing that was harmful to dolphins—outside of its own territory was necessary to protect dolphins. The GATT panel did not fully accept this argument, and ruled that Article XX was not applicable in this case. The U.S. use of trade restrictions against tuna from countries that the United States considered not to be dolphin-safe were seen by the GATT panel as unilateral and extrajurisdictional (i.e., application of national laws outside of the territory of the United States) such that they discriminated against "like products," thus violating the national-treatment clause of GATT.[66] Further, the panel pointed out that the United States could have pursued less trade-restrictive, bilateral or multilateral efforts to reduce dolphin kill. It did say, however, that dolphin-safe labels were allowed.[67]

Various interpretations of the Tuna-Dolphin ruling have been put forward. Some have argued that the way the MMPA operated was flawed and thus went against GATT rules. In particular, it determined permissible dolphin kill based on what American fishermen had achieved that same season—something Mexican fishermen could not know in advance. Yet many were critical of the GATT ruling against "extraterritoriality," which was viewed as artificial and inappropriate to contemporary cross-border environmental problems.[68] In the second Tuna-Dolphin dispute of 1994, the European Union challenged the secondary embargo imposed by the United States on tuna from third-party sellers that did not meet U.S. regulations. The GATT dispute panel again ruled against the United States, claiming that the application of the MMPA in this case was unilateral and arbitrary, and thus did not qualify for an exemption under Article XX.

The WTO has ruled on other environment-related disputes as well. It ruled, for example, in favor of Brazil's and Venezuela's complaints against U.S. gasoline standards that put more stringent requirements on foreign imports than on domestic production. In this case, the United States revised its Clean Air Act to remove the element of discrimination.[69] Another example is the 1996 Shrimp–Sea Turtle case. Initially, many environmentalists saw the ruling in this case as proof that the WTO was "anti-environmental." In this case, Thailand, Malaysia, Pakistan, the Philippines, and India all complained about a U.S. law discriminating against imports of shrimp that did not use turtle-excluding devices in their nets, which U.S. law requires. These countries claimed they were taking other measures to protect turtles. The United States tried to claim an exemption to WTO rules under Article XX. The WTO dispute panel ruled, again, that the U.S. requirement that other countries observe its domestic laws was not legal under WTO rules. This behavior was deemed to be discriminatory toward a product based on its production methods, and ineligible for exemption under Article XX because it was a unilateral measure. The United States appealed the case, and the Appellate Body altered the ruling. It ruled that in principle Article XX could cover the U.S. measures, but it added that these were nevertheless applied in an unjustified and arbitrary manner because it assumed that all countries could meet the requirements of the U.S. law equally.[70] The Appellate Body also ruled that the United States did not make enough efforts to negotiate a sea turtle conservation agreement with the complaining countries. The WTO prefers multilateral efforts to deal with such problems, not unilateral measures.

These rulings appear to put trade ahead of the environment. But some argue that a closer inspection shows that the WTO is in fact accommodating many environmental concerns.[71] The explicit reference to "sustainable development" in the WTO preamble has made it possible for dispute panels to consider environmental concerns as part of a trade dispute, something that the GATT could not do. In the case of the Shrimp–Sea Turtle ruling, for example, the decision that the U.S. measures in principle could be taken under Article XX(g), provided they were not applied in an unjustified and arbitrary manner, was very different from the Tuna-Dolphin ruling, which held that the measures to protect dolphins were not allowed at all under Article XX because they applied outside the jurisdiction of the United States. This ruling also indicates that the WTO is somewhat tolerant of the idea of discrimination based on production and processing methods. Some analysts have argued further that the recent rulings by the WTO Appellate Body show that the WTO has taken environmental issues on board, and the fact that in some of these cases it reversed earlier panels shows it is open to more environmental interpretations of Article XX. These rulings have set precedents that will make it easier in the future for countries to use Article XX to restrict trade on environmental grounds.[72]

Other WTO Agreements and the Environment

A number of the other agreements under the WTO from the Uruguay Round also mention the environment in one way or another.[73] The Agreement on the Application of Sanitary and Phytosanitary Measures (SPS Agreement) relates to food safety and human, animal, and plant health and safety regulations. The purpose of this agreement is to ensure that states do not adopt safety measures that create unfair obstacles to trade. The agreement includes exceptions that allow a country to adopt an SPS measure in order to protect the environment, as well as to block trade for a limited period of time in the case of scientific uncertainty. In these cases, measures must be subjected to risk assessment and scientific verification. This has been interpreted as a very limited use of the precautionary principle at the WTO. The WTO encourages countries to harmonize their SPS standards or to adopt internationally recognized standards, such as those developed by the Codex Alimentarius Commission of the Food and Agriculture Organization and World Health Organization. Countries are allowed, however, to have more stringent measures than those set by the Codex. SPS measures must also meet other criteria, in particular that they not be applied in an arbitrary or unjustified way.

Several disputes have been brought forward to the WTO claiming violation of the SPS Agreement. In 1996 the United States and Canada, for example, challenged the ban by the European Union on imports of beef treated with hormones. The WTO dispute panel ruled against the Europeans on the grounds that the ban was not based on a valid risk assessment.[74] Although the European Union lost this case, it maintained the ban and faced the trade sanctions the WTO allows as retaliation for violating the agreement. In a separate case, the European Union's 1998 rules suspending the import of GMOs were undertaken on the grounds that such measures were justified under the SPS Agreement, with the European Union declaring that its ban on the import of such products was temporary while it conducted scientific assessments. The United States, Canada, and Argentina initiated a challenge to the European Union at the WTO in May 2003, claiming that such measures were inconsistent with the SPS Agreement, in part because the suspension of imports lasted for an unreasonable amount of time.[75] The European Union had already begun to bring into place tough new laws on the labeling and traceability of GMOs in July 2003, enabling the European Union to lift its moratorium on the import of GMOs in 2004. The complainants, however, did not drop the challenge, as they saw these laws as still violating the SPS Agreement. The panel ruling, which comprised over three thousand pages including its appendices, was released in September 2006 after considerable delay. The outcome was mixed. The panel found that the European Union's de facto moratorium on GMO imports from 1998 to 2004 was inconsistent with some, but not all, of its obligations under the SPS Agreement. It also found undue delay in the approval of GMO products for import. In response to the ruling, the European Union agreed to amend those regulations that had not already been changed. Many critics, however, were not satisfied with this outcome, with some arguing that the WTO's ruling paid insufficient attention to the role of the Cartagena Protocol in reaching its decision.[76]

The Agreement on Technical Barriers to Trade (TBT Agreement), which seeks to ensure that countries do not use technical regulations and standards to block trade, also allows for countries to adopt technical regulations to protect human, animal, or plant life or health, or the environment. Again, these rules are subject to qualifications. In this case, such measures must be transparent and nondiscriminatory. Like the SPS Agreement, the TBT Agreement encourages countries to harmonize standards or to adopt internationally set standards where they exist, but again allows for exceptions where necessary. One dispute has been brought to the WTO that relates to the TBT Agreement and the environment. This is the Canadian challenge against the European Union regarding a French law banning the import of asbestos and asbestos products. Initially, the dispute panel ruled that the ban was not a technical regulation under the TBT Agreement and that it was allowed under Article XX(b) of the GATT. Following an appeal by Canada, however, the Appellate Body ruled that it was indeed a technical regulation, but nevertheless upheld the ruling that the ban was allowed under Article XX(b).[77]

The General Agreement on Trade in Services (GATS) also incorporates an article similar to Article XX of the GATT, which allows for exemptions if necessary to protect human, animal, or plant life or health, but not relating to the conservation of exhaustible natural resources. But again, such measures must not be arbitrary or unjustifiable. The Trade-Related Intellectual Property Rights Agreement (TRIPS)—which sets rules on intellectual property, such as patents and copyright, and on trade—also has significant environmental implications. It calls on countries to make patents available for inventions—involving both products and processes. This includes microorganisms and biological processes for the production of plants or animals. Plant varieties should be given protection either by patents or by an effective sui generis (designed for a specific purpose) system. But it allows WTO members to exclude from patentability any inventions that may endanger the environment, provided it can be shown that doing so is necessary for environmental protection. Finally, the Agreement on Agriculture (AoA), which seeks to reduce subsidies on agriculture, allows countries to continue to subsidize agriculture in ways that do not distort trade (such subsidies are referred to as the "green box"). Included in the subsidies allowed under the agreement are those that relate to protection of the environment, although these must be part of clearly defined government environmental or conservation programs. No environmentally related challenges have yet been brought forward regarding the AoA, TRIPS, or GATS.

The Committee on Trade and Environment and Potential MEA-Trade Conflicts

The WTO has made other explicit efforts to improve its image on the environmental front over the past decade. In 1994 the General Council of the WTO reconstituted the Committee on Trade and Environment (CTE; a group that was set up in the 1970s but had not met in years) as a permanent body to discuss the relationship of trade and the environment. This committee, which now meets periodically, discusses how the WTO should handle issues such as the link between trade agreements and multilateral environmental agreements (MEAs), ecolabeling, and the ways in which trade and the environment relate to trade rules regarding intellectual property and services. The CTE is charged with making recommendations on these items, including whether to amend the WTO agreements to more explicitly spell out rules on cases where environmental goals can supersede trade goals.

So far there has been a great deal of talk in the CTE, but not much concrete action on these fronts. The slow progress is partly due to opposition from developing countries, and partly due to a lack of leadership by the United States.[78] Developing countries are especially worried about border carbon adjustments now under discussion, because they fear these will allow developed countries to discriminate against them even though the Kyoto Protocol does not require them to reduce emissions. The issue of trade and the environment did not even appear on the official agenda at the failed launch of trade talks in Seattle in November–December

1999. It did, however, make it onto the agenda at the WTO ministerial meeting in Doha in 2001, which launched the Doha Round of trade negotiations. At that meeting, ministers agreed to undertake negotiations for the purpose of clarifying the relationship between MEAs and WTO rules.[79]

The question of the relationship between MEAs and WTO rules warrants further elaboration here because this is an area where there is potential for a trade dispute and the question has sparked considerable debate. There are more than four hundred MEAs, and about twenty of them include trade provisions. In other words, the environmental agreements restrict trade in some way or other as a means to meet an environmental goal. Some of the more prominent MEAs with trade provisions include the Basel Convention, the Convention on International Trade in Endangered Species, the Montreal Protocol, the Kyoto Protocol, the Cartagena Protocol, the Stockholm Convention, and the Rotterdam Convention (for details, see table 3.2). In some cases, an MEA might restrict trade in a risky or endangered product that it is seeking to control. In other cases, it might restrict trade between parties and nonparties in certain products as a means of encouraging countries to join the agreement, and because it would prevent nonparties from exploiting markets for products the agreement is seeking to regulate.[80] Many MEAs include other control measures that may also affect trade, such as provisions for technology transfers and prior informed consent before goods can be traded.

MEAs use such measures in an effort to ensure compliance and meet environmental goals. But some analysts argue that these have the potential to be WTO-inconsistent. That is, they could be interpreted as being incompatible with WTO rules on most favored nation and national-treatment issues, as well as with WTO requirements to eliminate quantitative restrictions. Such inconsistencies are seen to be most problematic when one country that is party to a multilateral environmental agreement applies trade restrictions against a country not party to the MEA, yet both countries are WTO members. An example is the Montreal Protocol, which imposes more stringent restrictions on trade in ozone-depleting substances, or products produced using them, between a party and a nonparty than between two parties. In cases where both countries are party to an MEA (and thus have voluntarily endorsed the provisions of the agreement), most analysts believe it is unlikely that a conflict over trade and environment rules would reach the WTO. But problems could arise in these cases, too. The Basel Convention, for example, bans the trade in hazardous waste between parties listed in the convention's annex VII (mainly non-OECD countries) and those that are not. Some parties to the convention could, in theory, argue that this is discriminatory because it discriminates based on a country's membership in the OECD. Other types of MEA trade-related measures, such as those that deal with the precautionary principle and intellectual property rights, also have the potential to create tensions.[81]

Thus far, no cases have come forward that officially challenge MEAs as being WTO-inconsistent. But environmental activists are concerned about the growing number of environmental agreements that certain countries have refused to ratify. For example, key environmental agreements, including the Basel Convention, the Cartagena Protocol, and the Kyoto Protocol, still lack ratification from a number of countries, most notably the United States. In such cases where there are significant economic players who are not parties to MEAs with trade provisions, the potential for a challenge at the WTO is heightened. The outcome of any such dispute would be binding under the WTO rules, and would set a precedent for future cases. Some argue, too, that the reach of various WTO procedures and agreements puts a "chill" on the capacity of MEAs to use trade restrictions effectively as legitimate environmental measures. The dominance and power of the WTO's dispute resolution process compared to the enforcement mechanisms of MEAs, and the fact that the United States is not a member of most MEAs, means that trade objectives tend to trump environmental ones.[82]

The WTO could address the potential inconsistencies between the WTO rules and MEA rules in several ways. One would be for the WTO to waive certain trade obligations when trade restrictions are imposed for MEA purposes. Another would be to amend the WTO rules to list MEAs exempted from trade challenges.[83] The WTO could also draw up rules that trade measures in MEAs would have to follow to be consistent with WTO rules.[84] Changing the WTO rules, however, can be done only by consensus, and not all countries agree on the need for change. Short of amending the WTO rules, the clarification of the relationship could by default be determined by dispute panel resolutions.

The relationship between MEAs and the WTO was debated at the Johannesburg World Summit on Sustainable Development in 2002. The Plan of Implementation from the Summit calls on signatories to "promote mutual supportiveness between the multilateral trading system and the multilateral environmental agreements, consistent with sustainable development goals, in support of the work program agreed through the WTO, while recognizing the importance of maintaining the integrity of both sets of instruments."[85] As of July 2010, these discussions were still under way.[86]

Reforming the WTO?

Although it could be argued that these steps taken within the WTO indicate that some progress is being made in that institution regarding environmental considerations, there still remains disagreement about whether, and the extent to which, such change should be taking place. Most market liberals argue that the WTO rules as currently constituted are sufficient for dealing with environmental concerns, and extreme market liberals argue that environmental issues have little or no place in trade policy. More moderate market liberals do acknowledge that some minor reforms to the WTO could help spell out its relationship to the environment more explicitly. Most market liberals see the merits of clarifying the relationship between MEAs and WTO rules, for example, but some are still skeptical, and want to be sure that environmental agreements represent genuine environmental efforts and are not just protectionism in disguise. Increased transparency and clear rules for ecolabeling are also sought.[87]

Such changes to the WTO are deemed minor, and overall most market liberals argue that WTO rules are green enough. The liberalization of agricultural trade is a good example of this point. The aim of the WTO's AoA is to reduce tariffs on agricultural products and to eliminate or reduce agricultural subsidies. Market liberals argue that such measures are key not just for facilitating trade and promoting growth, but also for environmental protection, because the removal of trade-distorting tariffs and subsidies will make land use much more efficient.[88] Working toward these original aims of trade liberalization, then, is the best policy for the environment, according to these thinkers. Market liberals are also wary of incorporating a precautionary principle into the WTO's rules, which they see as an open invitation for trade protectionism.

Institutionalists have tended to argue for more substantial reforms to the WTO. They see the merits of a WTO, and with some substantial retuning see it as essential in the quest for sustainable development. Institutionalists agree with market liberals on the need for a clearer delineation of the way MEAs and the WTO should interface, though they are stronger advocates of the use of trade sanctions in MEAs for achieving environmental goals. They also agree on the need for explicit rules for ecolabeling and measures such as border carbon adjustments, and on the necessity of transparency, accountability, and participation from civil society in the WTO processes. Going further than market liberals, most institutionalists back the limited use of the precautionary principle in dispute resolutions at the WTO and accept discrimination based on production and processing methods.[89] Some go so far as to argue for a new organization to counter the power of the WTO, one perhaps called a World Environment Organization (WEO).[90] Along with nongovernmental actors, such an organization could, these institutionalists believe, help embed environmental norms into the global political economy (see chapter 8).

Most social greens and some bioenvironmentalists are far more critical of the WTO, often calling for a complete overhaul of the institution or sometimes even its dissolution. Many social greens see the WTO/GATT system as an instrument used by capitalist states and corporations to liberalize trade.[91] Moreover, they argue that it is secretive, undemocratic, discriminatory, and antienvironmental. For this reason, they see genuine reforms using its existing framework as nearly impossible. And they praise the few international environmental agreements that include possible trade sanctions.[92] But they also argue that these treaties could go much further to restrict environmentally harmful trade. In addition, many social greens and bioenvironmentalists justify unilateral trade actions as a sovereign right of states to protect their own (and the global) environment,[93] although as noted earlier, there is some disagreement over the use of measures such as border carbon adjustments that could be particularly harmful to poor countries. Some also propose building alternative global institutions. Colin Hines of the International Forum on Globalization, for example, advocates a General Agreement on Sustainable Trade (GAST). A GAST would not discourage trade; rather, it would encourage states to treat domestic industry favorably, as well as allow states to give preferential treatment to goods from other states that respect human rights, labor rights, and the environment. It would also allow citizen groups to sue companies that violate the agreement.[94] Other proposals include an expansion of "fair trade"—that is, trade with communities that pay fairer prices for goods and that can ensure environmentally sound production to small scale producers. The fair-trade movement is growing rapidly for items such as coffee, handicrafts, and bananas. For example, the world's biggest coffee chain, Starbucks, which started serving fair trade coffee in 2000, increased its purchases of Fair Trade Certified coffee from just under 650,000 pounds in 2001 to 40 million pounds in 2009, making it the world's largest purchaser of Fair Trade Certified coffee.[95] Despite increasing steadily over the last decade, however, fair trade products continue to comprise a small proportion of global trade as well as overall natural resource consumption in developing countries.[96]

Regional Trade Agreements—Opportunity for Greener Models?

If there is some difficulty reconciling the tensions between trade and the environment within the WTO, what are the prospects for regional trade agreements? Here we look at NAFTA and the European Union as examples of regional trade agreements to examine how each grapples with environmental issues.

NAFTA, negotiated in the early 1990s, is arguably a more environmentally friendly trade treaty than the GATT/WTO. It explicitly deals with environmental issues to a much greater extent than the GATT, specifically mentioning the environment in the preamble and main text of the treaty (in five of twenty chapters). Chapter 1 of the NAFTA text stipulates that the trade provisions embodied in the environmental treaties named in the trade agreement take precedence (it mentions the multilateral treaties CITES, Basel, and the Montreal Protocol, as well as four bilateral treaties). This exemption for these agreements is qualified, however. NAFTA parties must choose the least trade-inconsistent measure to meet their obligations under these treaties, regardless of whether that measure is more politically or economically feasible.[97] Chapter 11 of the NAFTA text also explicitly discourages pollution havens. That is, it states that it is inappropriate for parties to encourage investment by relaxing domestic standards for health, safety, or the environment. (The next chapter of this book examines the concept of pollution havens and the investment provisions of NAFTA more closely.) Other parts of the NAFTA agreement also mention the environment in more flexible and explicit ways than does the GATT, incorporating a more precautionary approach.[98]

There is in addition an environmental side agreement parallel to the measures in the NAFTA text: the North American Agreement on Environmental Cooperation (NAAEC). The purposes of the side deal are to foster environmental cooperation among NAFTA parties, to ensure that states comply with and enforce their own national environmental laws, and to establish a mechanism for settling environmental disputes. Parties are required under the side agreement to report periodically on the state of their environment. A new institution, the Commission on Environmental Cooperation (CEC), was created to oversee the NAAEC. It conducts studies on environmental performance and policies in the NAFTA countries; it also hears environmental complaints and settles disputes. If a party feels that one of the other parties to the agreement consistently fails to enforce its own environmental laws, it can bring a complaint forward to the CEC. The CEC is much more open than the WTO to citizen and NGO participation, and these groups can individually launch such complaints.[99] The side agreement also established the Border Environmental Cooperation Committee (BECC) and a North American Development Bank.

Serious disagreements arose over the environmental dimensions of the NAFTA treaty during negotiations, including among environmental groups—particularly those in the United States. Groups drawing on social green and bioenvironmentalist analysis, including Friends of the Earth, Sierra Club, Public Citizen, and Greenpeace, were highly critical of most aspects of the agreement. In particular, these groups opposed the agreement for what they foresaw as its negative environmental and development impact. They were critical of the negotiation process, too, which they felt corporate interests had effectively co-opted. A more moderate position, taken by groups following more of the reasoning of institutionalists and market liberals, including the National Resources Defense Council, the WWF, and Environmental Defense, more or less endorsed the way environmental concerns were brought into the treaty.

They basically agreed that trade liberalization was a potentially positive force, or was at least going to occur regardless, and wanted to ensure that environmental safeguards were built into the treaty.[100]

Has NAFTA been "greener" in practice than other trade agreements? The answer is mixed. On the positive side, no major decline in standards or enforcement has occurred in the United States.[101] Meanwhile, Mexico's environmental laws have improved since NAFTA came into force, with levels of enforcement stable.[102] Serious environmental violations in the industrial region of Mexico near the U.S. border have also decreased considerably, with more than 400 firms signing environmental-compliance action plans.[103] And the share of dirty industry in Mexico's overall economy has declined.[104] On the negative side, Canada's post-NAFTA record has been less impressive.[105] So far, the CEC has also not accomplished much. Critics further argue that NAFTA has devastated Mexican smallholder agriculture, altering land-use practices and polluting the countryside with toxic chemicals. Mexican and Canadian imports of toxic waste from the United States have increased since NAFTA. Mexico alone saw a doubling of hazardous-waste imports in the second half of the 1990s. A report published by the Texas Center for Policy Studies found that exports of hazardous waste from the United States to Mexico increased from 158,543 tons in 1995 to 254,537 tons in 1999. Similarly, Canada saw its hazardous-waste imports rise from 383,134 to 660,000 tons in the same period.[106]

The case of the European Union's handling of the trade and environment tensions is somewhat different from that of NAFTA. The origins of the European Union go back to 1957 when the Treaty of Rome—an agreement that sought to promote economic and political integration among European states (rather than simply trade liberalization)—was signed.[107] The Single European Act (1986), the Treaty on European Union (1992), the Treaty of Amsterdam (1997), and the Treaty of Lisbon (2007) have reshaped the European Union in ways that promote further political and economic integration, including the harmonization of environmental laws.[108] Over this period, the European Union has developed and adopted EU-wide policies for environmental protection that are not seen as secondary to economic goals.[109] A number of policy directives have been adopted regarding the environment, in an attempt to harmonize regulation across the European Union. These include policies regarding waste, air pollution, water, nature protection, and climate change, for example.[110]

The European Union's goal of political as well as economic integration has meant that the implementation of harmonized environmental laws has taken place parallel to, but separately from, economic integration. For this reason, the economic impact of the harmonization of environmental laws across Europe has not generated much controversy. In fact, harmonization of standards has been considered important in facilitating trade in the region. Certain EU states, however—primarily those with weaker environmental standards to begin with, such as Greece, Portugal, Spain, and Italy—were more hesitant than others to adopt stringent EU-wide environmental policies. For this reason, the European Union's environmental directives have some flexibility built into them, to allow for some degree of difference in requirements according to member states' economic and environmental situations.[111] Despite these differentials, however, the European Union is widely seen to be a successful case of upward harmonization of environmental laws within the context of enhanced economic integration.

Conclusion

A consensus is growing within the global policy community that trade and environment are mutually compatible—indeed, that both are essential for the health of the other. The institutionalist position on trade has grown stronger over the last decade, and today, the global policy goal adopted by most governments is to "manage" trade to foster *both* economic growth and better environmental conditions. This requires stronger national environmental policies and international environmental laws and norms. It also requires states to integrate environmental concerns into free-trade agreements and to use trade tools for environmental protection, provided that they are well targeted and do not distort trade more broadly. The collective constraints of these rules must allow trade to still remain "free enough" that it can continue to power economic growth in all corners of the globe.

This does not mean the "fight" over trade is over. Market liberals still see the potential to go too far—and some feel this has already happened. That is, global environmental regimes will become yet another barrier to the accumulation of global wealth—and that, in turn, will impoverish all. Many bioenvironmentalists and social greens are far more skeptical than market liberals about the compromise of managed trade. They argue that it does not address the overconsumption or gross inequalities intrinsic to the global trading regime. It does little to restrain the destructive activities of multinational firms or to internalize the full social and ecological value into the prices of traded goods. It also does little to raise global environmental standards, and in many cases allows (and even encourages) states to keep domestic standards low (and can even create incentives for states to lower standards further). In short, social greens and bioenvironmentalists see global trade and free-trade agreements as little more than engines of unsustainable macroeconomic growth, creating ever-greater stress on the global ecological system.

Even many institutionalists feel there is a long way to go, although these concerns are more about particulars and less about the theoretical foundations of trade and the environment. Many institutionalists, for example, argue that under the WTO and NAFTA in particular, it is still far too difficult to ban trade in harmful products, especially on the basis of production or processing methods. Many note further that global trade rules do not set any minimum standards for environmental regulation—only ceilings, not floors. It is much easier under international trade law to contest another country's high environmental standards as unfair; it is far harder to contest weak environmental standards as unfair. Thus, the safer position for states is to weaken standards, reinforcing the race to the bottom. Finally, many institutionalists point out that the overlap of trade agreements with environmental clauses and environmental agreements with trade provisions creates unclear and at times ambiguous global trade and environment rules.

The norms of free trade and environmental trade clauses steer much of the environmental conduct of transnational corporations. In the next chapter, we explore the environmental consequences of TNCs in more detail.

6

Global Investment and the Environment

The public often associates corporations with the world's highest-profile transnational firms, such as General Electric, Royal Dutch Shell, Exxon-Mobil, Wal-Mart Stores, Microsoft, Toyota, Procter & Gamble, Johnson & Johnson, and ChevronTexaco.[1] There are, however, tens of thousands of parent TNCs and hundreds of thousands of affiliates. Sales and profits drive these firms. Economic globalization, which is extending free trade, privatization, deregulation, and the amount and speed of global financial flows, is also enhancing the power of these firms. There are increasing opportunities for higher sales and higher profits in a borderless world market. A quick glance at the opportunities in China—with over 1.3 billion people and with rapid increases in domestic consumption—is illustrative. Since China became a member of the WTO in 2001, TNCs have been eager to invest there, seeing great potential for immense profits from such investments.[2]

Market liberals applaud these developments. Successful TNCs mean a healthy global economy. A healthy global economy means strong growth in national economies—in both rich and poor countries—which in turn translates into more state and corporate funds for better environmental management. TNCs, moreover, have the financial resources for sizable investments in developing countries, bringing new technologies (which are cleaner and more efficient than their predecessors) as well as higher management standards. TNCs are, in other words, the engines of sustainable development for both developed and developing countries. Institutionalists agree here, although they add that the profit imperative of firms means that in some cases the international community (through policies and financial incentives) needs to guide the actions of firms to ensure sustainable development.

As with trade, bioenvironmentalists and social greens hold very different views from market liberals and institutionalists on the environmental impacts of TNCs. Bioenvironmentalists see TNCs as engines of overproduction and overconsumption. TNCs extract, process, and export to rich countries the bulk of the globe's natural resources, creating deforestation, overfishing, overmining, and desertification. They manufacture, brand, and sell the bulk of the world's products, constructing a culture of consumerism with clever advertising. They pollute the earth's air and water with dangerous chemicals like dioxins, furans, polychlorinated biphenyls (PCBs), and DDT. And they transport and improperly dispose of garbage and hazardous waste, sometimes shifting it from rich to poor countries. Social greens largely accept the bioenvironmentalist evaluation of TNCs with respect to the environment. They add, however, that in the quest to maximize profits TNCs are responsible also for the inequality and exploitation of much of humanity. Billions of people now scramble for small sums of currency to survive, tilling increasingly unproductive land for export crops (e.g., coffee, bananas, and soybeans) and slaving away in perilous factories (e.g., to sew cheap clothes or make inexpensive shoes). The nutritional balance of subsistence farming is lost. So is the community well-being and cooperative environmental management of the past. TNCs are, in other words, the engines of injustice and exploitation of both nature and humanity.

This chapter begins with an overview of recent trends in the globalization of TNCs and foreign direct investment. It then examines the debate over environmental standards, addressing in particular the question of whether firms relocate to take advantage of lax environmental rules in developing countries. Following this is an analysis of the site practices of TNCs in the developing world. The chapter also examines the reaction of TNCs to pressures for environmental reforms, dividing the discussion into those who argue the corporate sector is "greening" and those who argue it is basically "greenwash." It ends with a discussion of the role of TNCs in global environmental and investment governance.

Globalization and Transnational Corporations

Foreign direct investment (FDI) by TNCs is one of the main consequences and drivers of economic globalization. FDI entails ownership or investment in overseas enterprises in which the investor plays a direct managerial role. A TNC by definition invests in overseas operations. Generally, a parent firm of the TNC holds minority or majority shares in numerous branches, subsidiaries, or affiliates in more than one country, creating complex managerial structures.[3]

The number of TNCs has increased markedly since the 1970s, growing from 7,000 TNC parent firms in 1970 to over 79,000 in 2007, with some 790,000 foreign affiliates around the world (see figures 6.1 and 6.2). Foreign affiliates in 2007 accounted for 11 percent of global GDP. That year these firms employed eighty-two million people worldwide, with sales of US$31 trillion (a 21 percent increase from 2006).[4]

World FDI net inflows have risen significantly since the 1970s, although much of the acceleration has occurred over the last two decades. In 1982 FDI net inflows were valued at US$58 billion, growing to US$207 billion by 1990, and reaching US$1.4 trillion in 2000. Following a sharp drop in investment in 2001 after the bursting of the Internet investment bubble and the September 11, 2001, terrorist attacks in the United States, FDI rebounded to a new record high of US$2.3 trillion in 2007, only to fall again to US$1.8 trillion in 2008 following the global financial crisis (see figure 6.3). Although this growth in FDI has been particularly rapid over the past two decades, like trade it has not been evenly distributed.

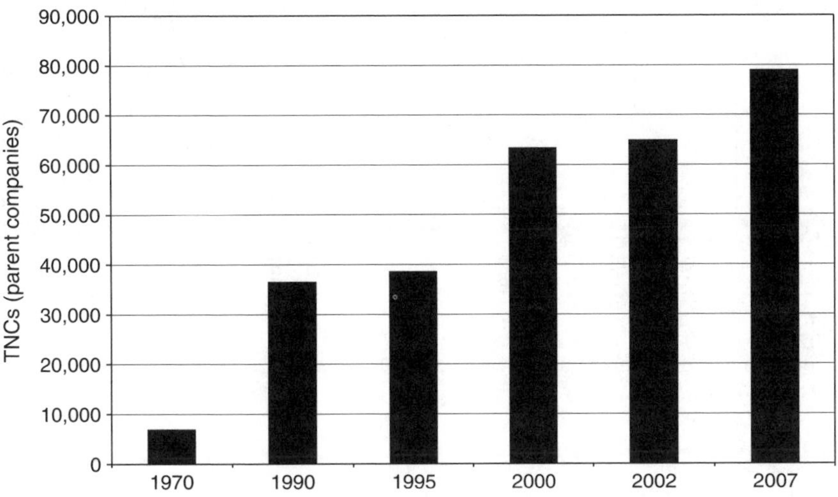

Figure 6.1
World total number of TNCs (parent companies). *Source*: UNCTAD World Investment Report 2008, <http://www.unctad.org/en/docs/wir2008_en.pdf pg xvi>.

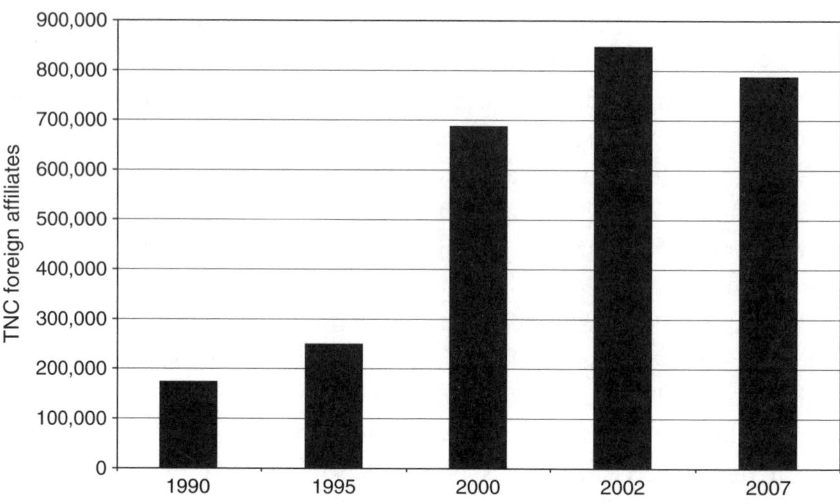

Figure 6.2

World total number of TNC affiliates. *Source*: UNCTAD World Investment Report 2008, <http://www.unctad.org/en/docs/wir2008_en.pdf pg xvi>.

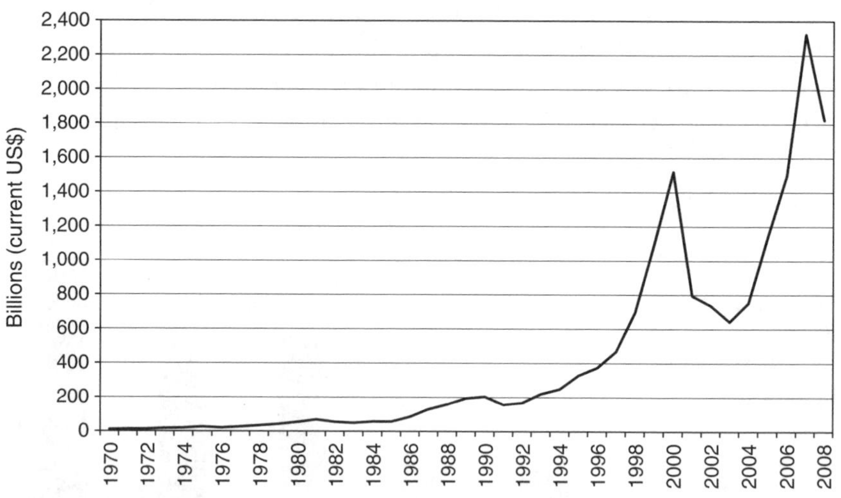

Figure 6.3

World foreign direct investments, net inflows, 1970–2008. *Source*: OECD, <http://stats .oecd.org/index.aspx>.

In 2007 developed countries accounted for 68 percent of FDI, with the European Union, the United States, and Canada attracting 43.9 percent, 12.7 percent, and 5.9 percent, respectively. Developing countries accounted for only 27.3 percent of FDI inflows in 2007, with 46 percent of those flows going to just six developing countries.[5] Of the developing country regions, Asia-Oceania (which includes China) and Latin America attracted respectively 17.5 percent and 6.9 percent of global FDI inflows in 2007, while Africa attracted a mere 2.9 percent.[6] The liberalization of investment rules since the 1980s has had much to do with the rapid pace of growth in FDI and TNCs in that period. Between 1992 and 2007, of some 2,540 regulatory changes to domestic FDI rules in countries around the world, 2,292 (90 percent) were changes that created a more favorable environment for investment.[7]

Many transnational corporations have expanded their power through mergers or acquisitions over the past two decades. The value of cross-border merger and acquisition activities was approximately US$200 billion in 1990, rising to a record US$1.6 trillion in 2007. From 1986 and 1990 cross-border mergers and acquisitions between TNCs grew at an average annual rate of 26.6 percent—in 2007 alone the rate was 46.4 percent.[8] The result has been increasingly large corporations, with sales figures that surpass the levels of GDP in many countries. According to the ETC Group, for example, Wal-Mart's revenues in 2007 exceeded the gross national income of both Greece and Denmark, while Toyota's revenues were greater than the gross national income of Venezuela.[9] Together, their impact is enormous. In 2007 the top 100 TNCs held US$9.2 trillion in assets, had US$7.2 trillion in sales, and employed over fifteen million people.[10]

Globalization has also concentrated corporate power by facilitating control over specific markets. In 2006, for example, the top 10 drug companies controlled 55 percent of global pharmaceutical sales, and within this group, the top four vaccine companies controlled 91.5 percent of the global vaccine market. Similarly, the top three seed companies in 2007 accounted for 47 percent of the global proprietary seed market. One company alone, Monsanto, accounted for 23 percent of the proprietary seed market, as well as 87 percent of the global area planted with genetically engineered seeds.[11]

Differential Standards: Pollution Havens, Industrial Flight, Double Standards?

Critics of TNCs often accuse them of intentionally locating their operations in areas with weak or poorly enforced environmental regulations. A core debate here is over whether highly polluting industries migrate to countries with lax environmental regulations. Is there an "industrial flight" of firms from countries that impose stringent environmental regulations? Do firms seek "pollution havens" in countries with lower environmental standards?[12] To follow these debates, it is essential to distinguish carefully between industrial flight and pollution havens. Industrial flight occurs when the *raising of environmental standards* in one country triggers the relocation of that industry to another country with lower standards. Pollution havens exist when two criteria are met: first, a country (usually, but not necessarily, a developing country) *sets its environmental standards at an inefficiently low level* (in economic terms) in order to attract foreign investment, and second, industry relocates to this country in order to save on pollution abatement costs. A pollution haven is not simply a state with low environmental standards. Nor is it simply an area with lots of pollution. These situations, all agree, are common throughout the developing world. World Bank economist David Wheeler explains:

> To qualify, the region has only to use weak environmental regulation as an inducement to investment by cost-conscious polluters. These polluting firms may be owned by either domestic or foreign investors; they may use new or re-located facilities; and they may produce for either domestic or foreign markets. What really counts for the pollution havens debate is neither ownership nor market location, but the willingness of the host government to "play the environment card" to promote growth.[13]

Pollution havens and industrial flight are thus two sides of the same coin: importantly, both concepts seek to capture the implications of differential environmental standards (and hence, costs) between different countries.

The debate over whether pollution havens and industrial flight are real phenomena began in the 1970s. It arose out of the fear that higher environmental regulations in the United States might drive industries, and with them jobs, to developing countries with lower environmental regulations. This concern remains strong even today. But it is paralleled by a concern about the environmental and health implications for countries that host dirty industries.[14] There is particular concern with the state of the environment in the developing world because, at least in some cases, conditions there continue to deteriorate as those in the industrialized world improve. Again, though, it is critical to note that this does not confirm the presence of industrial flight or pollution havens. We now turn to the evidence and arguments for and against the existence of these phenomena.

Differential Standards Are Inevitable

A number of studies in the 1970s and 1980s tested the underlying hypothesis connecting pollution havens and industrial flight—that is, the idea that firms relocate in response to changes in environmental regulations. Coming mainly from a market liberal perspective, these studies were primarily econometric and drew on pollution emissions and emission-control cost data for U.S. manufacturing firms. Most of the studies found that the probability of firms relocating to developing countries in order to avoid more stringent environmental regulations in rich countries was low.[15] One of the main reasons cited for the "elusiveness" of pollution havens and industrial flight is often simple: costs like labor and technology are far higher than environmental costs; therefore, a shift in environmental regulations is not enough to encourage firms to relocate on those grounds alone. Pollution control costs rarely constitute more than 2 percent of sales.[16] There are other factors as well. Some industries—such as electricity generation—need to remain close to their market, and it may be impractical to relocate. Also, for some industries stable and predictable standards matter more than compliance costs.[17]

The intensity of pollution is undeniably rising in some poor countries. Yet market liberals like David Wheeler do not believe this is connected to TNC reactions to regulatory differences between rich and poor countries. Wheeler in fact found that over the last decade air quality had improved in three of the larger recipients of FDI in the developing world—Brazil, China, and Mexico.[18] The decline in investment in "dirty" industries in industrialized countries since 1960 has no doubt corresponded with increasingly tight environmental regulation. And there has no doubt been a rise in many of the same pollution-intensive industries in the industrializing world. Yet market liberals maintain that the cause is shifts in *domestic* production and consumption patterns—often the products from "dirty methods" are for home use, not for export—and not shifts in TNC investment to take advantage of regulatory differences. There is, in other words, no "race to the bottom" as a result of industrial flight from rich countries or the pull of pollution havens. Moreover, in the rare instance when pollution havens arise, these are relatively short-lived phenomena. Economic growth soon provides the means for a more adequate environmental response such as improved regulation and a rise in investment in cleaner production processes.[19]

For market liberals, differential standards are thus not causing an influx of foreign investment into developing countries. Some have even argued that lax standards deter FDI, and that higher standards actually attract investment.[20] In some instances, as in the case of Japan, high domestic standards act as a deterrent to investment in Association of Southeast Asian Nations (ASEAN) countries with weaker standards.[21] Environmental problems arise mainly as part of domestic industrial growth (as seen in the environmental Kuznets curve predictions in chapter 4), not differential standards. For most market liberals, differences in environmental standards among countries are normal, or at least inevitable. In their view, as mentioned in chapter 5, it is unrealistic to expect all countries to maintain identical standards and ecological conditions. Different countries quite naturally have different pollution absorption capacities, both politically and physically.[22] Such differences can even give countries useful absolute or comparative advantages for production and trade. A 1991 World Bank memo, under the signature of Larry H. Summers, former vice president and chief economist of the World Bank, former president of Harvard University, former director of the National Economic Council in the Obama administration, and current Harvard professor, bluntly explained the logic:

> The measurement of the costs of health-impairing pollution depends on the foregone earnings from increased morbidity and mortality. From this point of view a given amount of health-impairing pollution should be done in the country with the lowest cost, which will be the country with the lowest wages. I think the economic logic behind dumping a load of toxic waste in the lowest-wage country is impeccable and we should face up to that.[23]

Market liberals contend that differential standards occur alongside the transfer of clean technologies and better management to developing countries. The transfer of these technologies and environmental management skills results in what some call a "pollution halo," where pollution levels drop as a result of FDI.[24] Some studies, for example, have found environmental improvements resulting from FDI in China.[25] In developing countries, market liberals argue, even TNCs in dirty industries are more likely than domestic firms to use more sustainable technologies and more elaborate environmental policies.[26] Why do TNCs perform better than local firms (which are often much smaller)? Most importantly, perhaps, TNCs have the financial and organizational resources to access more sophisticated technologies and managers. There are other reinforcing factors as well. TNCs sometimes keep overseas standards relatively high as part of a risk-management strategy, so sudden environmental reforms or lawsuits do not disrupt corporate activities. They sometimes raise environmental standards overseas to "voluntarily" satisfy NGOs, international organizations, and consumers. TNCs may also employ higher standards and cleaner technologies to gain a competitive advantage in production (as new technologies are generally more efficient and produce higher-quality products), and this efficiency is reinforced when they use the same practices in all of their manufacturing plants, regardless of location or differential standards.[27]

Market liberals tend to argue that smaller and local firms operating in developing countries are in fact more polluting than larger and multinational-affiliated firms. This occurs in large part because they lack funds to install cleaner production technologies. Domestically owned firms also tend to be more secretive and less accountable to both public and "outside" pressures. Further, FDI originating from industrialized countries tends to be "cleaner" than FDI from developing countries.[28] In addition, smaller firms are more dispersed, and for this reason it may be harder to ensure their compliance with environmental regulations. Locally owned firms may also be more polluting because incentives to be "cleaner" are lower where land ownership and rights are not clearly delineated, as is the case in much of Southeast Asia.[29] Market liberals see state-owned local firms in the developing world as particularly prone to be heavy polluters.[30]

Although they point out these differences between the environmental performance of TNCs and domestic firms, market liberals *do not* argue that the overseas standards of TNCs are always the same as those in their home country. Rather, they stress that the environmental performance of TNCs is often better than that of both domestic firms and the minimum standards of host-country laws. To further improve environmental performance of both domestic firms and TNCs, market liberals call for market-friendly measures like public disclosure of pollution emissions and pollution fines. They highlight cases in the developing world, such as in Indonesia and the Philippines, where public disclosure of emissions has had an impact on the environmental practices of firms.[31] The promotion of voluntary measures under the policy of corporate social responsibility (CSR) is also seen as positive, as TNCs with strong CSR policies are less likely to locate in poor countries with weak environmental regulations.[32]

Differential Standards Cause Environmental Harm

Others are far more skeptical of the impact of foreign direct investment on the natural environment. Social greens are perhaps the most critical, although bioenvironmentalists and institutionalists share some of their concerns. Most see pollution havens and industrial flight as real or potential threats. Indeed, the most environmentally damaging industries tend to have a high concentration of TNCs as the main investors and operators. Some of the worst offenders are global extractive TNCs that mine natural resources—loggers, oil drillers, and miners. This pattern of foreign direct investment in high-environmental-impact industries has afflicted both developed and developing countries, although the latter are now receiving more and more of this investment. The evidence social greens and bioenvironmentalists present, though, relies less on the statistical data of market liberals and more on in-depth case studies. Much of it, market liberals retort, is mere anecdotes. But for social greens in particular, this evidence allows for analysis of the impact of TNCs on real people—on individual lives and the well-being of communities, notions not easily captured by econometric surveys of data.

Social greens and bioenvironmentalists are concerned about the rise in the share of highly polluting industry in developing countries, and accuse TNCs of following what they call "double standards"—that is, when a firm applies one set of standards in their operations in one country, and a different set of standards in another.[33] Usually this means more lax standards are applied in TNC operations in developing countries, while higher standards for the same production processes are adhered to in the home country. Several UN studies have confirmed that TNCs operating in developing countries conform to standards lower than home-country standards.[34] As mentioned earlier, market liberals do not deny that this occurs—they just argue that this is still a better situation than following host-country standards. Social greens in particular see this practice as unjust, exploitative, and dangerous.

Social greens highlight the case of the American TNC Union Carbide in Bhopal, India, the site of the worst industrial accident in history, as a clear case of double standards and environmental injustice.[35] The Carbide Bhopal plant produced pesticides. Union Carbide owned 51 percent; the Indian government owned the rest. Many claim that Union Carbide was aware in the early 1980s that this factory had much more lax environmental, health, and safety standards than a similar pesticide plant owned by the same company in West Virginia (which produced an identical pesticide).[36] Little was done to upgrade the Indian plant. In the middle of the night on December 4, 1984, a leak at the Bhopal plant spread poisonous gas over the town. Eight thousand people were instantly killed, and several hundred thousand more were injured.[37] Union Carbide blamed the tragedy on the Indian government for not conducting better checks on the plant, and argued that sabotage likely caused the accident. But critics blamed it on the negligence of Union Carbide, arguing that it was not an accident but rather a "disaster waiting to happen."[38] The company was taken to court and was forced to pay compensation to many of the victims, although many others received nothing for their suffering and those that were given funds only received a small settlement. Union Carbide pulled out of India ten years later, and the company was bought out by Dow Chemical in 2001. An investigation by *New Scientist Magazine* claims the U.S. headquarters was responsible for the plant's design, and that Union Carbide had scaled back its investment in safety technologies at Bhopal in the early 1980s.[39]

The maquiladora manufacturing plants (factories that assemble goods explicitly for export) in Mexico are another case of TNC environmental abuse in a developing country. About 2,000 American companies, including General Electric, Ford, General Motors, and Westinghouse, moved their assembly plants over the border to northern Mexico in the 1980s and 1990s, avoiding laws in California regulating toxic emissions. Reasonable environmental laws are on the books in Mexico. But the Mexican government has not enforced them effectively. In the 1960s and 1970s most of the maquiladora factories were in sectors such as garment assembly and were not highly polluting. Over the course of the 1980s, however, the composition of these factories changed dramatically. By the 1990s the main sectors were chemicals, electronics, and furniture, all of which are highly polluting.[40] By the early 1990s nearly 90 percent of maquiladoras used toxic materials in their production. One reason for these changes was the increase in the quantity of investment in these sectors since the 1970s.[41] For example, from 1982 to 1990 investment rose twentyfold in maquiladora chemical factories after a tightening of environmental regulations on the industry in the United States.[42] Environmental studies of conditions near these plants in this period revealed polluted local rivers, soils, and water supplies. Workers have acute health problems, and there has been an increase in birth defects. At fault, critics claim, are corporations that are ignoring environmental regulations.[43]

For social greens and bioenvironmentalists, the ecological performance of TNCs in these cases is typical of how TNCs generally exploit the peoples and environments of developing countries.[44] Similar types of TNC practices have been found in other sectors, such as mining, oil extraction, and logging (as discussed shortly). Social greens stress the injustice and environmental harms TNCs cause; bioenvironmentalists tend to focus mainly on their ecological impacts. Both argue that economic globalization means that the smallest cost differences are increasingly influencing corporate decisions on where to locate, and on how these corporations will operate once they have relocated. It also means that governments increasingly hesitate to regulate strictly, for fear this will drive firms away. Globalization, then, is not diffusing cleaner technology globally as the market liberals argue, but allowing more and more TNCs to exploit remote habitats and peoples.

Bioenvironmentalists see all corporations—usually in the context of capitalism, profit maximization, and overconsumption—as potential sources of ecological problems. Social greens see TNCs as unconnected to local communities, which makes them more likely to exploit workers and the environment. For them, local firms are innately better environmental managers. Local firms are more oriented toward local needs.

They tend to be labor- rather than capital-intensive, and generally function on a smaller scale than TNCs. They also tend to produce goods more appropriate to the needs of the people living in the nearby region. Of course there are many exceptions, as in China, where small-scale manufacturing enterprises account for a significant proportion of pollution. Nevertheless, in many developing countries, TNCs produce the bulk of goods and thus pollution.

Expanding the Debate

Many critics of the market liberal view of TNCs argue that the debate over differential environmental standards has been too narrow and too polarized, making it difficult to move forward with concrete policy steps. The economic literature on pollution havens has dominated the debate. It defines "dirty industry" in narrow terms and in doing so fails to adequately encompass location decisions of all polluting firms that operate globally. It tends to rely heavily on data on pollution emissions or on expenditures for emission controls by manufacturers to determine which sectors are highly polluting and dirty. This focuses too much on manufacturers, critics argue, and ignores some of the world's worst polluters and degraders, such as the hazardous waste and natural resource firms.[45] Several recent econometric studies using more complete datasets and controlling for a number of factors have found some evidence of pollution haven effects, and have argued that certain datasets and methods are more likely to underestimate these effects.[46] Further, there may be havens of environmental degradation, rather than simply industrial pollution, but such a narrow framework ignores them.[47] Another problem with many of the older econometric studies of pollution havens is their focus on expenditure on emission controls as the only environmental cost faced by firms. This tends to underestimate total environmental costs.[48] Other environmental costs, though perhaps harder to measure, are equally important. These include costs related to public relations, liability, insurance, and certification of voluntary standards such as Responsible Care or ISO 14000 (discussed later in this chapter).

Institutionalists in particular are somewhat critical of the market liberal search for statistical proof of pollution havens (defined narrowly) in order to justify any concern over the environmental impact of FDI. The global community, they argue, needs to take a broader approach—

one that focuses on policies to remedy the inequities and detrimental impacts of differential standards. Such an approach needs to explore incentives—real or perceived—that affect state policies and corporate behavior. These incentives must be as much political as economic. They need to address issues like the possibility of a "regulatory chill"—a problem related to pollution havens but one most market liberal analyses of pollution havens ignore. The phrase "regulatory chill" describes instances where countries fail to raise environmental standards above the status quo, for fear that investment will either leave or not enter the country. Whether firms actually respond to this failure to raise standards is not the core issue. Rather, the point is that the fear that firms will act can affect the stringency of regulations in both rich and poor countries.[49] A regulatory chill is almost impossible to quantify. Some studies do indicate, however, that a fear of possible corporate reactions does affect government decisions on whether to strengthen environmental regulations. In both the United States and the European Union, for example, at various times industry groups have managed to convince governments to not impose stricter climate-change regulations by stressing competitiveness concerns.[50]

Those who call for a broader debate tend to advocate policies that are "win-win" even in the absence of hard evidence that pollution havens exist. All groups can agree on this broad point, but it is far harder to reach a consensus on the nature and extent of restrictions on corporate behavior. Those leaning toward market liberal and institutionalist views primarily advocate aid from rich to poor countries, support for public access to corporate information, stronger regulations, and more cost-effective ways to reduce pollution. Some also argue that the IMF and the World Bank should account for risks of more intensive pollution from structural adjustment programs.[51] Others stress the need for even stricter actions, such as a global investment agreement that would put rigorous environmental controls on foreign direct investment by TNCs (discussed later in this chapter).

TNCs and Site Practices

A second major strand of literature on corporate investment and the environment centers not so much on whether firms relocate to take advantage of lower environmental costs, but instead evaluates the actual environmental practices of transnational corporations. Often, authors concentrate on a simple question: do corporate operations harm the environment? Critics of corporations have dominated much of the literature here, often describing illegal and destructive activities of specific industries in specific regions.[52] Research institutes like the World Resources Institute and NGOs like the Rainforest Action Network, Mining Watch in Canada, Oil Watch in Southeast Asia, and Global Witness in Cambodia document these activities, with their findings then distributed worldwide through the Internet and free publications.[53] The literature on natural resource firms and manufacturers tends to generate somewhat different criticisms and debates. So does the literature on the site practices of firms in developed countries (which often have better environmental practices than in the past) versus ones in poor countries (which often have lower operating standards and less transparency).

To illustrate this body of literature, we briefly examine the logging, mining, oil, and electronics sectors in the developing world. This literature tends to reflect the views of bioenvironmentalists and social greens, because market liberals and institutionalists tend to presume that correct markets or correct institutions will solve these problems. Social greens and bioenvironmentalists see the problem as more entrenched and as arising from the nature of capitalism and corporations themselves. For them, capitalism—the enshrining doctrine of profits and greed—creates marauders and plunderers.[54] It is little more than a flashing neon sign: "Earth for Sale."[55] Globalization is enabling corporations to "Rule the World."[56] It is creating a "Corporate Planet."[57] Firms are becoming larger, more powerful, and less accountable, combing the world for cheap labor and "limitless" environmental resources.

Social greens and bioenvironmentalists see irresponsible and illegal corporate loggers as one of the main causes of forest degradation in the tropics (which opens the forest to fires and farmers and triggers the process of deforestation). Although bioenvironmentalists also stress population growth in conjunction with corporate behavior as key causes, social greens focus almost exclusively on the role of corporations in the global capitalist framework. Loggers routinely skirt environmental rules, often relying on state officials for access to sites and protection from prosecution. It is common—indeed, it is often necessary in order to stay in business—to bribe enforcement officers and customs officials. Numerous studies have found that few firms, if any, actually practice sustainable logging (i.e., in a way that the forest can regenerate so loggers can harvest the same amount again sometime in the future—say in fifty to eighty years).[58]

A glance at the Asia-Pacific region confirms the terrible consequences of logging. TNCs from Japan, Malaysia, and Indonesia have been at the center of the tropical logging industry over the last four decades. Japanese general trading firms like Mitsubishi Corporation have supplied funds and "guaranteed" markets.[59] Malaysian and Indonesian firms have often conducted or supervised actual logging operations. Natural forests still blanketed most of the Asia-Pacific area in the middle of the twentieth century. Now, forests cover around three-quarters of Melanesia and less than half of countries like Indonesia, Laos, Cambodia, Malaysia, and Burma. The situation is even worse in Thailand (one-quarter) and the Philippines (one-fifth). The extent of forest degradation is an even greater problem. More than 95 percent of Asia's frontier forests—areas large and pristine enough to still retain full biodiversity—are already lost.[60] The Philippines is effectively logged out (it has been a net importer of tropical timber since the 1990s). The rest of the Asia-Pacific region will, at recent rates and under current logging practices, follow the Philippines within a few decades.[61]

Transnational mining companies have had a similarly poor environmental record, and have been targeted by environmental groups seeking to raise public awareness of the problem as well as to raise the environmental and social standards in the industry. Because most of the easily accessible minerals in the world have already been mined, mining TNCs are forced to search far and wide for new deposits, and are increasingly setting up operations in developing countries to extract minerals that are more and more remote and difficult to obtain. Because of this, risks in the industry are growing. And because of these risks, mining companies are increasingly seeking investment insurance and guarantees from multilateral development banks and export credit agencies (see chapter 7), although until recently these agencies required little by way of environmental accountability.[62] There have been numerous recorded accidents and scores of cases of environmental pollution, as well as ecological and social disruption, including negative economic, health, and social impacts on women, resulting from mining operations over the past few decades.[63] Deforestation and toxic leaching of mine tailings and dangerous chemicals including cyanide into waterways and the soil are just a few of the many environmental problems associated with mining. Mining accidents and environmental destruction at mine sites—such as the burst tailings ponds in Guyana in 1995, the Philippines in 1996, and Romania in 2000—commonly occur because corporations do not take adequate precautions.[64] Often, miners displace indigenous peoples, and in some cases there are serious human rights abuses as TNCs make deals with host governments to suppress resistance. Activists have highlighted major ongoing environmental and human rights problems—for example, around the Grasberg mine in Indonesia and the Bougainville mine in Papua New Guinea.[65] Though the mining industry has taken some steps to improve its environmental performance through voluntary measures (discussed shortly), activists remain skeptical of their motives and the overall effectiveness of their actions.

Oil exploration and extraction have similar negative environmental consequences. Oil is at the core of the global energy system, and thus of global development. As demand rises and reserves dwindle, oil companies have expanded their exploration and extraction operations over a larger and larger territory, including in the developing world. As with mining, TNCs are the main players in oil and gas exploration globally, and they often seek funds to support their ventures from export credit agencies. The oil TNCs with markets in Europe, Japan, and the United States can swamp economies in the developing world. In Ecuador, for example, the corporate push for construction of oil pipelines has led to a dependence on oil revenues. Because oil prices on world markets are highly unstable, the overreliance on oil revenues drives a boom and bust cycle in the domestic economy, contributing to increasing levels of external debt, rises in poverty rates, and growing resistance.[66]

Activists have charged transnational oil companies operating in the developing world—from the Western Amazon to West and Central Africa to Southeast Asia—with human rights and environmental abuses. The environmental impacts of oil exploration are many, and include contamination of waterways from oily water wastes and toxic drilling muds that are stored in ponds similar to tailings ponds. Oil spills and gas flares also pose a constant threat to communities near oil drilling as well as during the cross-country and maritime transportation of oil.[67] As with mining, roads and settlements near exploration and drilling sites also trigger deforestation. Human rights abuses of local peoples are also common. Many were outraged when in 1995 the Nigerian government executed Ken Saro-Wiwa and nine other activists following a sham trial on trumped-up charges resulting from protests against the environmental impact of Shell's oil extraction in that country.[68] Despite widespread attention to this case, activists contend that environmental abuses such as pipeline spills have continued in Nigeria. In June 2009, Shell finally reached a settlement with the families of some of the victims, but did not admit any liability.[69] Social greens see these cases as clear examples of enclosure and the use of state violence to suppress indigenous groups (as discussed in chapter 4).

The activities of TNCs in the high-tech electronics sector, too, have been subject to scrutiny and criticism by activist groups. TNCs in this sector have increased their investment in Asia and Latin America, but their environmental performance has not been as state of the art as the technology they are producing.[70] The production of electronic goods and components—such as silicon chips, semiconductors, and computers—involves hazardous materials, such as arsenic, benzene, cadmium, lead, and mercury—all known carcinogens. These toxic materials are not only a threat to workers in this industry, but often end up in waste that pollutes land and waterways (local waste contractors hired by TNCs commonly dispose of this waste improperly). The manufacture of silicon chips and semiconductors also requires an enormous amount of water and energy, used for cleaning the products and keeping production sites clean.[71] In addition, the disposal of high-tech products at the end of their life is also problematic, especially because of their toxic components. Although there are some recycling programs for old computers, cell phones, and other high-tech items, a significant proportion of those collected in North America for recycling end up being exported to developing countries, particularly China, India, and Pakistan. Serious environmental harm can occur when countries without strict regulations recycle such items. Not only is there exposure to the heavy metals in these items, but the burning of PVC (polyvinyl chloride)-coated cables to reclaim the copper is highly polluting to the air and waterways.[72] Because the high-tech sector is one of the fastest growing in the global economy, critics argue that these problems are likely only to increase in the future unless action is taken now.

Greening or Greenwash?

In response to criticism of their environmental practices in the late 1980s and early 1990s, particularly in developing countries, many global firms began to take measures to become more "green."[73] With a lack of outside forces to regulate them, they have embraced self-regulation via voluntary environmental initiatives. This enthusiasm by industry for self-regulation is driven in part by an attempt to improve their public image, and in part by economic factors. Market liberals have hailed the economic benefits of greening as proof that the market is self-correcting and that firms can deal with environmental problems. Social greens and bioenvironmentalists, however, are skeptical of the sincerity and efficacy of these efforts. Institutionalists, on the other hand, take the middle ground in this debate.

Greener

Throughout the 1980s—in particular, in the runup to the Rio Earth Summit in 1992—many firms began to take environmental concerns more seriously.[74] Responsible Care, an environmental and safety code delineating practices for the chemical industry, was established in 1985 following the accident at Bhopal. In 1990 the Global Environmental Management Initiative (GEMI), set up by the International Chamber of Commerce (ICC), was charged with implementing the ICC business charter for sustainable development.[75] Industry also played a key role at the Rio Earth Summit in 1992 (see following discussion). Indeed, Agenda 21 sanctioned the use of voluntary initiatives as a key strategy to promote sustainable development. At the 2002 Johannesburg Summit, industry again played a key role in promoting voluntary initiatives (see following discussion). The result has been a flurry of voluntary codes established by industry, which today are diverse and wide-ranging. Through these efforts, industry has tried to keep the focus on voluntary corporate responsibility that includes environmental awareness—their preferred way of addressing environmental issues. Some NGOs that are critical of the slow-moving and bureaucratic nature of state-based agreements have begun to support voluntary initiatives, especially those involving multiple stakeholders.

There are four key types of international voluntary environmental initiatives.[76] The broadest category is principles and codes of conduct. Such measures typically involve firms signing onto broad statements that are usually set out in conjunction with the UN or with NGOs in order to give them more legitimacy. In this category is the UN's Global Compact (GC) with industry, launched in 2000, which encourages corporations to follow ten principles of social and ecological responsibility.[77] By 2010 more than 5,300 firms were participating in the GC. Other codes of conduct and sets of principles include the Principles of Sustainable Development in the mining sector, and the Carbon Principles for bank lending.

A second common type of voluntary measure is reporting and disclosure schemes.[78] By the second half of the 1990s, most major corporations began to issue CSR reports, the idea being that self-disclosure of information would put pressure on firms to improve their performance.[79] The Global Reporting Initiative emerged in 2000 to provide a set of reporting guidelines to provide a degree of standardization and comparability across CSR reports. Today, more than 1,000 firms use these guidelines.[80] Other reporting schemes include the Carbon Disclosure Project, which sets a standard measure to determine carbon emissions of major firms that investors can then use to guide decisions.[81]

A third common type of voluntary measure is environmental management systems standards, such as the ISO 14000 standards (see box 6.1). The European Union's Eco-Management and Audit Scheme (EMAS) also sets voluntary standards for environmental management. A fourth type of voluntary initiative popular with industry is the market-based mechanism, which includes a range of measures such as ecocertification of products like lumber, with the goal of encouraging firms to sign on in order to benefit from growing markets for such products. NGOs often play a role in these initiatives—what Ben Cashore at Yale University refers to as "non-state market driven" governance.[82] Participation in voluntary carbon markets, such as the Chicago Climate Exchange (CCX), has also been seen as a market-based approach by which to encourage firms to reduce their carbon emissions.[83]

For market liberals, adherence to voluntary initiatives promises only positive outcomes for the environment. Industry is seen as the best actor

Box 6.1
ISO 14000 environmental management standards

The ISO 14000 environmental management standards are perhaps the most widely recognized global-level voluntary initiative on the part of industry. These standards were developed in the early 1990s under the auspices of the International Organization for Standardization (ISO), directly following promises made by industry to establish voluntary initiatives at Rio. The ISO 14000 standards are management standards, meaning that they encourage firms to establish a management system that improves their awareness by setting their own goals for environmental improvement. The ISO 14001 standard, published in 1996, is the only certifiable standard in the series. To seek certification, a firm must provide an environmental statement showing that it intends to comply with all applicable environmental laws in the jurisdiction in which it is located. It must commit to the prevention of pollution as well as to continually improve environmental management. It must adopt a management system that ensures that it conforms to its own environmental policy statement. Further, firms should encourage suppliers and contractors to establish their own environmental management systems to conform to the ISO 14001 standard. By 2007 some 154,600 firms in 148 countries had gained certification to the ISO 14001 standard.[1] With a growing number of firms in both developed and developing countries adopting these standards, adherence to the standards will increasingly become a de facto condition for conducting business in the global marketplace.

Critics are concerned about whether the ISO 14000 standards will really make a difference in terms of environmental performance.[2] ISO 14001, for example, stresses that TNCs should comply with all environmental laws in the country in which they are operating. This differs substantially from Agenda 21, which calls on TNCs to follow home-country standards. In this way, ISO 14001 allows for differences in standards between countries, but it may not do much to improve standards in developing countries. At the same time, some businesses are attempting to use ISO 14000 and other voluntary industry codes as a deliberate attempt to head off more stringent regulation. Industries in a number of countries are pressing their governments for some form of regulatory relief, such as more lenience for ISO 14001–certified firms during monitoring of environmental regulations. For example, the United States, Argentina, South Korea, and Mexico have adopted measures that take ISO 14001 certification into account in the monitoring and enforcement of regulations.[3]

Notes

1. ISO 2008.
2. Krut and Gleckman 1998; Clapp 1998; Kimerling 2001.
3. Speer 1997, 227–228; Finger and Tamiotti 1999.

to set up such standards because it is most familiar with corporate possibilities and constraints. Market liberals prefer such an approach because it shifts the burden of regulation from the state to the firm, which can monitor environmental performance much more efficiently (important for "eco-efficiency"; see following discussions). Moreover, adherence to global standards helps ensure that even when state enforcement is weak (common throughout the developing world), firms will still abide by local regulations. Global standards like ISO 14000 also encourage firms to adopt cleaner technologies, rather than focusing on cleaning up after the fact, because firms consider environmental impacts throughout the decision-making process. Such standards are intended to spread good environmental practices up and down the supply chain, as well as across borders, because certified firms are required to encourage suppliers to become certified.[84] Indeed, some research has shown that adherence to ISO 14001 among TNCs has encouraged a "race to the top" wherein developing country firms adopt voluntary regulations in response to pressure from other firms higher up the supply chain.[85]

Industry coined the term "eco-efficiency" to refer to its new environmental awareness and the benefits this awareness would bring. By integrating environmental concerns into business practices, firms can expect economic as well as environmental benefits. These include lower costs for environmental cleanups, reduced capital costs due to lower risks, and increased market share secured by an enhanced image.[86] At the same time, economic efficiency is thought to have environmental benefits. Firms that are efficient generate less waste and incur fewer environmental costs. Efficiency and environmental protection then are seen to be synergistically beneficial. The push for eco-efficiency is a way to gain a competitive advantage in the marketplace, and firms that fail to adopt such principles could ultimately be pushed out of business. In this way, environmental stewardship is viewed as making sound business sense.

Today, analysts like the ecological modernization theorists in the institutionalist camp argue that environmental concerns are being institutionalized—that these are becoming part of institutional structures and decision-making processes.[87] Drawing on experience from Europe, these thinkers see a restructuring of the capitalist political economy as taking place so that both firms and governments will view environmental management not as a cost, but as a business opportunity. In this atmosphere, firms are increasingly going "beyond compliance," internalizing environmental norms and practices and thus actively seeking to do more than the law requires with respect to environmental practices.[88] Some prominent European oil companies, for example, came out in favor of the Kyoto Protocol, beginning the process of moving their energy interests away from fossil fuels and toward alternative and renewable energy sources.[89] Some TNCs in the chemical industry are also seen to be "exporting environmentalism" by taking their new internalized environmental practices and codes of conduct abroad to their operations in developing countries.[90]

Greenwash

Undeniably, firms everywhere now incorporate environmental language into public statements and official corporate policies. Critics of corporations, including most social greens and many bioenvironmentalists, label this *greenwash*, a phenomenon in which a company tries to convince consumers and shareholders that it is environmentally responsible, where the purpose is more about image than substance.[91] An example is a company that was instrumental in producing chemicals that destroyed the ozone layer now taking credit for "protecting" the ozone because it no longer produces these chemicals. Another is a company that takes waste from one polluting process and uses it as a raw material in another, labeling it "recycling." Or a company that claims to be "green" but at the same time lobbies (or funds lobby groups) against environmental regulations.[92] Some observers further argue that companies are working hard to weaken environmental agreements or subvert environmental campaigns or that firms are trying to "hijack" environmentalism to serve their own purposes.[93] Greenpeace took up the issue of greenwashing just before the Rio Earth Summit in 1992 and published the *Greenpeace Book of Greenwash* to distribute at the event.[94] Groups like Greenpeace and CorpWatch target the large industry "leaders" whom they feel are advertising falsely and regularly issue "Greenwash Awards" to such companies. Greenpeace's Stop Greenwash campaign web site not only updates instances of greenwashing in corporate advertising, but also reveals when corporations backtrack on green promises or fund antienvironment groups, including the recent closure of renewable energy programs by major oil corporations, and the funding of climate skeptics by auto and oil industries.[95] Activists Jed Greer and Kenny Bruno, in a book published by the NGO Third World Network, argue, "The reality is that TNCs are not saviors of the environment or of the world's poor, but remain the primary creators and peddlers of dirty, dangerous and unsuitable technologies."[96]

Social greens in particular argue that corporations are able to take the green high road mainly because they have managed to redefine sustainable development in a way that fits their own existing practices. Sociologist Leslie Sklair explains that "sustainable development was seen as a prize that everyone involved in these arguments wanted to win. The winner, of course, gets to redefine the concept."[97] As Sklair sees it, corporations won that argument in the 1990s and have redefined the concept of sustainable development in a way that suits their own interests. In doing so, companies have felt free to advertise themselves as being proenvironment, even when some of their practices are not. In some cases, major corporations spend more on publicizing their green credentials than they do on the actual green programs they are advertising.[98] Some critics argue that voluntary corporate responsibility (the industry term) should give way to corporate accountability, which would not just rely on promises to be responsible, but would also hold corporations accountable for their actions.[99]

Some of the more specific arguments made on this theme are that market liberals and institutionalists are conflating economic efficiency and environmental protection, which critics see as setting a dangerous precedent. In fact, at least in the early stages, nearly all of the examples of businesses adopting cleaner production strategies were a result of government regulatory pressure, and did not emerge from within business itself as an efficiency-gaining measure.[100] Some have also pointed out that "green investment" touted by business is merely a "cleanup" rather than "clean." In short, critics claim that industry is just changing labels rather than truly changing course.

Critics have also analyzed many pitfalls common to voluntary corporate initiatives. Typically, voluntary measures are weak with respect to implementation and enforcement. Because they are voluntary, the requirements for participation are generally minimal. Moreover, in order to encourage firms to sign on, such measures commonly prioritize those environmental measures that save firms money, rather than those that are more costly but equally, if not more, important for environmental protection. Voluntary measures such as management standards tend as well to focus on encouraging firms to improve their procedures, rather than actual environmental performance. Furthermore, many initiatives have poor participation rates in practice. When one considers that more than 700,000 foreign affiliate firms operate worldwide, the 150,000 firms certified to ISO 14001, the 5,300 participants in the Global Compact, and the 1,000 businesses using the GRI guidelines are just a small slice of that larger group. Complexity within and across various initiatives also makes it difficult for the average consumer or investor to discern what actual measures firms are taking. Finally, voluntary initiatives have the potential to exacerbate inequalities between rich and poor countries by raising costs for poor country firms and closing them out of international markets.[101]

TNCs and Global Governance for Investment and the Environment

At the same time that a fierce debate is raging over the environmental performance of TNCs, linked debates are intensifying over the role of TNCs in global environmental governance and the environmental consequences of the global rules that seek to regulate TNC activities.[102]

TNCs, through many different channels, have played an important role in the formation of global environmental governance. As mentioned in chapter 3, TNCs traditionally lobbied domestic governments in an attempt to influence the creation of global environmental rules, as a way to protect their own interests. Increasingly, however, industry players are also lobbying policy makers directly at the international level, attending negotiations over specific environmental treaties as well as larger environmental conferences. Compared to the fifteen-minute intervention made by industry at the Stockholm Conference in 1972, industry groups at both the 1992 Rio Earth Summit and the 2002 Johannesburg Summit put an enormous effort into their involvement.[103] Maurice Strong, the secretary-general of the Earth Summit, had encouraged his friend Stephan Schmidheiny of the Business Council on Sustainable Development (a corporate lobby group founded in 1990 that comprised some forty-eight TNCs) to take an active role.[104] The ICC was also active at the Rio

Summit; it formed the World Industry Council on the Environment (WICE) in 1992 to provide industry followup on Rio. In 1995 the two groups merged to form the World Business Council for Sustainable Development (WBCSD). The Business Action on Sustainable Development (BASD) was formed in 2001 as a coalition of business groups (primarily a joining of forces of the ICC and the WBCSD, and comprising some 161 TNCs) for the chief purpose of lobbying at the Johannesburg Summit.[105] The main message put forward by industry at both of these gatherings, clearly within the market liberal camp, was that voluntary self-regulation organized by industry, rather than outside regulation of TNCs, was the most effective and efficient way to promote sustainable development.[106]

The "structural power" of TNCs is another significant influence over global environmental governance. This refers to the ability of TNCs to influence the formation and functioning of governance through their dominant position in the global economy, which allows them to shape mainstream ideology and state policy formation. It is extremely difficult to measure this type of corporate influence over states and global institutions, but many scholars stress its importance for understanding global policy outcomes. This type of analysis comes mainly from social greens who draw on a historical materialist perspective—in particular on the ideas of Gramsci. These scholars outline the ways a "bloc" of actors—consisting not just of transnational corporations, but also of certain states and key intellectuals who serve the interests of industry—influence the dominant ideology and discourse on sustainable development.[107] This influence is present in the role corporate actors play in defining sustainable development (as mentioned earlier) as well as in the negotiation of global environmental agreements.[108] The current era of increasing global economic competition, for these analysts, has meant that many states pursue domestic policy outcomes acceptable to corporations in order to keep or attract investment in their country.

Some institutionalists have critiqued this notion of structural power among business actors in global environmental governance. Using the "business conflict model," analysts such as Robert Falkner at the London School of Economics have argued that business actors are rarely united into a single bloc in the context of global environmental negotiations. Divisions among corporate actors can give rise to political alliances between some more progressive business actors and environmental groups. The result can be stronger provisions in environmental agreements, which may benefit some business actors while disadvantaging others. Such divisions among business actors occurred in the negotiation of international agreements on ozone, climate, and agricultural biotechnology.[109]

Transnational corporate actors also influence global environmental governance through their participation in other international forums important to global environmental policy, such as through the setting of industry-based initiatives, as discussed previously. The ISO 14000 environmental management standards, for example—while initially designed as a voluntary set of standards—are now recognized as legitimate standards by the WTO. In this way, they have become part of global structures of environmental governance. Some critics have expressed concern about this development because TNCs dominated the drafting process with only minor input from governments and environmental groups.[110]

At the same time that industry players may influence the process of global environmental governance, intergovernmental investment rules also influence the actions of these firms, including ones not specifically designed to address environmental issues. Global and regional trade agreements with investment provisions, as well as agreements within the OECD, have had important implications for investment and the environment. Here we discuss the significance for TNCs of some of these investment rules.

One of the best-known sets of investment rules with ramifications for the environment is chapter 11 of NAFTA, which sets out investors' rights under this trade agreement. The main defining features of this chapter are that it calls for governments to provide national treatment to foreign investors, and it allows corporations to sue governments for compensation when they believe their investments have been expropriated or when actions "tantamount to expropriation" harm their profits.[111] It sets up an investor-to-state dispute settlement mechanism to handle such suits when they are brought forward. NAFTA is the first trade agreement to include such explicit rights for corporations. Although chapter 11 was originally designed as a general protection for corporations against nationalization of industries as a measure to encourage more FDI, in the end it has had enormous significance for the environment. Of the forty-nine suits filed under chapter 11 of NAFTA between 1994 and 2008, twenty-three are cases with clear environmental or natural resource implications. A number of these have been suits dealing with disputes over environmental regulations where firms have claimed that the imposition of certain environmental regulations violates chapter 11, in that new laws were deemed to have caused significant losses in profits for the complaining firm.

Social greens and bioenvironmentalists have lambasted chapter 11 of NAFTA. To them, it is a clear attempt to undermine environmental regulations and it encourages not just a regulatory chill but also a "race to the bottom." This is because they see the investment rules as restricting the ability of governments to protect the environment. For example, the suit by Ethyl Corporation against Canada in 1997 over its ban on imports of the chemical MMT (a gasoline additive) led the Canadian government to repeal its ban on that chemical. Moreover, the Canadian government had to pay US$13 million in damages to the corporation in an out-of-court settlement. In effect, critics say, with chapter 11 the way it is, governments will be afraid to implement new laws for fear of being sued.[112] For this reason, they would like to see the investment provisions removed or rewritten. Institutionalists and market liberals have tended not to see these cases as ones that threaten environmental regulations in a blanket fashion. Rather, they argue that in each case the tribunal had solid and specific reasons for each decision, and maintain that there is no bias in chapter 11 against environmental regulations.[113]

Another set of investment rules through the OECD, the Multilateral Agreement on Investment (MAI), was negotiated in the mid to late 1990s. This agreement planned to enshrine NAFTA chapter 11–type rules, in particular the investor rights' provisions, on a much larger scale. There was an attempt at the time the Uruguay Round was negotiated to include such measures in an agreement under the WTO, but many developing countries resisted this. For this reason, the OECD sought to develop its own agreement, with a view to eventually including it in the WTO once it gained broader support.[114] The MAI fell through in the late 1990s, however, after a broad coalition of activists publicized the secretive nature of the talks and revealed the problems with the agreement, including its lack of environmental safeguards.[115] Though OECD governments abandoned the MAI, many activists are still concerned that governments will slip such rules into other global agreements, because investment issues are under consideration in the Doha Round of trade negotiations at the WTO. The victory of activist groups in opposing this effort gave much steam to the antiglobalization movement.

Following the defeat of the MAI, the OECD revised its voluntary *Guidelines for Multinational Enterprises*. Originally adopted in 1976, the OECD has periodically reviewed and revised this set of guidelines, most recently in 2000, although consultations are now ongoing on a *possible* update. The 2000 revisions include recommendations that bring the guidelines into line with international environmental standards and treaties and with the Rio Principles. For example, the newly revised guidelines promote existing environmental-management standards such as the ISO 14000. But going further than International Organization for Standardization (ISO) standards, they also call for TNCs to adopt "measurable objectives, and where appropriate, targets for improved environmental performance." The guidelines also call for better information on the environmental activities of firms, as well as more consultation with affected communities.[116] Critics, however, remain skeptical.

The OECD *Guidelines for Multinational Enterprises* apply only to the OECD. In 1977, the United Nations Center for Transnational Corporations (UNCTC) launched negotiations on a globally applicable voluntary code of conduct for TNCs that included provisions on environmental conduct and outlined rights and responsibilities of TNCs.[117] The UNCTC, which was set up in the early 1970s, was mandated to monitor the economic, social, and environmental impacts of TNCs, particularly those operating in developing countries. The code of conduct aimed to ensure that foreign direct investment did not have adverse consequences in these areas. Talks on this code were ongoing until the early 1990s, but it was never finalized or adopted. Under pressure from the United States, the UNCTC was dismantled just prior to the Rio Earth Summit, and the United Nations Conference on Trade and Development (UNCTAD) took over its remaining activities. Instead of implementing the UNCTC code, the Earth Summit promoted voluntary initiatives developed by corporate actors themselves, as discussed earlier.[118]

Social greens and bioenvironmentalists are critical of the current investment rules and instead call for harmonized standards globally, preferably ones that discourage unsustainable investment and, in the case of social greens, ones that promote local production. Greenpeace and Friends of the Earth (FOE), for example, have both advocated the adoption of a binding set of rules to ensure better performance of TNCs on environmental, social, and human rights. Their proposals call for firms to adhere to the highest standards, for corporate liability for damage, and for implementation of the precautionary principle.[119] (Chapter 8 discusses these proposals more fully.) Institutionalists would prefer an agreement that prevents a race to the bottom, but many do not support harmonizing standards. They do not want to discourage global investment (under the proper conditions), because it is crucial for sustainable development. Although market liberals advocate a global investment agreement, they would like to see it focus on protection of corporate rights. Specific environmental-protection provisions in such an agreement, from their point of view, are unnecessary.

Conclusion

Underlying assumptions about the nature of firms shape much of the debate over investment and the environment. Market liberals see firms as efficient managers and as engines of economic growth. Most accept that negative environmental impacts of firms are a problem on occasion, but they nevertheless argue that the best approach is to let market incentives encourage firms to *voluntarily* work for better environmental management. Interfering with the natural pursuit of corporate profits is, as with global trade, unlikely to actually conserve nature. In fact, in most cases it simply perverts ecological, social, and—most significantly—financial goals. Institutionalists have sympathies with the market liberals, but they are more willing to see corporate activity as potentially negative more of the time. They argue that a strong state with a solid regulatory capacity and a vibrant civil society is necessary to reduce the impact of externalities and maintain a stable investment climate to ensure that firms have the proper incentives for sustainable development. For many institutionalists, globalization is also opening more possibilities for the global community to hold firms accountable—even those working in remote and weak states. This includes global agreements and standards to help evaluate corporate environmental performance. For institutionalists, these may be voluntary or binding.

Bioenvironmentalists and social greens have long condemned the ecological and labor practices of TNCs, particularly those operating in developing countries. In recent years, globalization has bolstered the economic might of TNCs relative to small states and, unsurprisingly, activists have begun to campaign even more vigorously against TNCs. The rather pejorative term "corporate globalization" signifies their concern that globalization is really about constructing a global political economy in the commercial interests of TNCs. Critics of corporations are suspicious of corporate motives as well as their structural capacity to improve environmental performance. Both bioenvironmentalists and social greens call for strict controls on TNCs. They point to the structures within TNCs that drive firms to pollute the world's air and water and deplete the world's oil, minerals, timber, and fish, especially in the developing world. Weak states, corruption, and greed allow firms to get away with reckless environmental abuses. Relying on green markets and voluntary corporate initiatives might do some good, but ultimately, it is critical to simply prevent some of the corporate exploitation of the global environment. For bioenvironmentalists, it is essential to control *both* TNCs *and* local firms—both are equally capable of plundering. Social greens take a slightly different tack. They call for measures to dismantle TNCs and replace them with smaller firms accountable to local people and local environments. This must occur alongside a broader reform of the global political economy, including, as the next chapter discusses, an overhaul of global financing.

7

Global Financing and the Environment

Global financing is increasingly at the heart of the economies of the developing world. This involves governments constantly borrowing and repaying public and private loans, a steady stream of multilateral and bilateral grants and technical assistance, as well as staggering private financial flows to corporations via investment funds. Each of these has profound and somewhat different environmental implications. While the broad split in views again falls along the now-familiar coalitions between social greens and bioenvironmentalists on the one hand and the market liberals and institutionalists on the other, each perspective has its own unique take on financing and the environment.

Market liberals and institutionalists both see international financing in a broadly positive light. For market liberals, it fosters global investment and economic growth, which benefits all countries by promoting global economic stability. It also specifically helps support efforts by poorer countries to develop, which enables them to better manage some of the inevitable environmental costs of early industrialization and natural resource exports. Market liberals, though, are wary of endorsing too much development aid, which can interfere with market incentives, they argue. Institutionalists agree with market liberals on the benefits of global financing, but add that it is also one of the most vital financial and technical mechanisms to sustain environmental institutions and norms. They see financing as the primary means of building the institutional capacity of developing countries, and ultimately of the international community, to manage local and global environmental affairs.

Bioenvironmentalists and social greens are highly critical of current patterns of global financing. Both perceive it as yet another global driver of economic growth, and as a source of funds and technical advice that allow developing countries to overexploit natural resources, urbanize, and industrialize. Development assistance in particular wrongly assumes that elevating the living standards of poor countries to the level of the rich countries is a sustainable goal, while doing little to address the dangerous fact that the ecological footprint of the world's wealthy is already well beyond the earth's carrying capacity. Some extreme bioenvironmentalists add that development assistance is utterly misguided, merely compounding the stress on the global ecosystem by supporting the overpopulation of the poor. To survive, in the words of Garrett Hardin, the rich must live "by the ethics of a lifeboat" and not perversely toss aid to the hordes who threaten to swamp the boat of wealth and prosperity.[1] Most social greens, on the other hand, see such extreme arguments as callous and immoral. Still, they agree that development financing is failing. It *does not* serve the interests of communities and individual well-being in the developing world, but instead serves the interests of international agencies, powerful states, and TNCs intent on exploiting the resources and peoples of the developing world. Far too much global financing supports environmentally unsound industrialization and intensive agriculture in poor countries, and far too little supports healthy communities and sustainable livelihoods.

This chapter explores the environmental implications of some of the main channels of global financing. It focuses in particular on development assistance and developing-country debt, as well as assistance geared toward improving environmental management. It also covers the environmental consequences of bilateral aid, export credit agencies, and private financial flows. It begins with an overview of various kinds of financing as well as recent trends in international finance.

Scope and Trends in International Finance

The globalization of financing is increasingly connecting the world in terms of money flows, tied to both trade and corporate investment. International financing occurs in a variety of ways. Development aid consists of grants (aid that does not have to be repaid), technical assistance (expert advice), and preferential loans (loans with lower interest rates and longer payback periods than standard public and bank loans). Multilateral aid is channeled through institutions like the World Bank and UN agencies. Bilateral aid is from one government to another. Multilateral and bilateral loans are referred to as *official debt*. Such loans may or may not be counted as official development assistance (ODA), which is assistance to multilateral institutions and developing-country governments that meets the guidelines of the twenty-three-member OECD Development Assistance Committee (DAC). To be counted as ODA, such assistance must be given explicitly for the promotion of economic development and welfare in developing countries, and it must also include a grant element of at least 25 percent. Much of the globe's international development assistance is ODA, but much of the world's official debt arises from loans borrowed on terms just outside of this definition as well (although these terms are still generally better than a private bank loan). Governments and private interests also borrow directly from commercial banks (referred to as *commercial debt*). Global financing comprises other private financial flows as well, such as markets for securities (stocks, bonds, and derivatives contracts), and foreign-exchange trading.

Development aid—that is, grants and the subsidized portion of loans—is an important source of financing for poor countries. In 1970 the UN adopted a goal that donors should contribute 0.7 percent of their gross national product to developing countries. This was reiterated in the call for a New International Economic Order (NIEO) in 1974, as well as at the Rio Earth Summit in 1992 and again in the past decade as part of the UN's Millennium Development Goals. This goal was reinforced in 2002 at both the Monterrey Financing for Development Conference and the Johannesburg World Summit on Sustainable Development. Yet throughout this time, donor governments have fallen far short of this goal. Only five donor countries meet that target (though several others are poised to meet it soon),[2] and the average for the OECD DAC now stands at just 0.30 percent of gross national income (figures 7.1 and 7.2). This percentage is significantly lower than it was in the early 1960s when it stood as high as 0.5 percent. In absolute terms, however, the total amount of money flowing into the developing world in the form of aid rose over that same period, despite dropping in the early 1990s and remaining stagnant until 2004–2005, when it began to rise again following the Asian tsunami. In 2008, ODA from DAC countries amounted to about US$120 billion (figure 7.3). The United States, however,

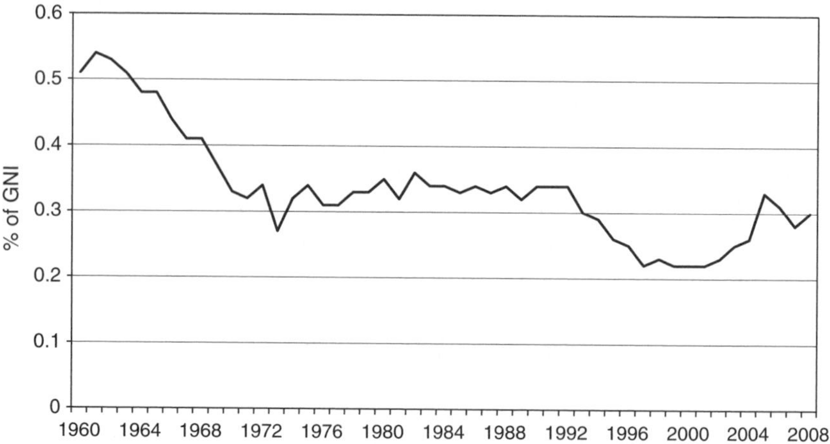

Figure 7.1
OECD DAC: Official development assistance (percentage of GNI), 1960–2008. *Source*: OECD, <http://stats.oecd.org/index.aspx>.

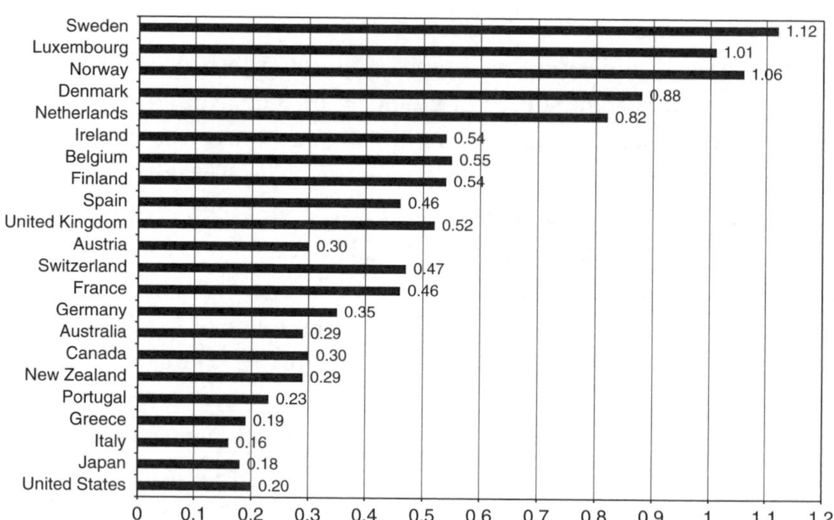

Figure 7.2
Official development assistance by donor as a percentage of GNI, 2009. *Source*: OECD, <http://stats.oecd.org/index.aspx>.

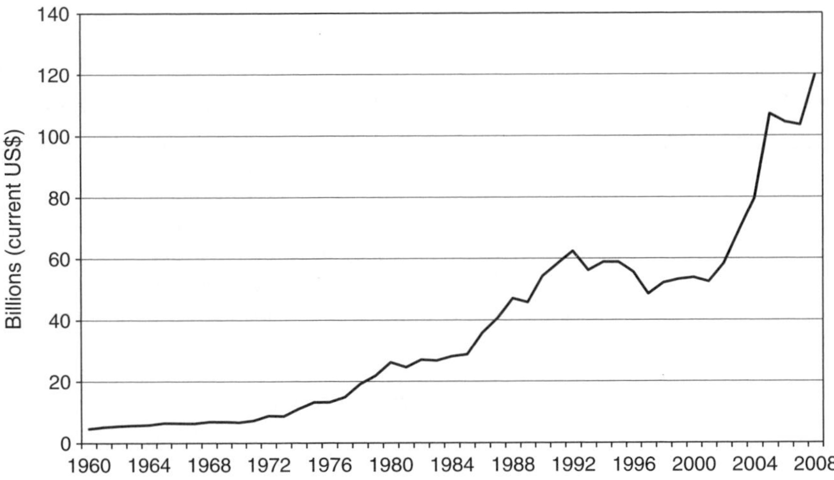

Figure 7.3
OECD DAC: Total official development assistance, 1960–2008. *Source*: OECD, <http://stats.oecd.org/index.aspx>.

contributed only 0.20 percent of gross national income in 2009, among the lowest of the major donors (figure 7.2).[3] Although in absolute terms aid has been rising, in per capita terms in donor countries, it has nearly halved since 1995.[4] In addition to DAC country aid, emerging donors such as China and India are becoming more important (despite also being recipients of aid), as are private philanthropic foundations as sources of financial assistance to developing countries. Some estimates put the aid of China, Brazil, and India at around US$4 billion in 2005, and the aid from the Bill and Melinda Gates Foundation alone at US$3 billion in 2009.[5]

External debt—another major source of international financial flows—has become a defining feature of most developing-country economies over the past 40 years (see box 7.1). In 1970 the total debt stocks of low and middle income countries stood at US$68.8 billion. It was US$494 billion by 1980, US$1.21 trillion by 1990, and US$3.72 trillion in 2008 (figure 7.4). Governments must make regular payments on both the principal and the interest on this debt. These payments, referred to as *debt service payments*, are a significant drain on the economies of most developing countries. In 2007 total debt service paid by developing countries was US$540 billion.[6] The past decade has seen an increase in debt relief measures for developing countries, although it is important to note that this increased forgiveness of debt has been counted as "aid."

Box 7.1
The developing-country debt crisis

The ballooning of developing-country debt over the past thirty years is the product of a number of factors, both within developing countries and internationally. In the early 1970s, as a result of an excess of dollars in international markets linked to the OPEC oil price rises and rising inflation, developing nations were encouraged to borrow at flexible interest rates with few conditions from banks and governments. High inflation meant that real interest rates (the rate of interest adjusted for inflation) were quite low, making borrowing appear to be cost free for many countries. Commercial debt—that is, loans from private banking institutions—grew mainly in Latin America and Asia. Official debt—in other words, loans from governments and multilateral institutions—affected all parts of the developing world and formed the main source of loans for sub-Saharan Africa. Lenders at this time made the assumption that countries would not default and were aggressive in offering them loans.

After these debts began to accumulate, however, changes in the global economy made the loans increasingly difficult for poor countries to repay. In 1979, OPEC nations doubled the price of oil, and in response in 1981 the U.S. Federal Reserve raised U.S. interest rates above 20 percent to fight inflation. International interest rates rose as well, sparking a worldwide recession. Demand for the products sold internationally by developing countries plummeted, and the prices of raw commodities, which made up a significant proportion of developing-country exports, did not rise in step with the prices of the industrial products they were importing. The result was a double blow for developing countries with commercial debt. They were faced with higher real interest rates (because the rate of inflation fell as interest rates rose) and, because of the recession, they were not earning enough income to repay the loans. Countries holding official debt were also affected, because the global recession made it extremely difficult for them to earn enough from their exports to repay their loans. In 1982 Mexico announced it was unable to pay interest on its foreign debt. Other countries followed, and by the mid-1980s the world was immersed in a full-blown debt crisis that has become more and more entrenched, particularly in the world's poorest countries. In return for loan refinancing by the IMF and the World Bank, indebted countries agreed to adopt strict structural adjustment policies dictated by these institutions (discussed later in the chapter). Despite the emergence in recent decades of a variety of debt relief and cancellation programs, the overall level of debt in the developing world has continued to climb.

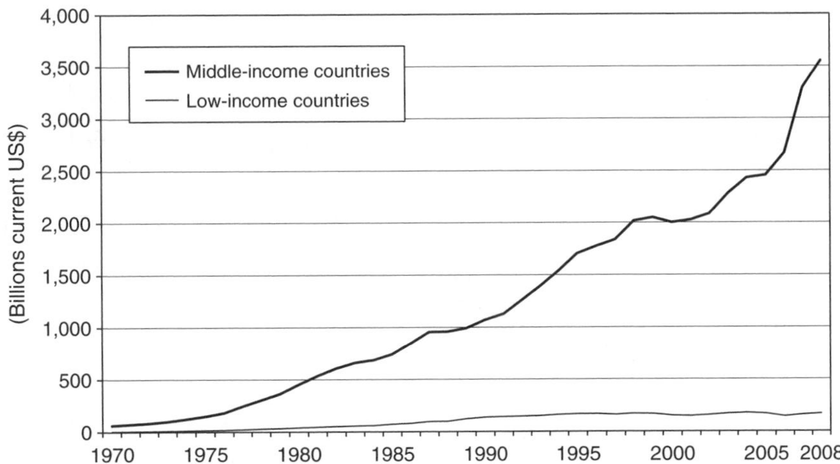

Figure 7.4
External debt stocks, total 1970–2008. *Source*: World Bank World Development Indicators, <http://data.worldbank.org>.

The liberalization of controls on financial flows over the last two decades has expanded global private financial flows well beyond international bank lending. Over three trillion U.S. dollars worth of national currencies is now traded across borders every day. As figure 7.5 shows, in April 1992, total average global foreign exchange market turnover was US$820 billion per day. This figure had more than tripled to US$3.08 trillion per day by April 2007 (despite declining from 1998 and 2001).[7] Private financial flows across borders now far outweigh trade in goods and services. Trade in stocks and bonds has also ballooned in the past decade.

Multilateral Lending: The World Bank and the IMF

Multilateral lending is a defining feature of international finance and the environment. The World Bank, as the main multilateral lending agency, is of particular significance. The World Bank plays a crucial role not only in lending to developing countries, but also in establishing the adjustment programs designed to ensure repayments of debts—both commercial and official. The World Bank provides a significant proportion of all multilateral assistance to developing countries, and its policies are extremely influential over the policies of other development banks and donor states. This influence affects some regions more than others, and is especially

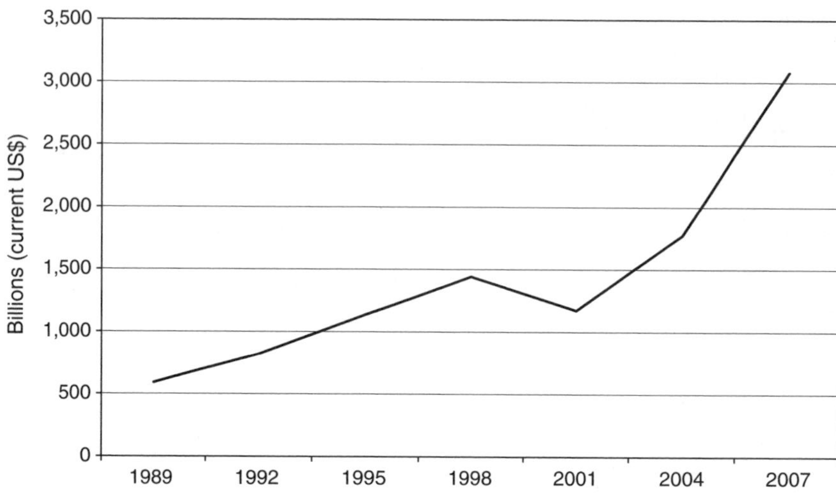

Figure 7.5
Global foreign exchange market turnover, 1989–2007. *Source*: Bank for International Settlements 2007.

important in areas like sub-Saharan Africa, where multilateral aid comprises the bulk of foreign assistance.

The World Bank and the IMF were established in 1944 after delegates from forty-five countries negotiated a monetary agreement at the UN Monetary and Financial Conference in Bretton Woods, New Hampshire. The Bretton Woods Agreement—which established a system of convertible currencies, fixed exchange rates, and free trade—was designed to support a liberal international economic order. The IMF was to be the overseer of exchange rates and provider of funds to meet balance-of-payments deficits. The original purpose of the World Bank, officially known as the International Bank for Reconstruction and Development (IBRD), was to finance projects to reconstruct a devastated Europe after World War II. The original task of the World Bank was largely complete by the late 1950s and early 1960s. By then, a host of newly independent countries were on the world stage. The World Bank therefore turned its attention to granting loans to the developing world, partly to stimulate global economic growth, and partly to ensure global economic stability. In 1960 the World Bank established a separate lending arm, the International Development Association (IDA), which specifically lends to the poorest developing countries on "soft" terms (low interest rates).[8]

Today the World Bank is the largest development lending agency in the world, recently lending approximately US$25 billion to developing countries a year.[9] The World Bank lends to support both specific projects and for structural adjustment. It disburses lending for structural adjustment more quickly, supporting shortfalls in balance of payments in return for policy changes in the borrowing country. The World Bank introduced structural adjustment lending in 1980, as a response to the economic stagnation of poor countries with high debt loads. This type of lending has become a significant feature of the institution's lending portfolio since then, with a climbing percentage of World Bank loans going to support adjustment programs. In the 1980s adjustment lending averaged around 17 percent of its loans. Today, around one-quarter of its loans are for adjustment or "policy lending," with around three-quarters supporting projects.[10]

A market liberal framework shaped the views within the World Bank on the environmental impact of its lending for most of the postwar period. Over the last few decades, however, the World Bank has slowly ebbed into a more institutionalist perspective, advocating, for example, global institutions for environmental protection.[11] Indeed, the World Bank was the first multilateral development agency to establish an Office of Environmental and Health Affairs (later renamed the Office of Environmental Affairs) as far back as the early 1970s,[12] although it did not begin to seriously address the environmental impacts of its loans until the 1980s and 1990s, in the face of mounting criticism from environmentalists.

Critique of Project Lending and the World Bank's Response

Environmental activists from around the world focused their campaigns on the World Bank's mega infrastructure projects and large-scale migration and resettlement schemes, especially in the 1980s and 1990s.[13] Critics—mainly social greens and bioenvironmentalists—attacked the Bank's market liberal assumption that economic growth would cure environmental problems. At the time the Bank, like most other international lenders, paid little attention to environmental issues in project design. This was not surprising, because most lenders—including the World Bank—did not have the staff to analyze every project to assess environmental impacts, let alone to incorporate environmental goals into the design of every project. The Office of the Environment had few staff members and was separate from the project design office. In large part this neglect was natural, because Bank officials were mostly traditional liberal economists who worked on the assumption that good economic policy was good for the environment.[14] Any environmental problems in the wake of projects were seen to be negative externalities inherent in the development process.[15]

The Bank's large infrastructure projects—such as roads, dams, and power plants, as well as industrial agricultural projects and migration schemes—did not account for environmental concerns. Many projects, as a result, contributed directly to higher rates of soil erosion and deforestation, a loss of native genetic material for agriculture, and the disruption of the lives of indigenous peoples. Even where the World Bank did have environmental policies in place in the 1980s, these were generally ignored.[16] Environmental groups were able to track the environmental performance of a number of World Bank projects, assisted by a growing network of local groups on the ground. Two World Bank–funded schemes drew particular media attention. The first was the Polonoroeste project, a colonization project in Brazil that included the construction of a major highway and large-scale resettlement into the Amazon. Critics have labeled this project a disaster because it caused negative social and health impacts on indigenous peoples, increased rates of deforestation, and damaged land unsuitable for development or agriculture.[17] The second scheme was the Narmada River Valley Projects in India, which aimed to build a series of dams along the Narmada River for irrigation and power generation. It involved significant resettlement, and was criticized for bringing social disruption, deforestation, and damage to ecosystems and watersheds from the flooding.[18] In both cases, critics accused the World Bank of not following its own policies regarding resettlement and consultation.

Armed with detailed information about environmental abuses in these and a growing number of other projects, environmental NGOs launched a concerted and sustained campaign to "green" the World Bank in the 1980s. Environmental groups had long tried to educate the public about the poor environmental practices at the World Bank. But their strategy took on a new angle at this time, as they began to lobby donor-government legislators responsible for approving government donations to the IDA of the World Bank. The U.S. government in particular was targeted because it has traditionally been skeptical of multilateral organizations (and thus frequently withholds funds from them). Environmental groups—such as the Environmental Defense Fund (now Environmental Defense)—testified before the U.S. Congress on the negative environmental impact of the Bank's various projects, most notably the Polonoroeste project. This prompted the United States to question the World Bank's environmental record. The United States did agree in the end to give its donation to the IDA replenishment—but with qualifications for the Bank to improve its environmental practices.[19] This led to a wider NGO campaign in North America, Europe, and the Pacific to reform the World Bank's environmental policies.

The World Bank halted the Polonoroeste project in 1985—the first time an ongoing project loan was halted on environmental grounds.[20] Environmental groups were able to pressure the World Bank to incorporate environmental issues more broadly in its project lending. The World Bank president at the time, Barber Conable, admitted in 1987 that the Bank had made mistakes in the past when it came to the environment and the displacement of people, and he promised to make the bank more "green."[21] The Bank conducted an independent review of the Narmada Valley projects in the early 1990s. The results, released in 1992, highlighted the severe problems associated with the project.[22] A year later the Indian government and the World Bank decided to discontinue the loan for the Narmada Dam scheme, opting for Indian self-funding rather than facing continuous international environmental opposition.[23]

The World Bank undertook a major restructuring in 1987, which included the reorganization of the environment staff from the Office of the Environment to a new Environment Department as well as into Regional Environment Divisions. The Office of the Environment at the time had just five specialists.[24] The new Environment Department was charged with researching the relationship between the environment and development, as well as assessing projects with potential environmental implications. After 1991 the World Bank also began to require environmental impact assessments of all projects. By 1990 the Bank had increased the environment staff by about 60 people; by 1994 nearly 200 had been added.[25] The Bank in addition pledged to design more projects with positive environmental benefits, and to consult with affected peoples and NGOs before embarking on projects with the potential to cause environmental harm.

The World Bank also helped set up a pilot project for the Global Environment Facility (GEF) in 1991. The purpose of the Facility, run cooperatively by the World Bank, UNEP, and UNDP, is to finance projects with global environmental benefits in developing countries with a per capita income of US$4,000 or less (we discuss the GEF more fully later). In 1992, to coincide with the Rio Earth Summit and to advertise its new green credentials, the Bank devoted the *World Development Report* to the theme of Development and the Environment.[26] This initial process of greening continued into the mid-1990s.[27] In 1993 the World Bank established an independent Inspection Panel to review claims by peoples affected by World Bank projects—a body seen by some observers as "remarkably" autonomous.[28]

Collectively, these changes amount to a significant shift in World Bank policies with respect to the environment in a relatively short period. The World Bank, increasingly drawing on the language of institutionalists, claimed it was "mainstreaming the environment" into its everyday practices. This meant far more funds for environmental projects.[29] Not everyone is convinced of the Bank's newfound environmental awareness and responsibility. Critics accuse the Bank of "greenspeak." No doubt, these critics note, there is a new *rhetoric* and every year the Bank is releasing many new *reports* on its environmental activities. Yet environment personnel at the Bank are still marginal players, and in practice most decisions on whether to lend funds for projects are made independently of the environment department. Projects with environmental benefits, critics note, tend to be small-scale and are of little interest to the Bank lenders because of the time required to design and manage them. Such projects are sidelined in part because the lending culture at the Bank tends to reward employees who lend large amounts, and not on the outcome of the loan itself (which is far harder to measure). The culture thus favors larger projects, the very ones with the greatest potential to cause environmental harm.[30] Critics note further that the concern with environmental matters in the first half of the 1990s seemed to fade over the next decade, only picking up again after 2003. In 1994 environmental lending made up 3.6 percent of total lending by the World Bank; loans for projects with environmental benefits increased some thirtyfold from 1990 to 1995.[31] Bank environmental lending as a percent of its overall lending then stagnated between 1995 and 2003, in some years even falling.[32] Between 2003 and 2008, however, loans directed specifically to improving the environment increased from 5 percent to 11 percent of its overall lending.[33]

Critics also complain that the Bank does not adequately consult NGOs regarding the potential environmental impacts of project loans. NGOs are now calling for more far-reaching reforms both to the consultation process as well as to environmental lending, and have pointed out the contradictions in the Bank's own messages. A prime example of this is the attention NGOs recently raised to the amount of World Bank lending supporting the construction of coal-fired power plants just as the Bank recommended a reduction in reliance on fossil fuels and warned that the world's poor suffer disproportionately from climate change.[34] The Bank's *World Development Report* for 2010, which focused on the theme of climate, made no mention of the billions it planned to spend on fossil fuel projects in a host of developing countries, including India, South Africa, and Botswana. A study by the Bank Information Centre indicates that the World Bank's lending for fossil fuel–based projects nearly tripled between 2005 and 2008, to over US$4 billion. This occurred despite a 2004 recommendation by the World Bank-commissioned Extractive Industries Review (EIR) that the Bank phase out lending for coal and oil by 2008.[35] The World Bank's chief economist defended its investment choices, declaring: "The [bank's] policy is to continue funding coal to the extent that there is no alternative and to push for the most efficient coal plants possible. Frankly, it would be immoral at this stage to say, "We want to have clean hands, therefore we are not going to touch coal."[36] This argument has not convinced critics, who are upset that money that they think should be spent on renewable energy projects is going instead into projects that rely on technologies that contribute to climate change.

The process of opening up and changing environmental policies of the Bank has, in the view of many NGOs, only just begun. Thus far, for example, there is little evidence of internalization of environmental costs, and environmental issues are not yet "fully integrated into the core logic of the World Bank's development strategy."[37] A recent independent evaluation of the World Bank's environment lending has found its performance weak and has recommended further reform.[38]

Structural Adjustment Lending and the Environment

One-quarter of World Bank lending currently supports structural adjustment programs (SAPs), more recently referred to by the Bank as "policy lending." This type of loan (also known as a structural adjustment loan, or SAL) provides balance-of-payments support to indebted governments. In return for these funds, governments agree to reform economic policies following the advice of the World Bank and the IMF. The intended purpose of this economic restructuring is to promote growth and eventually enable indebted governments to repay their debts. The World Bank began to undertake this type of lending in the early 1980s after many developing-country governments began to experience difficulties repaying external debts. Structural adjustment loans enabled the IMF and World Bank to wield considerable influence over economic policies and outcomes in poor regions of the world. Not only are these agencies themselves a significant source of funds to developing countries, but their policy advice also influences other development banks as well as donor states. Sub-Saharan African and many Latin American countries have been most profoundly affected by structural adjustment lending. The policy reforms called for under such loans have considerable environmental implications. Do these reforms help or harm the environment?

Not surprisingly, market liberals and institutionalists perceive the reforms under structural adjustment programs as critical to sustainable development. The main types of policy changes called for are those that open up developing-country economies to the world economy. The purpose is to facilitate trade and investment and ultimately economic growth. The first condition on a country's first adjustment loan is usually currency devaluation and the implementation of a floating exchange-rate system. The intent of devaluing the local currency is to boost the income of local producers and exporters. This has several further benefits for the local economy. It can lead to increases in production, which can lower prices and raise the volume of exports, allowing developing countries to recapture lost markets.

Other reforms called for in adjustment programs include a liberalization of domestic price policies and a removal of restrictions on trade.

These measures are aimed to help "get the prices right." The World Bank sees excessive government intervention in these areas as the primary explanation for why actual prices are lower than market prices. The loan conditions also call for the privatization of state-owned firms, reductions in government spending, and a relaxation of restrictions on foreign investment, all designed to reduce inefficiencies. Together, the adjustment reforms should make it easier for a country to repay debts because they encourage activities that earn foreign exchange and reduce government inefficiencies. SAPs were supposed to last three to five years only, and were seen as a one-time adjustment. In fact, they have been in place in a number of countries for nearly thirty years. Most countries in sub-Saharan Africa, and many in Latin America, as well as a number in Southeast Asia, had adopted SAPs by the late 1990s. These reform programs have become a dominant feature of their economies.

The IMF and World Bank did not at first consider the environmental dimensions of their adjustment lending. This is understandable because, as noted, the Bank did little to assess the environmental impact of its lending in the early 1980s.[39] But it also did not think that such scrutiny was warranted, because it was assumed that the economic growth stimulated by market-oriented reforms was necessary to improve environmental management anyway. This line of reasoning has continued to characterize the World Bank's response to criticism of the ecological impact of its adjustment lending. The Bank has argued that the impact of SAPs on the natural environment is hard to generalize, but that in most cases it has probably been either neutral or positive.[40] A World Bank report notes that "65 percent of recent SALs included an explicit environmental section, generally consisting of a statement to the effect that no environmental effects could be expected."[41]

Market liberals at the World Bank see the potential for some direct ecological benefits from SAPs. They can promote conservation by raising prices of previously underpriced natural resources, such as timber and minerals, to levels closer to true market value.[42] Price liberalization and currency devaluation also encourage the production of export crops like coffee, rubber, palm oil, and cocoa. The root systems of these tree crops can help prevent soil erosion. The removal of government subsidies, for example on fuel, agricultural chemicals, and industries, can improve environmental performance because these subsidies support inefficient industries and practices. Removing subsidies should reduce inefficient consumption—such as single-occupancy vehicles—which means less pollution and less chemical use.[43] And because SAPs encourage private ownership of both businesses and land, it creates, in the World Bank's view, incentives for stewardship and conservation.[44] If there are unexpected adverse effects from adjustment programs, these generally arise from imperfect policies, markets, or institutions.[45]

Critics—social greens in particular—have attacked SAPs as significant sources of unsustainable development in adjusting countries. Many claim that poverty and inequality have increased under adjustment programs, in clear contrast to World Bank declarations. Higher rates of poverty only fuel environmental degradation, as people deplete natural resources to survive (although, as chapter 4 discusses, the poor are not to blame). More directly, critics have attacked SAPs for increasing unsustainable exports of natural resources and encouraging pollution-intensive foreign investment. Many claim that SAPs virtually ensure that heavily indebted countries need to export unsustainable amounts of natural resources in order to earn the foreign exchange to service and pay off debts.[46]

Some critics, for example, claim that currency devaluations and the removal of trade restrictions create strong incentives to export large quantities of timber, which in turn drives deforestation. Data show that some of the highest rates of deforestation are in the most highly indebted countries. It is difficult to make the direct link between SAPs and deforestation, but there are, critics assert, some striking patterns. Brazil is one of the most highly indebted countries, and it has one of the most alarming rates of deforestation. The rate of deforestation began to rise at the time of the debt-crisis outbreak and during the implementation of adjustment policies in the 1980s.[47] In Ghana deforestation rose dramatically after it adopted an SAP in the 1980s. Between 1983 and 1986, Ghana's exports of tropical timber rose by 42 percent and the volume of timber production tripled.[48] Commercial logging has contributed to the shrinking of the country's forest area to one-quarter of its original size. Critics also point to unsustainable increases in tropical timber exports from Cameroon, Côte D'Ivoire, the Philippines, Tanzania, Thailand, and Zambia following adoption of adjustment reforms.[49]

Export-oriented adjustment policies have also pushed countries to increase their agricultural exports, often requiring a change in crops and agricultural techniques. In many cases, but not all, this has led to a decrease in food production and an increase in export-oriented crops. In some cases, plantation owners and farmers have brought large areas of marginal land under cultivation, contributing to deforestation, flooding, high rates of soil erosion, and falling agricultural yields. One example is Senegal, where policies to encourage an increase in the production of groundnuts contributed to severe soil degradation and lower productivity.[50] Industrial agricultural methods are on the rise in many countries in a bid to increase exports. Production of crops like cotton, flowers, and tobacco—with particularly harmful ecological impacts because of the chemicals and fertilizers needed to maximize productivity—often rises under SAPs because these crops generate foreign exchange (in some cases, the World Bank expressly encourages the development of such crops). Production of all three of these crops, for example, increased in Tanzania under its adjustment program.[51]

Debate over the links between structural adjustment lending and the environment is closely tied to broader questions of the impact of economic openness on ecological systems. This is a heated debate, as this book discusses throughout. The World Bank, although it continues to see its adjustment lending as an overall positive force for the environment, has begun to engage its critics and undertake studies on ways to better understand the linkages.

Multilateral Environmental Aid: The GEF and Climate Funds

Consistent with the 1987 Brundtland Report and the message put forth at the 1992 Earth Summit in Rio de Janeiro, institutionalists continue to stress that much more aid from rich to poor countries is necessary to help finance environmental protection in developing countries. Such financing, institutionalists assert, is best directed toward specific environmental activities within those countries, as well as toward global institutions that promote sound ecological practices. To achieve this, it is essential to ensure that more of the rich countries live up to their commitments. Far too often, institutionalists lament, this does not occur. At the 1992 Earth Summit, for example, rich countries were willing to commit only to providing US$125 billion of the necessary US$625 billion required to implement the conference action program to promote sustainable development (Agenda 21).[52] Both rich and poor countries agree that more aid is needed for the promotion of sustainable development, but they disagree on where it should come from, how much is needed, and how it should be allocated.[53] Furthermore, disputes have arisen over the "conditionality" of environmental aid—that is, over the restrictions donors attach to the aid, sometimes broad policy reforms as in the case of structural adjustment loans, or sometimes specific measures to protect the environment.[54] Concerns have also been raised about the "additionality" of aid, which refers to whether aid for the environment is to be "new" aid that is over and above existing aid flows, or just existing aid that is redirected to other uses.

For social greens, the problem is less about the *amount* of aid and more about the *source* and *focus* of aid. It is essential, in their view, to restructure the global political economy to rectify global economic inequalities. This will require far more than just lump sums of aid and loans, but new trade, investment, and financial flows. It is critical to restructure the financial system so that it empowers local agencies, rather than global agencies, with the funds to support sustainability.[55] Social greens tend to be suspicious of global institutions responsible for funding sustainability, because too often these institutions function with hidden agendas that are more about the interest of capitalists and donors than about the interest of communities of peoples.

This debate between institutionalists and social greens has played out over the environmental funding by the Global Environment Facility (GEF). The GEF undertaking was very much an institutionalist project, because it foresaw that global cooperation and funding through such an institution was essential for long-term environmental protection. From its inception, it has been a highly politicized institution. The GEF was set up with a budget of about US$1.3 billion just prior to the 1992 Earth Summit in Rio de Janeiro at the urging of several EU countries. The initial idea was to create a global "green fund" to help finance sustainable development initiatives called for in Agenda 21. The impetus for the GEF was to provide funding in the form of grants for developing countries to help them pay the "incremental costs" of global environmental obligations under international environmental agreements. Incremental costs, difficult even for the GEF to define precisely, refer in a broad sense to the extra costs developing countries pay for projects with "global" benefits. Along these lines, the GEF funds only projects with "global" environmental benefits. Initially, these were restricted to projects that focused on international waters, climate change, ozone depletion, and biodiversity. In 2002 the GEF added land degradation and persistent organic pollutants to this list.[56]

The GEF is run by the World Bank, UNEP, and UNDP, each of which plays a role that mirrors its own expertise, along with seven other agencies that have joined the GEF since it was established.[57] The World Bank is the lead agency and handles the finances. The UNDP directs the technical assistance associated with GEF projects, and the UNEP coordinates between the GEF and the global environmental agreement process. It was foreseen from the beginning that NGOs would have a role in GEF activities, although they had no formal rights in the pilot phase of the project. Funding for GEF projects comes from the GEF Trust Fund, a fund to which countries pledge donations. By May 2010 the GEF had committed more than US$8.8 billion to over 2,400 projects in over 165 countries, and had leveraged over US$38 billion in cofinancing.[58]

During the early years, the GEF was subject to heavy criticism from both environmental NGOs and developing countries. The World Bank quickly came to dominate the new institution, and there was in fact little consultation with NGOs and local communities on project design and implementation. Developing countries were unhappy with the GEF's commitment to only fund the "incremental" costs of projects. There was also concern that the dominant role of the World Bank gave too much weight to the needs of donors over the recipients. Attorney Bruce Rich at the U.S. advocacy group Environmental Defense succinctly explains, "The formulation of the Global Environment Facility was a model of the Bank's preferred way of doing business: Top-down, secretive, with a basic contempt for public participation, access to information, involvement of democratically elected legislatures, and informed discussion of alternatives."[59] For these reasons many NGOs were wary of endorsing the GEF as the one and only fund connected to global environmental treaties.[60]

Discussions on how to restructure the GEF to ameliorate these problems began as early as the 1992 Rio Earth Summit and continued over the next few years. Environmental NGOs and developing countries argued for a more accountable and democratic institution.[61] They were in favor of a voting structure similar to the UN system of "one country, one vote." They also argued for the management of the GEF to become independent from the World Bank. Environmental NGOs called for greater participation of nonstate actors, including NGOs and local communities, in GEF decision-making procedures and implementation. The donor countries on the other hand were most concerned about ensuring an efficient and effective institution. For this reason, they preferred that the World Bank continue to manage the GEF, and preferred a voting system similar to the Bretton Woods system of "one dollar, one vote."[62] The result was a compromise solution. The reformed GEF incorporates a mixed voting system with elements of both the UN and Bretton Woods voting structures.[63] The management of the GEF is formally housed in the World Bank, but is now functionally independent. The GEF now consults more formally with NGOs on its projects, and NGOs are allowed to attend GEF Council meetings as observers, and in some cases as participants, which is unusual in global institutions. The GEF was formally adopted as a permanent body in 1994. As part of this restructuring, the GEF was made the financial mechanism for the UN Framework Convention on Climate Change and the Convention on Biological Diversity. Today it is also the financial mechanism for the Stockholm Convention on Persistent Organic Pollutants and the UN Convention to Combat Desertification.

Social greens remain skeptical of the GEF, even though it is now undeniably a bit more democratic, transparent, and accountable. There are broad concerns about the "bandage" approach of an institution like the GEF. It papers over a few particular problems with money, rather than addressing the underlying structural and systemic causes. The GEF largely ignores, for example, issues like consumption. Social greens have several specific concerns as well. GEF money is given in the form of grants. Social greens agree that this is appropriate, and indeed essential. Yet because the GEF pays only for the "environment component" and only the part with "global benefits," most of the project grants are in fact in some way tied to other World Bank loans. This arises almost inevitably, critics argue, from the Bank's continued dominance of the GEF process. A report sponsored by the NGO Environmental Defense and the Halifax Initiative notes that the World Bank "has used the GEF to externalize environmental costs and increase the indebtedness of

Southern countries by 'sweetening' proposed loans with green grants."[64] Social greens further accuse the GEF of focusing mainly on transferring technology to developing countries. Prior to 2008, some 90 percent of GEF projects for climate change, for example, funded greenhouse gas (GHG) emission mitigation projects rather than adaptation projects that would have more local benefits in developing countries.[65] Projects are, moreover, top-down and do not adequately incorporate the views and participation of the local people most affected by them.[66]

In addition to and in some cases in conjunction with the GEF, other multilateral funds have been set up in recent years, in particular to address climate change. The Climate Adaptation fund was approved at the 2007 Bali meeting of the United Nations Framework Convention on Climate Change (UNFCCC), although it had been on the agenda since 2001. This fund is financed from a small percentage of the proceeds (which is approximately 2 percent of certified emissions reductions) from project activities under the UNFCCC Clean Development Mechanism, as well as other sources. The GEF administers the Adaptation Fund. In 2008, shortly after the Adaptation Fund was approved, the World Bank set up a separate series of climate investment funds. The World Bank manages these funds, which are implemented in conjunction with the regional development banks: the African Development Bank (AfDB), the Asian Development Bank (ADB), the European Bank for Reconstruction and Development (EBRD), and the Inter-American Development Bank (IDB). According to the World Bank, these new funds were designed in consultation with donors, the GEF, the UN, the private sector, and civil society organizations. The funds include a Clean Technology Fund, a fund for low carbon energy projects, and a strategic climate fund, which finances three separate programs for climate resilience, forest investment, and the scaling-up of renewable energy. Together, these funds have already accumulated over US$6 billion in pledges from donors.[67]

The proliferation of funds to finance projects to address climate change over a relatively short period of time (most of these funds were set up between 2006 and 2008), with overlapping mandates among different initiatives, has quickly crowded the landscape of international environmental financing. Some argue that the overlap and confusion among these various programs will likely hamper their effectiveness.[68] Further, critics point out that a shift appears to be occurring away from the GEF and toward the World Bank as the main organization for channeling climate change financing. This shift has several potential problems. Critics claim that the new World Bank funds were set up too quickly, with little meaningful input from potential recipient countries, civil society groups, and indigenous peoples. It was also external to the UNFCCC process, which some perceive as an attempt to undermine the multilateral negotiation process over climate financing. A further problem is that the new funds will provide loans, not grants. The Halifax Initiative points out that "asking already indebted countries to borrow money to address a problem created by the North violates the UNFCCC principles."[69] Additionally, the fact that the World Bank already has a fairly poor record on climate (as discussed earlier with reference to its lending for fossil fuel–based projects), has made critics wary of these new funds. They point out that the Clean Technology Fund, for example, allows some investment in more efficient fossil fuel–based energy projects. Critics would like to see new funds for climate financing projects to be exclusively focused on renewable energy. Finally, critics emphasize that the new climate money will be counted as ODA, and as such will not necessarily be additional development funding.[70]

Bilateral Finance: Export Credit Agencies

Export credit agencies (ECAs) are another important source of international finance. ECAs are public agencies in developed countries that provide credit—in the form of government-backed loans, investment guarantees, and risk insurance—to support both investment and trade in developing countries. These credits are specifically tied to business contracts with companies based in the lending country.[71] Developing countries that accept projects financed by ECAs are in effect taking out a loan from the ECA, which then goes to the corporation providing the service. The developing-country borrower (sometimes a government, sometimes a private entity) must eventually repay the loan, not the company involved from the industrialized country. Most ECA loans to private entities in the developing world require that their governments back the loans, often turning what appears to be privately held debt into public debt.[72] Most industrialized countries have at least one ECA that provides such assistance. Examples of some of the more prominent ECAs are the Export-Import Bank of the United States (the Ex-Im Bank), the U.S. Overseas Private Investment Corporation (OPIC), the Japan Bank for International Cooperation (JBIC)/International Financial Corporation (formerly JEXIM), the Canadian Export Development Corporation, the German Hermes Kreditversicherung-AG (Hermes), the UK Export Credit Guarantee Department (ECGD), and the Australian Export Finance and Insurance Corporation (EFIC).[73]

In providing this financing as well as loan guarantees and insurance services, ECAs seek to assist corporations from their own country as well as finance development projects in developing countries. These agencies provide a large and growing amount of development finance for trade and investment, at interest rates higher than those of other bilateral and multilateral assistance. ECAs provide some US$400 billion in loans annually, with some US$55 billion going to developing countries each year, making them significant financial players.[74] In 2007 they supported about 10 percent of global exports.[75] ECAs account for roughly a quarter of developing country debt owed to official creditors.[76] In some countries, the ratio is higher. In 2003, for example, export credits accounted for 64 percent of Nigeria's external debt and 42 percent of the Democratic Republic of Congo's external debt.[77]

ECAs have come under attack by environmental groups in recent years. The nature of export credit lending means ECAs tend to finance and insure ventures with high risks, including oil, gas, logging, and mining projects, as well as large-scale dams, nuclear power, chemical facilities, and road building in remote areas. As Aaron Goldzimer of the NGO Environmental Defense notes, "When an ECA will take on most of the risk and provide nearly full compensation if something goes wrong, there is every incentive for corporations and banks to move ahead with any overseas transactions—even excessively risky ones."[78] In the mid-1990s critics began to target ECAs, arguing that they did not properly consider the environmental implications of these projects before funding them. ECA Watch, an international NGO campaign on export credit agencies, pointed out that "half of all new greenhouse gas-emitting industrial projects in developing countries have some form of ECA support."[79]

Unlike bilateral ODA and multilateral development agencies, most ECAs are not subject to any environmental or social standards. They are also highly secretive agencies that do not disclose much information about the projects they finance.[80] The U.S. ECAs are an exception in this regard, because the United States is the only industrialized country to subject projects funded by its ECAs to environmental assessments. Under pressure by U.S. environmental groups, in recent years the environmental impacts of ECA loans have been gaining more and more attention. The case for more transparency has been particularly strong in the United States, because the U.S. ECAs have links to the United States Agency for International Development (USAID), which on paper has strict environmental standards.[81] This type of pressure, however, is lacking in many other industrialized countries. As a result, when a U.S. ECA rejects a project on environmental grounds, there is still a high probability that another ECA will finance it. This is what happened with financing for the Three Gorges Dam in China. Construction began on the twenty-year project in 1993. The Canadian, German, Swedish, and Swiss ECAs have all helped finance this project, which will displace up to 2 million people and is projected to cause widespread ecological disruption, even though the World Bank and the U.S. Ex-Im Bank rejected it.[82]

The United States and the United Kingdom have pressured other G-8 countries since the late 1990s to work toward a set of environmental and social guidelines for ECAs. The OECD has followed this lead and also launched talks on common standards for ECAs in these areas. At the same time, environmental NGOs have kept up their campaigns on ECAs, and in 2000 launched the Jakarta Declaration for Reform of Official Export Credit and Investment Insurance Agencies.[83] This declaration calls for greater transparency of ECAs, binding social and environmental guidelines and standards that are at least equal to those of the World Bank and the OECD, and the adoption of criteria on human rights and corruption. It also calls for an end to nonproductive investments and the cancellation of ECA debt of the poorest countries. In 2001 the OECD adopted a voluntary framework for common approaches to environmental assessment and export credit finance. This framework was strengthened in 2003 and again in 2007.[84] It is nonbinding, however, and despite its gradual strengthening over the years, critics still see it as a weak set of standards, and in particular stress that the guidelines make no mention of climate. In 2009, ECAs still allocated over a third of their lending for long-term projects in climate-sensitive sectors, including power, mining, industry, and infrastructure. An Environmental Defense study noted that in the 15 years up to 2009, ECAs had committed some US$1.5 billion per year toward coal plant financing alone.[85]

Private Finance and the Environment

The environmental impact of private global financing on developing countries has received far less attention than concessional assistance. This is understandable. Public financing is easier to track, and in the case of agencies like the World Bank, there are official policies to hold it accountable. It is much more difficult to hold private financial agencies accountable, especially when environmental standards and guidelines do not exist, as in the case of ECAs. The volume of private finance is enormous. Every day, as we noted earlier, more than US$3 trillion changes hands in private financial markets, mainly currency speculation and investments in stocks and bonds and other financial products such as derivatives. Because of its sheer size, private financing will inevitably alter the global environment, although the effects may be less obvious than in the case of public global financing.

Currency speculation and securities trading in particular can contribute to great economic instabilities. The 2007–2009 global financial crisis, for example, saw banks fail and prices of securities tumble sharply. The impact of the crisis was far-reaching. Following major instability in the banking sector in late 2008, governments stepped in with bailout packages in the trillions of dollars for the world's major banks to prevent them from failing, as well as government spending packages in the billions designed to stimulate economies in order to stave off a deeper recession. The broader impact of this crisis hit most economies hard in 2009, a year that saw global output, trade, and employment drop in ways not seen since the Great Depression in the 1930s. Especially remarkable out of this crisis was the volatility in the financial system. The value of the U.S. dollar was low when the crisis hit, rose sharply as banks began to fail, and then fell again as the recession took hold. Prices for commodities and financial derivatives based on them swung sharply from record highs and back down again. Oil prices climbed some 36 percent in 2008, to a record US$150 per barrel, and down 37 percent in 2009 following the financial collapse.[86]

Sudden and unexpected crises like the global financial crisis of 2007–2009 only reinforce the short-term mentality among investors in currency markets. Similarly, the money invested in stocks and bonds through mutual funds and in other financial derivatives demands short-term gains as well, so most investment ends up with the firms that promise such gains. As the global economy begins to recover from this major financial crisis, investors have jumped right back into the markets, prompting concerns that they have already forgotten the risks associated with past behavior.

There is an increasing recognition both inside and outside of the financial industry that the short-term mentality of financial markets tends to go against the long-term vision needed by firms to promote environmental protection and sustainability. Critics worry that it increasingly makes more financial sense, for example, to harvest an old-growth forest and invest the proceeds in financial markets today than it does to harvest the forest sustainably over a number of years. The returns on the former are significantly higher than on the latter. Such realities prompt firms and the banks that back them to pursue investment projects that lead to environmental destruction in the short run, with little consideration for the long run. By operating in this way, financial markets naturally tend to discriminate against firms that promote sustainable practices.[87] Unless banks and individual investors demand environmental responsibility as a condition of receiving money, there is little prospect that it will be a serious consideration. Some, such as Swiss industrialist Stephan Schmidheiny, are working to convince actors within private financial markets that they must promote principles of sustainable development.[88] He argues, along the lines of market liberals, that changing the behavior of banks and private investors is the key to the promotion of sustainable development. If these actors do not make decisions now to promote environmental protection within the private financial system, he maintains, economic growth prospects will be harmed in the future.[89]

Over the past decade, agencies like UNEP have pushed for greater environmental accountability of banks and financial markets. The UNEP Finance Initiative (UNEP FI) was founded in 1992 with the aim of engaging financial institutions in the broad discussions on sustainable development.[90] Today, UNEP FI works closely with more than 170 financial institutions that have signed onto its various statements to promote linkages between financial performance and sustainable development. Financial institutions participate in part because of NGO pressure, but also out of fear that financial market investors and banks could be held liable for the environmental damage done by corporate borrowers. A 1990 legal case in the United States, for example, decided that a bank was partly responsible for the environmental damage done by one of its borrowers.[91] Such concerns have been a key factor in getting banks to even consider the need to evaluate the potential environmental implications of borrowers. UNEP FI organized the Statement by Financial Institutions on the Environment and Sustainable Development and the Statement of Environmental Commitment for the Insurance Industry.[92] These statements are market-friendly, yet still support a precautionary approach to environmental management. The statements are not binding, but rather request signatories to abide by the principles set out in them. This has led some to ask whether it is a genuinely green move, or just a strategy to minimize risk on the part of banks. For others, it shows the willingness of private financial institutions to take some responsibility for environmental management, knowing a healthy global environment means a healthy global economy.

Other initiatives aimed at greening global finance have also emerged. The Equator Principles were first launched in 2003 as a voluntary code of conduct for private banks that provide project financing to developing countries.[93] Formulated by private sector banks in consultation with NGOs and the World Bank's private lending arm, the International Finance Corporation, the principles aim to ensure that private financial institutions do not fund projects in developing countries that may be unsustainable either socially or environmentally. Although they are strictly voluntary for financial institutions, the principles have been widely adopted, with some 50 private banks now adhering to them.[94] In the face of critiques by some NGOs that the principles were too weak, the principles were revised and strengthened in 2006. As of 2009 some sixty-seven private banking institutions from twenty-eight countries had signed on to these principles, including some export credit agencies such as Export Development Canada.[95] Critics have noted that the Equator Principles lack transparency and apply only to project finance, a relatively small percentage of the activities of private banks.[96]

In addition to seeking to reform banks, a number of initiatives have also targeted investors in order to encourage them to put their funds in more sustainable financial instruments. Investor groups, known as institutional investors, include pension funds, hedge funds, and mutual funds as well as banks and insurance companies. Together, these institutional investors account for a significant portion of global investment finance. The Principles for Responsible Investment (PRI) is a set of voluntary principles for investors, adopted in April 2006 on the initiative of the UNEP FI and the UN Global Compact. These principles influence corporations indirectly, as investors who sign onto the principles ask their clients to report and provide information with respect to environmental, social, and governance issues. The PRI principles call for investors to incorporate these issues into their investment analysis and decision making. As of 2010 the PRI had more than seven hundred signatories, representing US$20 trillion in assets across thirty-six countries.[97] The speed with which the PRI has gained signatories is impressive, although critics note that it is very much a voluntary set of broad principles with no real enforcement mechanism.

The Carbon Disclosure Project (CDP), launched in 2000, is another initiative aimed at raising awareness among investors about carbon emissions associated with their investments. The CDP encourages firms to disclose their emissions, which is reported to its signatories. As of 2010, there were some 475 institutional investors who collectively held US$55 trillion in assets.[98] More than 5,000 firms provide data to the CDP, as investors are increasingly demanding this information to inform their decisions. In this way, investor pressure through such voluntary initiatives can encourage more consideration of environmental issues. Critics note, however, that the CDP does not have an enforcement mechanism for disclosure and, furthermore, the information it does disclose is sometimes inconsistent and often complex, leading some investors to discount it when making decisions.[99]

In addition to initiatives designed to "green" financial markets, there has also been a rapid growth in participation in carbon trading markets (such as cap and trade schemes) that use financial incentives as a means by which to encourage firms to reduce their carbon emissions. These schemes operate by capping total carbon emissions allowed by a group of firms and rationing rights to emit. Firms are issued a limited number of carbon credits that they can then trade with one another. This trade in carbon credits, then, effectively puts a price on carbon emissions, which previously did not have a price tag attached to it. Because they must pay for each ton of carbon they emit, firms participating in these carbon trading schemes have a strong financial incentive to reduce their emissions. At the same time, carbon trading markets have become an arena for financial investment and even speculation, as carbon-based derivatives operate much like any other financial derivative market. Carbon trading has become a booming business, with the adoption of the European Union's mandatory Emissions Trading Scheme (EU ETS), and the Clean Development Mechanism of the UNFCCC. There is also a voluntary carbon trading market that has grown dramatically in recent years.[100] As the United States debates its own cap and trade scheme, carbon trading markets are set to grow yet further (see box 7.2 for an overview of carbon markets).

Box 7.2
Carbon markets

In a relatively short period of time, carbon markets have come to represent one of the world's most significant international policy mechanisms to mitigate climate change.[1] In aggregate, these markets are currently worth over US$126 billion, but have the potential to reach US$1 trillion if a U.S. market is established.[2] Carbon markets are meant to reduce CO_2 emissions by establishing property rights to the environment, generally measured as a right to emit one metric ton of CO_2.

These property rights can be used to reduce GHG emissions using two different market-based approaches. One is the cap and trade market, which places a quantitative limit on the amount of carbon that can be emitted in a region, country, or sector over a period of time. The amount of carbon allowed in the system, or the cap, is then divided into permits and distributed to emitters within the system. By enforcing a scarcity on the total number of permits, a firm that emits over its allocated amount of permits must purchase or trade for additional permits to offset its liable pollution within the compliance period. If an emitter is able to reduce its emissions below the cap in the compliance period, it can sell or trade its surplus permits to over-emitters for a net gain. The second type of market is one with a baseline and credits, or offsets. In these markets, an emitter can purchase a credit from a project that reduces its emissions below a "baseline" usually based on its business-as-usual emissions. This credit "offsets" CO_2 emissions generated by the purchaser.

Box 7.2
(continued)

A variety of global, regional, national, subnational, and private authorities are now using carbon markets as a mechanism to reduce GHGs emissions.[3] The EU ETS was initiated in 2005 and is currently the world's largest carbon trading market worth over US$91 billion and representing 73 percent of global carbon trade. The EU ETS is a regional cap and trade market designed to reduce emissions within the European Union by 20 percent by 2020 from a 1990 baseline.

The Clean Development Mechanism (CDM) is a global baseline and credit market designed by the Conference of the Parties (COP) to the United Nations Framework Convention on Climate Change (UNFCCC) under Article 12 of the Kyoto Protocol. Trading on the CDM began in 2005 and is now worth US$33 billion, representing 26 percent of global GHG trading.

There are also voluntary carbon markets. Although these markets represent just 2.2 percent of global emissions trading, they are growing at double the rate of the EU ETS and CDM. The CCX, for example, is a voluntary and private cap and trade market that began trading in 2003.

In order to be effective in mitigating climate change, carbon markets must ensure that the price and trade of each property right to emit one metric ton of CO_2 emissions (known as tCO_2e) represents one real metric ton of carbon from the atmosphere. For cap and trade markets to be effective, the cap must be low enough to create scarcity, raising demand and prices for tradable permits. Many economists believe a price above US$35 is necessary to encourage the adoption of renewable energy and sustainable technologies.[4] Currently, in the EU ETS, the price of a permit to emit one metric ton of carbon remains below this threshold, but as the cap is strengthened in the next compliance period, the price is expected to increase enough to encourage renewable energy adoption.

For baseline and credit markets to be effective, market regulators must ensure that each project's GHG reduction from the baseline is verifiable, transparent, and ultimately, real. This level of oversight requires a regulatory structure that often increases the transaction costs for each credit generated. For this reason, baseline and credit markets are often criticized as more inefficient than cap and trade markets, where a reduction target is explicit. But, from an environmental perspective, these costs are justified because high levels of transparency are required to ensure the market actually reduces emissions.[5]

Notes

1. We are grateful to Jason Thistlethwaite for his input on this box.
2. Capoor and Ambrosi 2009.
3. Ibid.
4. Hamilton et al. 2007, 20.
5. Nordhaus 2007.
6. Chan 2009.

As initiatives for greening global finance have grown rapidly in the past decade, debate has increased over whether these sorts of measures will make a substantial difference in terms of the global financial sector's impact on global sustainability.[101] Advocates see emerging initiatives, such as responsible investment, carbon financing, and carbon markets, as some of the most important ways to reorient global capitalism toward more sustainability. Critics, on the other hand, stress the difficulty of actually tracking and evaluating the environmental consequences of complex and obscure financing arrangements, and see much of what advocates are measuring as progress as illusionary or temporary. Some critics go even further, arguing that the net result is merely reinforcing and extending a global system of ever-rising production and consumption without making the hard choices necessary to deal with the rising total environmental costs of global financing.

Conclusion

Market liberals see global financing as an effective means to promote growth and prosperity in developing countries. Grants, such as for education, food relief, and health services, can no doubt help countries adjust, reorient, and get back on the development track. But market liberals also argue that it is equally critical to recognize the value of concessional loans from organizations like the World Bank. These loans help stabilize developing economies, which in turn benefits the entire global economy. Large loans allow governments in developing countries to lay the foundational infrastructure (especially for transportation and energy) to compete in global markets. No doubt, market liberals concede, economic development often requires large quantities of natural resource exports and the production of energy, including carbon-based energy, to earn the necessary foreign exchange to finance loans. Yet the period of resource exports and carbon-based energy should be seen as the first stage of development, a necessary step toward processing and manufacturing, and eventually services and information. This ratcheting-up process will inevitably involve some irreversible environmental changes. But this is not a reason to condemn it. It is paternalistic to want to "deep-freeze" the developing world into a state of low economic growth. Why should the peoples of the developing world be denied the lifestyle of those in the developed world? Europe and North America drew heavily on their natural resources to develop. Market liberals also support voluntary initiatives to green private global finance because such measures use the market to create incentives for investors.

Institutionalists agree with the broad parameters of the market liberal interpretation of the environmental consequences of global financing. There are, however, some notable caveats. For them, at a bare minimum, governments need to strive to meet the international promise of 0.7 percent of their gross national product. Donors also need to target more aid at specific environmental problems, especially global ones that are quite naturally difficult to fund in countries struggling to feed and house their own citizens. Institutionalists call for more mechanisms like GEF—with innovative mandates to finance projects with global environmental benefits—because these are the only effective and practical way forward. There is a fine balance between the financial needs of development and the financial needs to protect the global environment. Notably, the World Bank's recent effort to mainstream the environment into lending is shifting it away from a market liberal approach and toward an institutionalist one. Further, voluntary initiatives for private finance can be a useful way to reorient global finance toward goals of sustainable development, especially when part of efforts that combine international institutions and civil society along with private sector actors.

Compared to market liberals or institutionalists, bioenvironmentalists have far less to say about global financing. Most bioenvironmentalists see much of development lending as potential fuel for out-of-control growth and consumption. Environmental grants certainly can do some good, but these are a mere ripple in the global political economy, and the amounts are unable to counteract the effects of economic growth, consumption, trade, and investment. Foreign debt is, moreover, a drag on the ability of developing states to shift economic course, pushing them to harvest the last of their natural heritage just to service and repay loans.

Social greens are the main critics of the market liberal and institutionalist views of global financing. Ideally, from their point of view, the globe needs a more just political economy, in which the rich make do with less and do not deprive the poor of sustainable livelihoods. Then there would be less need for international aid, because the poorest would have access to land and would no longer be poor. The global political economy needs to empower communities, so locals again control decisions and destinies. This will never occur within current global financing structures. Social greens see the GEF- and World Bank–run initiatives, such as the recently launched Climate Funds, as elitist and designed to serve the interests of the World Bank and donor states. The ECAs meanwhile are opaque and unaccountable, in many ways even worse than the World Bank, which is at least inching ever so slowly toward more environmental sensibility, even if it has many lapses. Public and private loans in their view primarily serve global capitalism and local elites. The common people rarely benefit from this form of financing. The debt burden is adding to the social and environmental woes in the most destitute countries. There is a critical need to waive foreign debts completely, and most social greens support debt cancellation initiatives (i.e., outright forgiveness of debts of the world's most heavily indebted countries).[102] Many are, at the same time, critical of World Bank efforts to ease debt burdens through initiatives that make limited debt relief dependent on sustained commitment to structural adjustment measures.[103] Social greens are also skeptical of the ability of voluntary environmental initiatives in private financial markets to support sustainable development, as such efforts are typically weak and lack any meaningful enforcement mechanisms.

All perspectives agree that global financing *alone* will not be enough to solve global environmental problems. What, then, is needed? The conclusion examines the heated and radically different answers to this deceptively simple question.

8

Paths to a Green World? Four Visions for a Healthy Global Environment

How can we ensure a healthy and thriving ecological future? Should governments and global organizations control the forces of globalization? If so, what are the most effective means? Should we rely on technology, regimes, a world government, or local communities? There are radically different visions of the best way forward: ones rooted in radically different explanations of the causes and consequences of global environmental change.

Market liberals call for reforms to facilitate a smooth functioning of markets. They want eco-efficiency, voluntary corporate responsibility, and more technological cooperation. Institutionalists call for reforms to facilitate global cooperation and stronger institutions. They call for new and better environmental regimes, changes to international organizations, and efforts to enhance state capacity to manage environmental change. Bioenvironmentalists call for reforms to protect nature from humanity. They call for lower rates of population growth and consumption as well as a new economy based on an ethic of sustainability, one that operates at a steady state, designed to preserve the globe's natural heritage. For some, this can be done cooperatively with new and far stronger institutions—a vision not much different from institutionalists. For others, conquering the human instinct to consume ecological space will require a coercive and supreme authority—perhaps a world government. Social greens call for reforms to reduce inequality and foster environmental justice. People must rise up and dismantle global economic institutions to reverse globalization. The new global political economy must empower communities and localize trade and production. And it should respect the rights of women, indigenous communities, and the poor.

This book aims to help readers better navigate the vast array of competing arguments about how, why, and to what extent the global political economy is causing environmental change—as well as evaluate the "solutions" and "reforms" that seem to follow logically from various "analyses." We are not aiming to convince you that one perspective—or one set of options—is "more accurate" or "better" than another. Thus, this concluding chapter highlights the key insights from each of the worldviews, and explains the proposed paths for change emanating from each. It also examines the extent to which each worldview accepts or rejects the proposals of others. Our goal is to leave you better able to decide which elements of the worldviews we have mapped out in this book you find most convincing and most useful.

Market Liberal Vision

Market liberals see a future of ever-greater prosperity. Even the most superficial glance through history, they argue, confirms the great strides of humanity over the last few centuries. The past was not the idyllic paradise of goodwill and well-being that social greens so often seem to assume. It was nasty and short, full of suffering, disease, tyranny, and paranoia of "outsiders." Globalization is breaking down artificial barriers and enhancing cooperation and sympathy, which in turn promotes tolerance, democracy, and prosperity. True environmentalists, market liberals argue, should embrace globalization in all its forms: cultural, political, and, most important, economic. They should strive for a world with less government and more consumption, investment, trade, and development assistance. The immediate aim is to raise global and national per capita incomes. But the *real* aim is to ensure enough production so *all* of humanity prospers: only then will societies have the political will and the funds to implement sustainable development effectively.

Market liberals believe sustainable development is in the natural interest of business. "Pursuing a mission of sustainable development," argues the World Business Council for Sustainable Development, "can make our firms more competitive, more resilient to shocks, nimbler in a fast-changing world, more unified in purpose, more likely to attract and hold customers and the best employees, and more at ease with regulators, banks, insurers, and financial markets." The goal, in other words, is "sustainable human progress."[1]

Without economic prosperity, market liberals contend, it is unrealistic—indeed, it is unfair—to ask nations or individuals to sacrifice in order to save the global environment. Ordinary people need jobs, wages, training, and pensions *before* they will worry about climate change and ozone depletion. People will also, quite naturally, deplete and degrade the surrounding resources to feed, house, and sustain their families. Prosperity and better environmental quality, for market liberals, go hand in hand. A prosperous world, moreover, is a more ethical world, where people have the freedom to make dreams come true and the freedom to lead productive and useful lives.

Specific reforms need to unleash the power of the invisible hand of the market to drive economic growth, foster innovation and efficiency, and correct distortions. This requires open and competitive markets that foster trade based on comparative advantage. It requires steady and predictable investment climates: that means stable intellectual and physical property rights, reliable contracts enforced by the rule of law, transparent and accountable standards (the elimination of state corruption), and tax reforms to avoid unfair penalties on income earners (reasonable taxes on waste and pollution are fine). It requires fewer command-and-control regulations and better market incentives to ensure "voluntary" compliance (if compliance is voluntary, but with proper incentives, it will be in the financial interest of firms to comply). It further requires global rules to prevent governments from implementing policies—like bans, subsidies, and tariff barriers—that distort the smooth functioning of markets. *The Economist* magazine explains the consequences of such policies: "Bans and rules . . . have a habit of attracting a more pernicious sort of price signal, corruption. A quicker way for a government to improve its environment is to examine its list of subsidies. Artificially cheap water and energy, as well as tax breaks for mining, are scourges of the environment worldwide."[2]

The WTO and the IMF, for market liberals, are good models of how global cooperation can enhance economic well-being by promoting the benefits of open markets and discouraging government intervention in the economy. Further efforts are still needed, however. As the World Business Council for Sustainable Development notes, governments that impede business from doing business "and that try to take the place of business in meeting people's needs keep their people poor." The Council then adds, "Denying poor people and countries access to markets is planet-destroying as well as people destroying." As evidence, the Council points to the correlation between the Index of Economic Freedom and the Human Development Index: "Roughly, the more economic freedom, the higher the levels of human development."[3]

Market liberals like Bjørn Lomborg and Julian Simon also warn against premature actions to correct what might well be fictitious or overstated environmental problems. Activists and academics, they argue, are just as prone to personal and professional biases as government and corporate leaders. Groups like Greenpeace also often exaggerate the extent of environmental problems—on occasion because of incompetence, but primarily because this is what they do: scare the masses, fundraise, scare the masses some more. "The often heard environmental exaggeration has serious consequences," Lomborg writes. "It makes us scared and it makes us more likely to spend our resources and attention solving phantom problems while ignoring real and pressing (possibly non-environmental) issues."[4] Global leaders must be careful to avoid dancing to the tune of misinformed civil societies. They must show courage and leadership, acting only when there is real scientific evidence of a problem. Market liberals do recognize that on occasion some environmental problems are severe and immediate actions are necessary. It is best, however, to minimize the problems in the first place, relying on measures like fostering eco-efficiency, voluntary corporate greening, and technological innovation. We now explore each of these in more detail.

The *eco-* in *eco-efficiency* stands for both "economic" and "ecological." By efficiency, market liberals mean we have to make optimal use of both economy and ecology. The basic idea is to achieve more from less energy and raw-material inputs. "Eco-efficiency," explain Livio DeSimone and Frank Popoff of the World Business Council for Sustainable Development, "harnesses the business concept of creating value and links it with environmental concerns. The goal is to create value for society, and for the company, by doing more with less over the entire life cycle, that is, from creation of raw materials to disposal of products at the end of their life." It is "a *management philosophy* that links environmental excellence to business excellence."[5] The bottom line here is simple: benefits will accrue to companies that follow the principles of eco-efficiency. This approach will enhance their image, save them money, and gain them markets. It will also do away with the need for command-and-control regulations. It will, most importantly, ensure future growth, which is essential for the health of societies, the health of business, and ultimately the health of the planet.

Market liberals also stress the need for voluntary corporate social and environmental responsibility to ensure sustainable development. Voluntary corporate initiatives, as discussed in chapter 6, include ISO 14000 environmental management standards, the UN's Global Compact, the public-private partnerships, corporate social responsibility, and corporate environmental stewardship. For market liberals, participation in these initiatives promises only positive outcomes for the environment. And industry is seen to be the best actor to set up such initiatives because it is most familiar with corporate possibilities and constraints.

Market liberals prefer such an approach because it shifts the burden of regulation from the state to the firm, which can monitor environmental performance much more efficiently (important for eco-efficiency). Moreover, adherence to voluntary standards such as ISO 14000 helps ensure that even when state enforcement is weak (common throughout the developing world), firms will still abide by local regulations. In addition, such measures encourage firms to adopt cleaner technologies, rather than focusing on cleaning up after the fact, because firms consider environmental impacts throughout all stages of their operations. Voluntary programs are also intended to spread good environmental practices up and down the supply chain, as well as across borders, because certified firms are required to encourage suppliers to become certified.[6]

Today, as noted in chapter 6, just about every transnational corporation frames its approach toward, and input into, national and international environmental management through a policy of corporate social responsibility (sometimes shortened to "corporate responsibility"). Advocates see CSR as a way for TNCs to raise environmental standards globally, such as the policy of the Swedish firm Electrolux to require suppliers and contractors in developing countries to follow its code of conduct. Meeting the triple bottom line also encourages corporations to voluntarily go "beyond compliance" with existing laws, particularly

TNCs investing in developing countries with relatively low standards. CSR is a way as well to capture new markets for green products.[7]

Initiatives like eco-efficiency and voluntary corporate responsibility alone will not suffice to solve all global environmental problems. A sustainable future, market liberals stress, will inevitably require scientific progress, imagination, and forward thinking—the same forces that led to so many of history's great discoveries and advances. The world's wealthiest nations must nurture and harness the greatest minds—the Isaac Newtons, Henry Fords, and Albert Einsteins—for the benefit of the entire world. Societies and governments, market liberals argue, must draw on these geniuses to think us out of puzzles like climate change, declining soil fertility, drought, and deforestation. There is no other way. How, for example, given the desires and needs of people in all corners of the world to drive automobiles, can the global community possibly "solve" urban air pollution, climate change, and the inevitable end of oil with policies alone? The only feasible path is new technologies. The World Business Council for Sustainable Development is blunt here: "Recent history suggests that those living in wealthier countries do not intend to consume and waste less. Given that the other 80% of the planet's people seek to emulate those consumption habits, the only hope for sustainability is to change *forms* of consumption. To do so, we must innovate."[8]

There are many grand visions and much ongoing research on future environmental technologies, like hydrogen fuel cells for automobiles.[9] For market liberals, solutions to problems like climate change that discount the potential of new discoveries and technologies—such as calls by radicals to limit or ban cars—are naive, wishful thinking, or downright harmful to the ultimate goal of a better world. Market liberals argue that innovation that brings cleaner production processes feeds into the goal of eco-efficiency, and is thus by its nature in the interest of firms. Where cleaner production is not possible, technological advances can help mitigate and remediate. Research into geoengineering and carbon capture and storage, for example, seek to counteract global warming.[10]

It is equally vital, market liberals contend, to transfer these technological advances to the developing world. The transfer of environmentally sound technology can help developing countries "leapfrog" past development stages that tend to rely on inefficient and polluting technologies.[11]

Transfer of innovative environmental technologies, according to market liberals, works best when done by transnational corporations, in a business-to-business setting. Governments and international organizations should, therefore, reduce tariff and nontariff barriers to trade as well as actively entice foreign direct investment that brings in more advanced technologies.[12]

Institutionalist Vision

Institutionalists agree with many of the core recommendations of market liberals, including the need to embrace globalization; raise per capita national incomes; and facilitate trade, investment, and financing. They agree, too, that free markets, technology transfers, and voluntary corporate greening are all potentially valuable strategies to promote better global environmental management. Yet institutionalists add an important caveat to these strategies. They argue that more concerted efforts to harness and channel globalization toward sustainable development, beyond relying on markets alone, *must* accompany these strategies.  Doing this will require stronger international organizations, norms, and regimes as well as stronger national and local capacities to manage environmental affairs.

Without such measures, problems like climate change *could* slow or even stall the steady progress toward a more civilized and prosperous world. The global community should not risk such a future. Like market liberals, institutionalists stress that modern scientific inquiry should guide all analyses, decisions, and agreements. At the same time, it is important to recognize that the scientific method is often a slow process, involving claims and counterclaims. Thus, great uncertainty often surrounds scientific inquiry, especially for problems as multifaceted as climate change. Institutionalists see the precautionary principle as a useful tool to help overcome the limitations of scientific investigation. Principle 15 of the 1992 Rio Declaration defines this principle: "In order to protect the environment, the precautionary approach shall be widely applied by States according to their capabilities. Where there are threats of serious or irreversible damage, lack of full scientific certainty shall not be used as a reason for postponing cost-effective measures to prevent environmental degradation."[13] The precautionary approach has been particularly important as a justification for creating and strengthening global environmental regimes.

Environmental regimes are at the core of the institutionalist vision of a sustainable future.[14] This emphasis arises from the belief that negotiated global agreements, principles, and norms offer the best way to address environmental problems in a world of sovereign states. Global "rules" embodied in regimes provide a common point around which states can coordinate their own environmental behavior while at the same time being assured that behavior change for the global common good does not compromise their national sovereignty. Regimes also allow states to establish common but differentiated responsibilities for developed- and developing-country governments, such that states with more capacity can make greater commitments, while those with fewer resources receive less ambitious targets and more time or financial assistance to meet their commitments. This is one way for the global community to more fairly divide up responsibilities in a highly unequal world. More than 400 multilateral treaties now deal with environmental matters. This encompasses a wide range of regimes, including, for example, the 1989 Basel Convention on the Transboundary Movement of Hazardous Waste and Its Disposal; the 1992 Convention on Biological Diversity; the 1994 UN Convention to Combat Desertification in Those Countries Experiencing Serious Drought and/or Desertification, Particularly in Africa; and the 1997 Kyoto Protocol (see table 3.1). Because of their belief in the power of international agreements as a key means to address global environmental problems, many institutionalists were deeply disappointed by the outcome at the 2009 Copenhagen climate summit, which failed to produce a strong new global agreement with binding targets for GHG emission reductions.

Although institutionalists believe strongly in the potential of international environmental cooperation through regimes, they also call for efforts to make those regimes more effective.[15] Some of the more recent international environmental agreements, such as the 2007 Nairobi International Convention on the Removal of Wrecks, still need additional signatories to ratify them before they will come into force.[16] Better monitoring and greater compliance with many of the agreements already in force is needed, too. And there is a need for better coordination between agreements that are related, such as between the Kyoto Protocol and any successor agreement on climate change and the Montreal Protocol that addresses ozone depletion. Similarly, better coordination could occur among the Basel Convention (hazardous waste trade), the Stockholm Convention (persistent organic pollutants), and the Rotterdam Convention (hazardous chemicals and pesticides). Institutionalists also point out that there is a need for more financing, especially for the implementation of agreements in developing countries, although the GEF, with commitments by 2010 of more than US$8 billion in grants in over 140 countries, is certainly helping. There is also a need to ensure that global agreements feed into national policies and actions.

One of the greatest concerns of institutionalists is whether regimes are in fact effective. In other words, are they actually contributing to "solving" the problem? It is extremely difficult to measure regime effectiveness, because multiple intervening factors can influence environmental change in unexpected ways. Over the past decade, institutionalist scholars have shifted from their initial focus on regime formation to issues of implementation, compliance, and enforcement, which more closely reflect a regime's effectiveness. The purpose is to analyze what works and what does not in a bid to design better agreements in the future.[17]

Institutionalists also call for broad reforms to global environmental institutions. UNEP, for example, argues:

> Many environmental institutions were originally set up under different conditions and to perform different functions from those they are expected to exercise today. . . . Many institutions are constrained by a lack of human capacity and funding, despite increased environmental challenges, and this limits their effectiveness. These are clearly issues that need to be addressed if institutions are to fulfill their present obligations and confront emerging environmental issues.[18]

Institutionalists also call for specific reforms to international economic organizations, as previous chapters discuss. Many people, for example, would like to see the World Bank improve the consistency of its environmental performance as well as its accountability by ensuring more participation of affected peoples and NGOs in both project and adjustment lending.[19] Some call for reforms to the WTO, like clarification of the relationship between multilateral environmental agreements and trade rules, as well as incorporation of the precautionary principle into the rules of the WTO.[20]

Some institutionalists, such as political scientist Frank Biermann, go further and call for the creation of a World Environment Organization (WEO) to counterbalance the economic power of organizations like the WTO.[21] The idea here is to effectively replace the relatively weak UNEP with a much stronger body. This could mean upgrading UNEP from a program to a specialized agency, or it could mean streamlining a variety of activities into a larger agency to better coordinate global environmental policies. Some would even like to see an organization with enforcement powers to ensure that states actually implement and comply with the environmental agreements they sign and ratify, something UNEP currently does not have. Others argue that such an organization will not necessarily improve global environmental governance and may harm the interests of developing countries. Instead, they would like to see the current global environmental treaty structures enhanced.[22]

A further broad set of institutionalist recommendations focuses on the need to enhance state capacity to manage environmental issues more efficiently and effectively. This call for capacity building is geared particularly but not exclusively toward developing countries.[23] Improved capacity at the level of the state as well as other key environmental actors can help a state better comply with its international environmental commitments, and to improve its domestic environmental performance.[24] Meeting this goal, institutionalists argue, will require more aid for developing countries—at a minimum, developed countries should strive to meet their promise to provide development assistance worth 0.7 percent of gross national product (see chapter 7 for details).

Capacity building is a core mandate of the UNEP and UN system as a whole. "Over the past several years," UNEP notes, "it has become clear that capacity building is central to the quest for sustainable development."[25] Institutional and state capacity building also figured prominently in the 2002 Johannesburg Plan of Implementation, as well as in the World Bank's 2003 *World Development Report* on the theme of sustainable development.[26] This focus on strengthening institutional capacity marked an important addition to the World Bank's analysis of environmental issues, incorporating a much more institutionalist approach than it had in the past. Building capacity requires more than just more development assistance in terms of grants and loans. It requires technology transfers as well. Today, computers and satellites no doubt provide the most effective means of monitoring the environmental quality of resources like forests, rivers, and air.

Bioenvironmentalist Vision

Bioenvironmentalists tend to see a future of environmental doom. For many, humans are like any other animal: they act and react to survive—what some label as self-interest and others label as instinct. The human species, then, is seen to be a primary problem for the earth's ecology. The genetics that allowed the species to conquer all others in an era of seemingly limitless land and resources now leaves the species overflowing into every possible ecological space.[27] Under these circumstances, bioenvironmentalists demand radical changes to confront today's ecological crisis: the earth is beyond its carrying capacity and it is the duty of intelligent leaders to act immediately. They call for a new global political economy that respects the biological limits of the planet earth. This requires limits on the growth of the global population and the global economy. It also requires new attitudes, norms, and policies that internalize the value of ecospheres, and in particular the value of nonhuman life.

A common theme for bioenvironmentalists is the need to stem global population growth. Early bioenvironmentalists—most notably, Stanford University professor Paul Ehrlich—called for stern measures to stop the "population bomb" from wiping out the planet. Strong governmental powers should be granted, he argued, to coerce people into having fewer children. He advised donors to withhold food aid to developing countries unless they took drastic measures, including forced sterilization, to bring their populations under control. "Coercion?" Ehrlich ponders, "Perhaps, but coercion in a good cause. . . . We must be relentless in pushing for population control around the world."[28] Garrett Hardin, writing around the same time, called for a supreme authority to control global population growth, to implement "mutual coercion, mutually agreed upon."[29] Herman Daly, professor of ecological economics at the University of Maryland, continues this theme, advocating the issuing of transferable "birth licenses."[30] These strategies, however, now sit on the more extreme end of bioenvironmentalist thought.

Despite demographic trends in the past thirty-five years that suggest a slowing down of the overall rate of world population growth, more contemporary bioenvironmentalists continue to stress the need to control population growth rates.[31] The total number of humans is still on the rise, these bioenvironmentalists argue, and the planet is already beyond its carrying capacity. Today the strategies of bioenvironmentalists have softened in the face of widespread criticism over the years that employing coercive measures to control population growth violates human rights. Most bioenvironmentalists now stress the need for education, health care, and distribution of contraceptive devices that give men and women a choice of various family-planning options.[32] For example, providing universal primary education for boys and girls as well as enhanced educational opportunities at higher levels for women could go a long way toward the goal of population reduction. The idea is that educated women are more likely to pursue careers that use that education, which delays marriage and reduces family size. Enhancing health care to reduce child mortality is another humane way to encourage parents to have fewer children.

Most bioenvironmentalists tie their call for population control tightly to their call for a reduction in economic growth and consumption. These measures, say bioenvironmentalists, are equally important and act to reinforce one another to deplete the earth of its natural resources. The calls to reduce economic growth and consumption have focused on bringing about fundamental changes to global consumption patterns, especially in rich countries. For this reason, some extreme bioenvironmentalists advocate controls on immigration from developing countries to rich countries. Many moderates, however, propose education as the best (effective and noncoercive) way to confront the emerging global culture of consumption. This involves questioning the implicit link between well-being and buying more and newer things. And it involves deconstructing the branding of desire to free good people from the brainwashing of advertisers.[33] One of the core educational goals is to teach citizens to *choose* to live within limits—to use less, waste less, recycle, and alter lifestyles.[34] This approach tends to stress the role of individuals in reducing the global impact of consumption and production.

An important strategy to achieve these population and consumption goals is to promote a new economy based on a new ethic of sustainability. Daly refers to this as a "steady-state" economy, in which the number of humans and the amount of capital are constant, and at levels sufficient for a good and sustainable life. In addition, a steady-state economy has a rate of throughput of matter-energy as low as is feasible.[35] Under such an economy, there is room for society to develop and improve human well-being, but not to grow. Daly is firm, though, in stating that a steady-state economy does not imply zero growth in GNP, because it is not defined in terms of GNP.[36]

One way to instill this new sustainability ethic and move toward a steady-state economy, some bioenvironmentalists argue, is to do away with our measures of "progress" and "well-being," in particular the misleading focus on GDP and GNP. Bioenvironmentalists propose a number of alternative economic indicators, some of which focus on adjusting GDP and GNP figures, and others that focus mainly on environmental indicators. Two of the more prominent indicators are the index of sustainable economic welfare (ISEW) and the genuine progress indicator (GPI). The ISEW, first proposed by Herman Daly and John Cobb,[37] measures real per capita personal consumption expenditure, and adjusts it to correct for income inequality, pollution, loss of natural capital, value of household labor, and a variety of other factors not counted in GNP or GDP. The GPI, as proposed by the U.S. NGO Redefining Progress, measures the financial transactions from GDP that relate to well-being and adjusts them for factors similar to those incorporated into the ISEW.[38] Both indicators are trying to measure roughly the same thing, human well-being beyond mere GDP, but the methodologies differ slightly. Comparing the ISEW and the GPI against GDP growth over the past fifty years shows clearly that although GDP has continued to grow, the other indices show far less improvement. The ISEW and GPI have gained important recognition among environmental groups such as Friends of the Earth, and they have now been calculated for a number of countries.[39]

Indicators promoted by bioenvironmentalists that focus mainly on environmental measures include the ecological footprint (see chapter 4) as well the living planet index (LPI). Organizations such as the WWF use the ecological footprint to demonstrate the extent of planetary resource depletion. The WWF also developed the LPI, which averages changes on three indices: forests, freshwater ecosystems, and marine ecosystems.[40] Such measures are not designed to replace GDP or GNP,

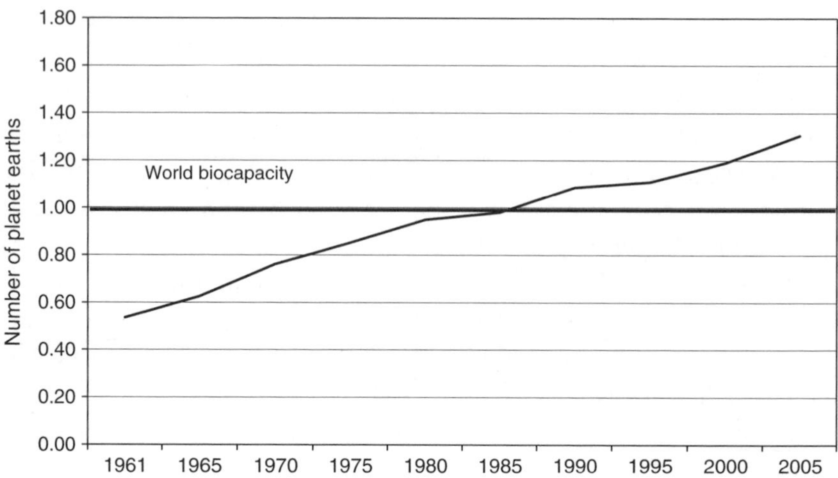

Figure 8.1
Humanity's ecological footprint, 1961–2005. *Source*: Global Footprint Network, <http://footprintnetwork.org>.

but rather to supplement them. The aim is to demonstrate not just to individuals but also to governments that we are living beyond the planet's biological capacity. The WWF's analysis of ecological footprints, for example, shows that human consumption of natural resources is about 30 percent over the earth's regenerative capacity (see figure 8.1).[41] The aim is to use this information to change human behavior in ways that are less consumptive and more conducive to conservation.

Bioenvironmentalists also commonly call for global restraints on production, trade, and especially transnational corporations as a way to move toward a more sustainable global economy. Some call for a world government or stronger international authority to ensure that such restraints are put in place. Others instead call for a constellation of strong institutions, norms, and rules—from global to local and public to private—to constrain production, trade, and transnational corporations so humanity again lives within the carrying capacity of the earth. Many recognize, however, that this will be a difficult challenge, given what is seen to be humanity's innate drive to plunder the planet.[42] Measures along these lines called for by bioenvironmentalists are similar to the institutionalist proposals discussed earlier, but bioenvironmentalists add (and stress) that a fundamental shift in human values must first occur.

Overall, then, bioenvironmentalists see a need to profoundly change Western values and lifestyles to solve global environmental problems. We must move, as environmental studies scholar Thomas Princen argues, from values based on efficiency and cooperation alone to ones focused on "sufficiency." In other words, accept "enough" and stop aiming for "more" through the growth paradigm.[43] There is no escaping this—not even with Nobel Prize–winning technological breakthroughs. How, these bioenvironmentalists ask, can technology ever replace an unidentified species? We cannot replace what we never knew we lost.

Social Green Vision

Social greens want a world that ensures both social and environmental justice. This is not just an ethical stance. Without this justice, social greens argue, the resulting global ecological crisis will swamp us all. Social greens largely agree with bioenvironmentalists on the need to address the culture of overconsumption and the need to adopt a new global ethic. Their focus, however, is much more squarely on rejecting globalization and relying on citizen action to promote justice in local communities. Political and environmental scientist Michael Maniates, for example, provocatively argues that the "individualization of responsibility" for environmental choices—such as green consumption and recycling—can feed into corporate goals and the culture of consumption. This can appease the guilt of overconsumption and diffuse the societal will to imagine collective ways to tackle environmental threats. He makes the important—and often overlooked—point that "individual consumption choices *are* environmentally important, but . . . control over these choices is constrained, shaped, and framed by institutions and political forces that can be remade only through collective citizen action, as opposed to individual consumer behavior."[44]

How can the global community bring about social and ecological justice in ways that reduce overall global consumption? Social greens call on the global community to dismantle or downgrade global economic institutions, change the nature of trade and production, cancel developing-country debts, and localize economies to empower local communities—in effect, to reverse globalization and industrial capitalism and alter the nature of the global political order. We now address each of these proposals.

In terms of global trade rules, most social greens argue that global leaders and ordinary citizens should reject the latest round of WTO trade negotiations, at least until a full environmental assessment of WTO activities is conducted.[45] Many, such as the members of the International Forum on Globalization (IFG), would like to see the WTO rules rolled back and the institution dismantled. Activist Colin Hines, for example, suggests a General Agreement on Sustainable Trade (GAST) to replace WTO rules, as well as a World Localization Organization (WLO).[46] He envisions that such organizations would encourage states to give preferential treatment to goods and services from other states that respect human rights, treat workers fairly, and protect the environment, and would favor domestic industry over foreign goods. Others, such as Greenpeace, take a more pragmatic approach and call for clear incorporation of the precautionary principle into WTO rules. They also call for precedence of multilateral environmental agreements over WTO rules, as well as greater WTO transparency and consultation. They also recommend rules to allow for discrimination based on production and processing methods and rules against patents on life forms (elements of what is sometimes called "safe trade").[47] Where trade is unavoidable, social greens also call for "fair trade"—that is, small-scale trade with communities that ensures fairer prices for goods and environmentally sound production.[48]

On global financing, social greens call on donors to forgive foreign debt as well as create a new organizational structure for global financing. Among social greens, there is a split between those who want to dismantle the existing global economic infrastructure and those who call for significant reforms. Most social greens, it is important to note, are not against global economic rules per se. The IFG, a group of more radical social greens, states, "Clearly, there is a need for international rules. To serve the whole of humanity, however, they must be based on the consent of the governed, and enforcement must be left primarily to democratically elected local and national governments."[49]

Many, both radicals and reformists, would like to see the IMF and World Bank decommissioned and favor a rollback of all structural adjustment programs. A new organization, providing financial advice but not loans, could work toward meeting goals of debt relief and local sustainability. Some would like to see international funding in the form of loans only available on a regional scale, and only for short-term emergencies.[50] Some envision an International Insolvency Court (IIC), which would oversee debt forgiveness. The New Economics Foundation and the Jubilee Coalition have proposed a Jubilee Framework, somewhat more reformist in nature, to oversee debt relief.[51] Many social greens also propose some sort of Tobin Tax—that is, a small tax of less than half a percent—on global financial transactions as a way to dampen global financial speculation, with the proceeds going to debt relief and/or local green initiatives.[52]

With respect to TNCs and global investment, most social greens advocate some sort of global agreement on corporate accountability to curtail what they see as systemic environmental and social abuses by TNCs. Most argue that voluntary measures are wholly inadequate, and that a legally binding agreement with teeth is required. Several proposals have been put forward by social green organizations. Greenpeace, for example, has proposed the "Bhopal Principles on Corporate Accountability." This includes stricter liability for corporations and rules to ensure the application of the highest standards regardless of location.[53] Although this proposal calls on firms to adopt these principles, their ultimate aim is for a legally binding treaty along these lines. Friends of the Earth International has also called for a legally binding treaty on corporate accountability.[54] Some social greens go so far as envisioning a UN organization to oversee such an agreement, as well as to collect data on TNC activities.[55]

The logical other side of the antiglobalization stance of social greens is the promotion of localization, also sometimes referred to as "subsidiarity." This means scaling down from the global where possible, and can refer to several different sublevels. As the IFG states, "Depending on the context, the local is defined as a subgroup within a nation-state; it may also be the nation-state itself or, occasionally, a regional grouping of nation-states. In all cases, the idea is for power to devolve to the lowest unit appropriate for a particular goal."[56]

All the institutional changes mentioned here are intended to work toward this goal of localization. Only those issues that require truly global coordination, such as global health and environmental issues, should have some level of international cooperative decision making, but others, like most economic, cultural, and political decisions, should be local. In practical terms, this means promotion of small-scale local agricultural systems and local production through small-scale enterprises. Corporations would be bound by "site here to sell here" policies.[57] It also means promoting local currencies, such as local exchange trading systems (LETS).[58] Some trade may still be necessary, but social greens stress that local communities should not be overly dependent on it. All of these localization measures would preserve diversity and reinvigorate local economies and communities. "Long-term solutions to today's social and environmental problems," writes social green Helena Norberg-Hodge, "require a range of small, local initiatives that are as diverse as the cultures and environments in which they take place."[59]

A final general set of proposals from social greens call for measures to re-empower voices marginalized in the process of economic globalization. Those who have been silenced include indigenous people, women, and the poor, who have been displaced from their traditional lands. These groups, social greens argue, are holders of critical ecological knowledge, and must be given the space and voice to apply it—that is, to practice sustainable livelihoods and educate others. Social greens call for local communities to reclaim the commons, including land, knowledge, water, biodiversity, and so on.[60] Reinstituting traditional and indigenous systems to protect and control common resources is, social greens contend, the most promising path to ensure a just world and a clean environment. This would involve the valuing of nonmonetized activities as legitimate economic activities, similar (but less formally measured) to the bioenvironmentalist measures of progress. Social greens see land reform as vital to enable this vision, because the world's poor and displaced peoples need fair and just access to land.[61] Social greens also lend their support to global social movements working to promote local and democratic movements to "reclaim the commons." This includes global movements for food security, land reform, indigenous rights, and sustainable local economies. Nicholas Hildyard, a former *Ecologist* editor and member of the Corner House, an NGO that works to support local communities in their environmental struggles, puts it succinctly: "Ultimately, it is only through the direct and decisive involvement of local peoples and communities in seeking solutions to the environmental crisis that the crisis will be resolved."[62]

Clashing Visions?

How compatible or incompatible are these four visions? Market liberals see potential pitfalls with the reform efforts of institutionalists, worrying their proposals to harness globalization with government and intergovernmental policies could harm economic growth. To meddle too much with the market, even if well intentioned, will result in inefficiencies and slow down progress toward better environmental conditions. In general, market liberals want less government, not more, and many are skeptical of global institutions such as the United Nations. Many market liberals are also critical of global environmental treaties, arguing that in too many cases the rules of these agreements contradict the rules of the WTO, which should guide the world economic order.

At the same time, market liberals consider the antiglobalization proposals of social greens and bioenvironmentalists as wrongheaded and downright dangerous to global well-being. Shutting down economic globalization will, in the view of market liberals, cause far more environmental harm (linked to increased poverty) than growth. Moreover, they argue that the bioenvironmentalists and social greens are far too pessimistic about the benefits of new technologies, which in the view of market liberals will reverse most growth-related environmental damage. They agree with social greens that population is not as grave an issue as bioenvironmentalists claim. But their reasons are different. Market liberals see economic growth as the key mechanism by which to ease population growth. Many market liberals also share some of the social green skepticism of global institutions. Yet they see the social green proposals for localism as nothing more than a recipe for protectionism, endemic poverty, and parochialism.

Social greens and bioenvironmentalists have strong alliances in some respects, because both see the so-called solutions of market liberals as little more than greenwash. They are opposed to large-scale global markets and are highly skeptical of market-based measures at this level. They have very little faith in voluntary corporate measures. According to these thinkers, standards such as ISO 14000, for example, have done little in practice to transfer clean technologies or improve the global environment. They also think it is dangerous to rely on Western science to generate new technologies, which in the long run could aggravate rather than solve global environmental problems. But although social greens oppose large, global markets, they do not oppose the use of small-scale markets in the local context (which market liberals also advocate).

Both social greens and bioenvironmentalists see institutionalist proposals for global environmental cooperation through regimes as little more than bailing out a few buckets of water as the *Titanic* sinks, because global environmental agreements have done little to improve the state of the global environment. Yet social greens and bioenvironmentalists interpret the reasons for this failure differently. Bioenvironmentalists think the authority of global regimes is too weak to be effective. Social greens think that focusing on global regimes is a misguided strategy because such efforts do not adequately include the voices of those most affected by environmental change, which means these agreements are inevitably ineffective.

Bioenvironmentalists and social greens disagree sharply at times over other issues. Social greens are often critical of the bioenvironmentalist calls for coercive measures to reduce population growth or to curb immigration to rich countries. A split within the U.S. Sierra Club (and as well between the U.S. and Canadian arms of the organization) illustrates this division nicely.[63] A slate of candidates ran in 2004 for election to the U.S. Sierra Club Board calling for a crackdown on immigration into the United States on ecological grounds. Ben Zuckerman, a member of the Sierra Club's national board of directors, and a professor of physics and astronomy at the University of California, Los Angeles, argues that immigrants cause great national, and ultimately global, ecological harm, because allowing a poor person into a rich country like the United States dramatically increases that person's ecological footprint.[64] Opponents within the Sierra Club called such views "racist," no different from the views of white supremacists.[65] Such views led the Canadian arm of the Sierra Club, which takes more of a social green perspective on population and immigration issues, to stress that it is a separate organization from its U.S. counterpart, and to develop its own population policy.[66]

Some social greens also take issue with bioenvironmentalists over the quantification of well-being in alternative measures of progress, arguing that they are merely replacing old numerical biases with new ones. Many bioenvironmentalists, while agreeing with many of the social green critiques of globalization, also still think that global cooperation, perhaps through a higher authority, is nevertheless essential (with the important caveat of a new global sustainability ethic). Thus, many bioenvironmentalists are not keen on a localization strategy that minimizes global institutional coordination.

Institutionalists, on the other hand, tend to see little harm in the solutions of market liberals—for them, these efforts will certainly help—but *only if* institutional safeguards are put in place to ensure that markets function in the best interest of the environment. They take issue, however, with those market liberals who criticize environmental treaties for contradicting the trade rules under the WTO, seeing such views as potentially dangerous for global environmental cooperation. For many institutionalists, international environmental laws should have as much standing as trade laws. Institutionalists also agree with bioenvironmentalists on the need to manage population growth more effectively. Their solutions, however, are more incremental and less coercive, focusing mainly on international cooperative efforts, through aid and technical advice, to enhance education, especially for women. Institutionalists also tend to support the general social green view on the need to rebuild the social structures of local communities. This should occur, however, through a globally coordinated effort, such as the local Agenda 21 action plans in communities tied to the global Agenda 21. Institutionalists also tend to see bioenvironmentalists as too pessimistic about the future, and see the social green vision of a world economic order of local communities as naive. They do, however, tend to see little harm in social greens' promotion of fair trade.

So what does all this mean for global environmental policy? Many would argue that what we have in practice today in both the global and national arenas are policies that largely follow the market liberal and institutionalist worldviews. The powerful global economic institutions and some key states—most notably the United States, but also including at times Australia, New Zealand, Japan, and Canada—have adopted a more or less market liberal worldview and are eager to see it influence global environmental governance. This is reflected in the current thrust from these players toward further economic globalization. The WTO's push for further liberalization of trade in agriculture and investment, and the continued push for programs of structural adjustment by the IMF and the World Bank, are prime examples of the market liberal influence. The global community is at the same time forming and extending more and more global environmental regimes, building a thick institutional structure of global environmental governance. These efforts are supported by organizations such as UNEP, as well as by some key states (most notably those in Europe, many developing countries, and sometimes Canada). In some cases, however, such as with the Kyoto Protocol, there are formidable stumbling blocks to the successful adoption and implementation of these agreements, as states with a more market liberal stance resist them. The social greens and bioenvironmentalists offer their proposals as an alternative to this reality.

During international environmental negotiations, strong alliances frequently exist between social greens and bioenvironmentalists on the one hand, and institutionalists and market liberals on the other as debates occur over the ecological impacts of globalization and economic growth. The proposals of the market liberals and institutionalists, however, clearly have the upper hand in the formal proceedings, although bioenvironmentalists and social greens do manage to influence the formal agenda, such as the proposal for a global treaty on TNCs at the 2002 Johannesburg Summit. At more recent negotiations, as at Johannesburg and at the Rio Conference in 1992, proposals for change adopted by the global policy community often converge around moderate institutionalist thought, because these views seem to offer a compromise between market liberal and more radical ecological views. Institutionalists, too, seem keen to broker this compromise. The recent publications of UNEP, for example, often canvass the arguments of market liberals, bioenvironmentalists, and social greens, drawing on different facets to promote their own agenda on the need to strengthen capacity and institutions.

The trend toward institutionalist thought within the global policy community *does not* mean institutionalists have "found" the answers and solutions. All it means is that they are managing, for the moment anyway, to muster the broadest base of political support by proposing "compromise solutions" to exceedingly difficult problems. Such small sacrifices, bioenvironmentalists and social greens continue to fret, may only slow down, but not reverse, the current crisis. More extreme market liberals, on the other hand, find these efforts to be constraining (although moderate market liberals are more likely to support modest institutional compromises). It remains to be seen how extensively and effectively the global community will implement the ideas and proposals of this institutionalist compromise.

Wading through the various worldviews on the political economy of the global environment was doubtless bewildering at times. There are no straightforward trends, explanations, or solutions. Our choices have multiple global impacts: some good, some bad, some comprehensible, others not. There is therefore no trouble-free path to a green world. Some global environmental change is inevitable. No one disputes that. Understanding such grand change means mistakes in analysis and policies and personal choices are inevitable, too. Some days, we must admit, we wish the world was not such a dense thicket of political, socioeconomic, and ecological complexity and uncertainty. On many days, we wish there were in fact simple solutions—predictable, fair, and painless. It would certainly make our own everyday choices easier. Should we eat a conventionally grown banana from Ecuador? Buy a shirt made in China? Drive our children to school? Turn on a lamp to read this book?

Simple answers and simple solutions would undoubtedly allow us all to feel less hypocritical about our everyday decisions. Yet from the knowledge of the complexity of the political economy of global environmental change comes perhaps the most important lesson of this book: the need to comprehend, tolerate, and yes, even respect the views of others as each of us develops our own vision of how to best move forward to create a healthy and prosperous planet. Embracing such diversity of knowledge, of knowledge with many "correct" answers and no "absolute" certainties, will not, we sincerely hope, paralyze environmentalists with perplexity. Instead, we hope it will empower environmentalists—market liberals, institutionalists, bioenvironmentalists, and social greens—to probe their own beliefs, so as to one day have a humility of mind that comes from a true understanding of the arguments and evidence of others.

Notes

Chapter 1

1. Other overviews of global environmental politics that complement this book include: Axelrod, VanDeveer, Downie 2010; Conca and Dabelko 2010; Kütting 2010; O'Neill 2009; Vig and Kraft 2009; Mitchell 2009; DeSombre 2007; Chasek, Downie, and Brown 2006; Speth and Haas 2006; Dryzek 2005; Dryzek and Schlosberg 2005; Switzer 2004; Elliot 2004; Maniates 2003; Lipschutz 2003; Paterson 2001c. See Dauvergne 2005 for a survey of research trends by more than thirty leading scholars of global environmental politics. See Dauvergne 2009 for an overview of environmentalism, including a short history, explanations of central concepts, ideas for student research projects, and a comprehensive bibliography organized by topics, themes, and regions. See Wapner 2010 for an analysis of the future of American environmentalism.

2. We use the term "political economy" to mean simply the interactions between political and economic processes. We are not implying any particular theoretical bias by using this term.

3. WCED 1987, 43.

4. "Our Durable Planet," 1999, 29.

5. World Bank, 1992, 30.

6. Simon 1981, 1996. The late Julian Simon was an economist and taught business administration at the University of Maryland. He became well known for his work on population growth and resource use, arguing that humanity is not facing a future of global scarcities of natural resources. He published widely on these themes, including books such as *The Ultimate Resource* (1981) and *Population Matters* (1990).

7. Easterbrook 1995. Gregg Easterbrook is a senior editor of the *New Republic* and a contributing editor to the *Atlantic Monthly* and the *Washington Monthly*. In his 1995 book *A Moment on Earth: The Coming Age of Environmental Optimism*, he argues that the global environmental situation is improving rather than deteriorating.

8. Lomborg 2001, 2007. Bjørn Lomborg is director of the Copenhagen Consensus Center, Adjunct Professor at the Copenhagen Business School, and the former

director of Denmark's National Environmental Assessment Institute. Lomborg's 2001 book *The Skeptical Environmentalist* draws on statistics on environmental quality to argue that, overall, the state of the global environment is in far better shape than ecologists and activists claim. This book sparked considerable controversy. The Danish Committee on Scientific Dishonesty, in response to complaints that Lomborg misused statistics and exaggerated points, ruled that the book was dishonest and contrary to standards of good scientific practice. The Danish Ministry of Science overturned this ruling in December 2003.

9. Bhagwati 2002. Jagdish N. Bhagwati is a professor of economics at Columbia University in New York. He has also served as Economic Policy Advisor to the Director-General, General Agreement on Tariffs and Trade (GATT), as Special Advisor to the UN on Globalization, and as External Advisor to the WTO. Bhagwati has published scores of books and articles arguing the case for free and open markets, including *In Defense of Globalization* (2004).

10. Schmidheiny 1992. Stephan Schmidheiny, a Swiss industrialist, was the founding head of the Business Council on Sustainable Development (which later became the World Business Council for Sustainable Development). His 1992 book *Changing Course: A Global Business Perspective on Development and the Environment* argues that it is in the economic interest of businesses to green themselves. Schmidheiny also played a central role in organizing business interests at the Rio Earth Summit in 1992.

11. Haas, Keohane, and Levy 1993; Chambers and Green 2005; Rechkemmer 2005.

12. Neumayer 2001, ix.

13. Paehlke 2003.

14. Young 1989, 1994, 1999, 2002; also see Young, Schroeder, and King 2008. Oran Young is a political scientist at the University of California, Santa Barbara. He has published many books and articles on the theme of institutional cooperation for the environment and is known especially for his work on international environmental regimes.

15. UNEP 2002a.

16. Keohane and Levy 1996; Hicks et al. 2008.

17. See, for example, Victor, Raustiala, and Skolnikoff 1998; Miles et al. 2001; Chambers 2008.

18. For a recent example, see Biermann and Siebenhüner 2009.

19. Lovelock 1979, 1995, 2009. James Lovelock is an independent scientist and environmentalist. Many see his concept of Gaia as one of the most imaginative and significant contributions to understanding the global environment.

20. Rees 2002, 249. William Rees is a professor of human ecology and environmental planning at the University of British Columbia in Canada. He has published widely on issues of consumption and has developed, with Mathis Wackernagel, the concept of the "ecological footprint," which measures human impact on the earth in terms of the amount of land needed to sustain lifestyles (see chapter 4 for details).

21. Rees 2002, 249.

22. Malthus 1798/1826.

23. Ehrlich 1968. Professor Paul Ehrlich is a biologist at Stanford University. He is best known for his work on population issues—many consider him a modern-day Thomas Malthus. In his 1968 book *The Population Bomb*, he argued for example that human overpopulation was destroying (or would soon destroy) the natural environment and create widespread famine and social instability. For this work he is sometimes referred to as the "prince of doom." Chapter 3 discusses his work in more detail.

24. Hardin 1968. Garrett Hardin, author of 27 books and more than 350 articles, was a professor of human ecology at the University of California, Santa Barbara. His work has been widely cited, and articles like "The Tragedy of the Commons" (1968) and "Living on a Lifeboat" (1974) have appeared in hundreds of anthologies. He was a member of the Hemlock Society, believing in the right to plan his own passing, and in 2003 he committed suicide at the age of eighty-eight. He was joined by his wife, Jane, who was suffering from Lou Gehrig's disease. His many books include *Exploring New Ethics for Survival: The Voyage of the Spaceship Beagle* (1972), *Filters against Folly: How to Survive Despite Economists, Ecologists, and the Merely Eloquent* (1985), *Creative Altruism: An Ecologist Questions Motives* (1995), and *Living within Limits: Ecology, Economics, and Population Taboos* (2000).

25. See Myers 1979; Princen 2005; and Ehrlich and Ehrlich 2004, 2008.

26. Myers 1997; Rees 2002, 2006.

27. Daly 1999, 34.

28. Herman E. Daly is a professor at the School of Public Policy at the University of Maryland. He served as Senior Economist at the Environment Department of the World Bank in the early 1990s, a post from which he resigned. Daly is best known for his promotion of ecological economics—economics that takes the physical limits of the planet into account. He argues that economies must attain a "steady state" that is in balance with the physical environment (for details, see chapter 4). He was a founder of the International Society for Ecological Economics and has published widely on the concepts of growth and environment-trade linkages. He received the Right Livelihood Award (alternative Nobel Prize) in 1996. Among his many books are *Toward a Steady-State Economy* (1973) and *For the Common Good: Redirecting the Economy toward Community, the Environment, and a Sustainable Future* (with John Cobb, 1989).

29. Hardin 1974; Ophuls 1973, 210.

30. For an overview of the origins of these views, see Helleiner 2000 and Laferrière 2001.

31. Schumacher 1973; Sachs 1999.

32. McMurtry 1999; see also Korten 1999.

33. Paterson 1996; Levy and Newell 2002. See also Stevis and Assetto 2001.

34. Shiva 1989; Mies and Shiva 1993. Vandana Shiva is a scientist, philosopher, and feminist whose activism and scholarly writings on the themes of global agriculture, the environment, and women have won her international acclaim. She is best known for her work on intellectual property rights over seeds, and the impact of the WTO on agriculture, women, and the environment in the developing world. She is director of the Research Foundation for Science, Technology, and Ecology, based in India, which she founded in 1982. She is also a leading member of the International Forum on Globalization and is advisor to many organizations, including the Third World Network. Her many books include *Staying Alive: Women, Ecology and Development* (1989), *The Violence of the Green Revolution* (1992), *Biopiracy: The Plunder of Nature and Knowledge* (1997), *Stolen Harvest: The Hijacking of the Global Food Supply* (2000), and *Soil not Oil: Environmental Justice in an Age of Climate Justice* (2008). She received the Right Livelihood Award in 1993.

35. Sachs 1999. Wolfgang Sachs is a scholar and researcher at the Wuppertal Institute for Climate, Environment, and Energy in Germany. He is also a former chair of Greenpeace Germany. Sachs is best known for his work on the interaction of globalization, development, and the environment, in which he focuses on the role of power in the global economy and its implications for the developing world and for the environment. His books include *Global Ecology: A New Arena of Political Conflict* (1993), *Planet Dialectics: Explorations in Environment and Development* (1999), and *Fair Future: Limited Resources and Global Justice* (with Tilman Santarius et al., 2007).

36. Goldsmith 1997. In 1969 Edward Goldsmith founded the magazine *The Ecologist*, which he edited for three decades. Today, under the editorship of his nephew Zac Goldsmith, it remains a leading outlet for social green analysis. Edward Goldsmith is coauthor of *The Social and Environmental Effects of Large Dams* (with Nicholas Hildyard, 1984), coeditor of *The Case against the Global Economy: And for a Turn towards Localization* (with Jerry Mander, 2001), and author of *The Way: An Ecological World-View* (1992). He received the Right Livelihood Award in 1991. He died in Italy in 2009.

37. Mies and Shiva 1993, 278; also see, for example, Princen, Maniates, and Conca 2002; Jha and Murthy 2006.

38. *The Ecologist* 1993, 140–150. See also, for example, Vandana Shiva's books.

39. See, for example, Mander and Goldsmith 2001; Cavanaugh and Mander 2004; Barlow 2008.

40. Shiva 1993b, 150.

41. International Forum on Globalization 2002. Also see the IFG website at http://www.ifg.org (accessed July 19, 2010).

42. Hines 2000, 2003. Colin Hines is an associate of the International Forum on Globalization and co-director of Finance for the Future, which encourages investments to reduce the use of fossil fuels. He also served as coordinator of the Economics Unit at Greenpeace International. He has published widely on the theme of promoting localization and argues against free international trade.

43. See, for example, Shiva 2008; Korten 2006, 2009; Bello 2009; McKibbon 2007, 2010.

44. Shiva 1997, 2008; Agyeman, Bullard, and Evans 2003; Agyeman 2005; Salleh 2009.

Chapter 2

1. For critical inquiry into the extent of contemporary globalization, see Hirst and Thompson 2000; Scholte 2005; Sassen 2006, 2007; Streeter, Weaver, and Coleman 2009.

2. Scholte 1997, 14.

3. A telephone mainline is one that connects a subscriber to the telephone exchange equipment (ITU 2009).

4. UNDP 1999.

5. ITU 2009.

6. UNEP 2002b, 20; the Size of the World Wide Web, <http://www.worldwidewebsize.com> (accessed July 19, 2010).

7. UNEP 2002b, 36.

8. Eberhart 2007. The International Air Transport Association (IATA 2007) predicts the number of air passengers will reach 2.75 billion by 2011.

9. Held et al. 1999, 170.

10. Scholte 1997, 17.

11. UNDP 1999, 25.

12. WTO 2001; World Bank 2006b; UNCTAD 2001, 9; UNCTAD 2002, xv, 272; UNCTAD 2008, xvi. See also World Bank World Development Indicators at <http://www.worldbank.org> (accessed July 19, 2010) and World Trade Organization Time Series data at <http://stat.wto.org/StatisticalProgram/WSDBStatProgramHome.aspx?Language=E> (accessed July 19, 2010).

13. Note, for example, the formal name of North Korea: Democratic People's Republic of Korea (DPRK).

14. Some of these democracies are considered weaker than others. Some organizations are devoted to tracking and ranking democracies around the world. See, for instance, World Audit at <http://www.worldaudit.org> (accessed July 19, 2010).

15. See, for example, International Forum on Globalization 2002.

16. UNEP 2002b, 35, 37.

17. See World Bank World Development Indicators at <http://www.worldbank.org> (accessed July 19, 2010).

18. Dauvergne 2003.

19. UNEP 2002b, 24.

20. Helliwell uses the terms *globaphiles* and *globaphobes* (from Burtless et al. 1998) to stress these contrasting views of globalization. He writes that "globaphiles are convinced that universal market openness is the single vital key to higher living standards. Globaphobes, by contrast, regard globalization as the tool multinational corporations are using to rob the world's poor by exploiting their labour, resources, and environments; destroying their culture; and commanding their vassal governments to implement whatever laws and trade agreements would make these transfers easier to achieve" (2002, 78).

21. UNCTAD 2000; World Bank 2006a. See also World Bank World Development Indicators at <http://data.worldbank.org> (accessed July 19, 2010).

22. UNEP 2002b, 329.

23. Lomborg 2001, 4, 7.

24. Ibid., 5, 61.

25. FAO 2008, 6; FAO 2009; for the 1969–1970 figure, see Committee on World Food Security 2001 at <http://www.fao.org/docrep/meeting/003/Y0147E/Y0147E00.htm#P79_3644> (accessed July 19, 2010).

26. Sen 1981.

27. Simon 1996, 12.

28. World Bank 1992, 29.

29. Runge and Senauer 2000.

30. Paarlberg 2000.

31. Today the thickness of ozone over the Antarctic is generally 40–55 percent of its pre-1980 level (UNEP 2000, 5). For details on environmental management and the environmental politics of the Antarctic, see Elliot 1994; Joyner 1998; Bargagli 2005; Herber 2007; Triggs and Riddell 2007.

32. UNEP 2000, chap. 2.

33. Soroos 1997, 169.

34. See Clapp 2009.

35. Goldsmith and Hildyard 1986.

36. For data and predictions for global cancer rates, see the web site for the International Agency for Research on Cancer at <http://www.iarc.fr> (accessed July 19, 2010).

37. See Davis and Webster 2002, 15–16.

38. See the web site for the American Cancer Society at <http://www.cancer.org> (accessed July 19, 2010).

39. The U.S. figures are from Wenz 2001, 5–6, the quote is on page 6. The global ones are from Green et al. 2005, 550.

40. Pojman 2000, 1.

41. Brown, Gardner, and Halweil 1999, 17.

42. Population Division of the Department of Economic and Social Affairs of the UN Secretariat, 2001.

43. For details see Dauvergne 2008, 49–56. See Paterson 2007 for an analysis of the environmental politics of the global culture of cars.

44. UNFPA 2001, 1, 6–7, 45.

45. Weisman 2007.

46. Iles 2004; Selin and VanDeveer 2006; Selin 2010.

47. Halweil 2009, 15.

48. WMO 1997, 9. See Gleick, Burns, and Chalecki 2002 for an overview of the technological, political, and economic forces behind the global freshwater crisis. See Conca 2006 for a comprehensive analysis of global water governance.

49. UNFPA 2001, 5; WHO 2008, 8.

50. UNEP 2001b.

51. Worldwatch Institute 2003; quotation from the summary at <http://www.worldwatch.org/press/news/2003/01/09> (accessed July 19, 2010).

52. Thomas et al. 2004, 145–148.

53. Pounds and Puschendorf 2004, 108. For more on the politics and science of biodiversity, see Steinberg 2001; Mushita and Thompson 2002; Cooney and Dickson 2005; Chester 2006; McManis 2007; Chivian and Bernstein 2008; Deke 2008; Jeffery, Firestone, and Bubna-Litic 2008.

54. See Stern 2006; FAO 2010.

55. Princen, Maniates, and Conca 2002, 6.

56. See, for example, Mander and Goldsmith 2001; Cavanaugh and Mander 2004.

57. Conca 2001; Dalby 2004.

58. IFG 2002, 5. The "consensus" authors with the IFG are like a "Who's Who" of social greens. They are Jerry Mander, John Cavanagh, Sarah Anderson, Debi Barker, Maude Barlow, Walden Bello, Robin Broad, Tony Clarke, Edward Goldsmith, Randy Hayes, Colin Hines, Andy Kimbrell, David Korten, Sarah Larrain, Helena Norberg-Hodge, Simon Retallack, Vandana Shiva, Vicky Tauli-Corpuz, and Lori Wallach.

59. UNEP 2002b, 62.

60. Notably, the rise in food prices in 2007-2008 caused the number of undernourished people to rise by 75 million in 2007 and 100 million in 2008, so that for the first time more than 1 billion people are now undernourished. See FAO 2008, 6; FAO 2009.

61. WHO 2009.

62. See the WHO Fact Sheet on "Obesity and Overweight" at <http://www.who.int/mediacentre/factsheets/fs311/en/index.html> (accessed July 19, 2010).

63. Lemonick and Bjerklie 2004, 58–69; also see information on obesity at the U.S. Surgeon General's web site at <http://www.surgeongeneral.gov/topics/obesity/calltoaction/toc.htm> (accessed July 19, 2010).

64. See the UNAIDS web site at <http://www.unaids.org> (accessed July 19, 2010).

65. Karliner 1997.

66. Shiva 2000; Kneen 1999.

67. Shiva 1997; Brac de la Perrière and Seuret 2000.

68. IFG 2002, 10.

69. See Falkner 2008, chap. 3. Two books (Grundmann 2001; Parson 2002) dispute the common claim that Dupont began to support global regulation of CFC production and emissions only because it had developed a substitute. Instead, these books argue that "by 1986 Dupont, concerned about its reputation, felt the weight of the scientific evidence rendered its position untenable and, furthermore, that industry leaders feared the prospect of litigation or unilateral action by the U.S." (Layzer 2002, 119).

70. These rankings rely on NASA data for annual average temperatures for the earth's surface (see <http://data.giss.nasa.gov/gistemp> [accessed July 19, 2010]). The average temperatures for 2009, 2005, and 1998 are close, although importantly 2005 did not experience the warming effect of an El Niño.

71. Paterson 2001a. See Dressler and Parson 2006 for an overview of climate change. For more on the global political economy of climate change, see Rowlands 1995, 2000; Paterson 1996, 2001b; Soroos 1997, 2001; Newell 2000; Skjærseth and Skodvin 2001; Cass 2005; Bäckstrand and Lövbrand 2006; Victor 2006; Depledge 2007; Bulkeley and Moser 2007; Volk 2008; Selin and VanDeveer 2009; Okereke, Bulkeley, and Schroeder 2009; Bulkeley and Newell 2010; Newell and Paterson 2010.

72. Rees and Westra 2003.

Chapter 3

1. Dryzek 2005; Dryzek and Schlosberg 2005.

2. For example, Hurrell and Kingsbury 1992; Brenton 1994; Caldwell 1996; Tolba and Rummel-Bulska 1998.

3. Such problems pushed some states to cooperate—for example, Canada and the United States signed the Migratory Birds Treaty in 1918.

4. Schreurs 2002, chap. 2. For an overview of environmental policy in Japan, see Imura and Schreurs 2005.

5. Crosby 1986.

6. Grove 1995.

7. Grove 1995; Fairhead and Leach 1995; Bryant 1997. Today, many ecologists recognize the logic of swidden farming in sparsely populated tropical areas. The burning of the degraded forest cover and vegetation supplies the soil with essential nutrients (acting like a natural fertilizer). After harvesting, the soil needs time to recover, so if there is opportunity it makes sense to move on and cultivate a new area.

8. Adams 1990, 16–20.

9. Grove 1995, 18–19.

10. Adams 1990, 20–21. The IUCN shortened its name to IUCN–The World Conservation Union in 1990.

11. The IUCN's web site is <http://www.iucn.org> (accessed July 19, 2010).

12. Lear 1997, 416–417.

13. Jasanoff 2001, 309–337.

14. World Bank World Development Indicators, <http://www.worldbank.org/data> (accessed July 19, 2010).

15. Rapley 2007.

16. See, for example, Frank 1967; Rodney 1972.

17. Helleiner 1994.

18. Ehrlich 1968, 132.

19. Hill 1969, 1–2.

20. Meadows et al. 1972. Twenty years later Donella Meadows, Dennis Meadows, and Jørgen Randers published *Beyond the Limits: Confronting Global Collapse, Envisioning a Sustainable Future*, which revisits many of the arguments in *Limits to Growth*. Donella Meadows, Jørgen Randers, and Dennis Meadows published *The Limits to Growth: The 30-Year Update* in 2004.

21. Meadows et al. 1972.

22. See, for example, Maddox 1972; Simon 1981; Commoner 1990.

23. Zacher 1993, 10.

24. Kumar 1996.

25. Schumacher 1973.

26. Rees argues that "a genetic predisposition for unsustainability is encoded in certain human physiological, social and behavioral traits that once conferred survival value but are now maladaptive. . . . Homo sapiens will either rise above mere animal instinct and become fully human or wink out ignominiously, a guttering candle in a violent storm of our own making" (2002, 249, 267). See also Brown, Gardner, and Halweil 1999; Brown 2005, 2008; Ehrlich and Ehrlich 2004, 2008; Rees and Westra 2003; Rees 2006.

27. UN Resolution 2398, quoted in Caldwell 1996, 58.

28. Rowland 1973, 49–51. The 1971 *Founex Report on Development and Environment* is available at <http://www.earthsummit2012.org/index.php?option=com_content&view=article&id=96:the-founex-report&catid=41:historical-ngo&Itemid=67> (accessed July 19, 2010).

29. Caldwell 1996, 68; also see Brenton 1994.

30. Cited in D'Amato and Engel 1997, 14.

31. Cited in Brenton 1994, 39.

32. Adams 1990, 38.

33. Murphy 1983; Hoogvelt 1982.

34. Taylor 1995.

35. Akula 1995; Rangan 1996; Guha 2000.

36. IUCN 1980.

37. Chapter 7 discusses the environmental effects of structural adjustment in greater detail. For a good overview of the economic situation in this period, see Rapley 2007, 2004.

38. WCED 1987, 8.

39. Bernstein 2001.

40. See Chasek, Downie, and Brown 2006.

41. See, for example, Krueger 1999; O'Neill 2000; Clapp 2001.

42. Rogers 1993, 238–239.

43. The Rio Declaration and Agenda 21 were published in UN, 1992. Also available at <http://www.un.org/esa/dsd/agenda21> (accessed July 19, 2010).

44. Brack, Calder, and Dolun 2001, 2. The full name of the document is the "Non-legally Binding Authoritative Statement of Principles for a Global Consensus on the Management, Conservation and Sustainable Development of All Types of Forests." For analysis of global forest politics, see Dimitrov 2005; Moran and Ostrom 2005; Humphreys 2006.

45. See Corell 1999 and Corell and Betsill 2001 for details of the global negotiations on desertification. For background on the causes of desertification, see Geist 2005.

46. Streck 2001, 75. The three implementing agencies of GEF are UNDP, UNEP, and the World Bank. It supports policies and programs in developing countries to address biodiversity, climate change, persistent organic pollutants, desertification, international water management, and ozone depletion.

47. Brack, Calder, and Dolun 2001, 2.

48. See the GEF web site at <http://www.gefweb.org> (accessed July 19, 2010).

49. See, for example, Chatterjee and Finger 1994; Sachs 1993.

50. Chatterjee and Finger 1994, 115–117; Connelly and Smith 1999, 206.

51. Shiva 1993b; Lohmann 1993.

52. For example, see Sachs 1993; *The Ecologist* 1993.

53. Osborn and Bigg 1998.

54. Chapter 5 discusses the structure of the WTO and its relationship to environmental issues more fully.

55. Davies et al. 2008.

56. Wade 2001, 2004.

57. Tabb 2000.

58. Mehta 2003, 122.

59. See Johannesburg Summit 2002 at <http://www.johannesburgsummit.org> (accessed July 19, 2010).

60. Mehta 2003, 122.

61. Johannesburg Declaration on Sustainable Development, <http://www.un.org/esa/sustdev/documents/WSSD_POI_PD/English/POI_PD.htm> (accessed July 19, 2010). Point 12 states: "The deep fault line that divides human society between the rich and the poor and the ever-increasing gap between the developed and developing worlds pose a major threat to global prosperity, security and stability." Point 14 states: "Globalization has added a new dimension to [global environmental problems]. The rapid integration of markets, mobility of capital and significant increases in investment flows around the world have opened new challenges and opportunities for the pursuit of sustainable development. But the benefits and costs of globalization are unevenly distributed, with developing countries facing special difficulties in meeting this challenge."

62. UN 2003.

63. Doran 2002, 2; See also Utting and Zammit 2006; Ivanova, Gorden, and Roy 2007.

64. Henderson 2006; Makower and Pike 2009.

65. Burg 2003, 116–118.

66. Wapner 2003.

67. Fisher 2010.

68. See Dimitrov 2010. Thanks to Matthew Paterson for providing us a detailed account of the Copenhagen summit and outcomes.

69. The literature on global environmental governance is expanding quickly. A few examples include Hempel 1996; Haas 1999; Lipschutz 1999; Conca 2000; Vogler 2000, 2003; Paterson, Humphreys, and Pettiford 2003; Newell 2008; Bretherton 2003; Jordan, Wurzel, and Zito 2003; Najam, Papa, and Taiyab 2006; Biermann and Pattberg 2008; Webster 2008; Biermann et al. 2009; Bernstein, Clapp, and Hoffmann 2009; Dauvergne and Lister 2010a.

70. Bernstein and Ivanova 2007; Biermann et al. 2009.

71. See, for example, Litfin 1998.

72. Principle 21 reads: "States have, in accordance with the Charter of the United Nations and the principles of international law, the sovereign right to exploit their own resources pursuant to their own environmental policies, and the responsibility to ensure that activities within their jurisdiction or control do not cause damage to the environment of other states or of areas beyond the limits of national jurisdiction." The Stockholm Declaration on the Human Environment, adopted June 16, 1972, is available at <http://www.unep.org/Documents/Default.asp?DocumentID=97&ArticleID=1503> (accessed July 19, 2010).

73. See Myers and Tucker 1987.

74. Clapp 1997, 129.

75. Andresen 2001a; Ivanova 2007, 2010.

76. Downie and Levy 2000.

77. Chasek 2000; Kassa 2007.

78. UNEP 2001a.

79. Political scientist Stephen Krasner defines international regimes as "sets of implicit or explicit principles, norms, rules, and decision-making procedures around which actors' expectations converge in a given area of international relations" (1983, 2).

80. Data from Ronald B. Mitchell, 2002–2007, *International Environmental Agreements Database Project (Version 2007.1)*. Available at <http://iea.uoregon.edu> (accessed July 19, 2010). Mitchell has documented over 900 multilateral environmental agreements, protocols, and amendments. Of these, 455 are multilateral environmental agreements.

81. Paterson 2001c; Kütting 2000.

82. There is a rich environmental literature on regime effectiveness. See Victor, Raustiala, and Skolnikoff 1998; Weiss and Jacobson 1998; Wettestad 1999; Young 1999, 2001, 2002; Miles et al. 2001; Hovi, Sprinz, and Underdal 2003; Mitchell 2002, 2006; Breitmeier, Young, and Zürn 2006; Mitchell et al. 2006; Stokke and Hønneland 2007; Vezirgiannidou 2009.

83. For research on the interplay of environmental institutions and the reform agenda for global environmental governance, see Rosendal 2001; Andersen 2002; Young 2002, 2008; Selin and VanDeveer 2003; Najam et al. 2004; Haas 2004; Skjærseth, Stokke, and Wettestad 2006.

84. Stokke and Thommessen 2003. See also Bill and Melinda Gates Foundation 2008.

85. Union of International Associations statistics, available at <http://www.uia.org/statistics/organizations/types-2004.pdf > (accessed July 19, 2010).

86. Betsill and Corell 2001, 2007.

87. Clapp 2001.

88. Wapner 2002.

89. Wapner 1996, 322. For an analysis of the global politics of whaling, see Peterson 1992; Stoett 1997; Andresen 2000, 2001b; Epstein 2008.

90. There is now an extensive literature on the role of NGOs in global environmental politics. See, for example, Princen 1994; Wapner 1996; Kolk 1996; Humphreys 1996; Jasanoff 1997; Najam 1998; Keck and Sikkink 1998; Auer 1998; Newell 2000; Tamiotti and Finger 2001; Bryner 2001; Corell and Betsill 2001; Hochstetler 2002; Ford 2003; Gunter 2004; Park 2005; Betsill and Corell 2001, 2007; Pellow 2007; Carmin and Bast 2009.

91. For details on the global political economy of persistent organic pollutants and the 2001 Stockholm Convention, see Lallas 2000–2001; Schafer 2002; Selin and Eckley 2003; Downie and Fenge 2003; Clapp 2003; Yoder 2003.

92. Background on the European Green Party can be found at <http://europeangreens.eu/menu/learn-about-egp> (accessed July 19, 2010). For the Charter of the Global Greens, see <http://www.globalgreens.org> (accessed July 19, 2010), under "Global Green Charter."

93. Susskind 1994; Gleckman 1995; Levy and Newell 2005; Fuchs 2005.

94. See, for example, Barnet and Cavanagh 1994; Nash and Ehrenfeld 1996; Korten 1999, 2001; Falkner 2003; Clapp 2005; Pattberg 2007; Sklair 2001; Finger and Kilcoyne 1997; Levy and Newell 2005; Fuchs 2005; Dauvergne 2008; Speth 2008; Clapp and Fuchs 2009.

95. Bulkeley and Betsill 2003; Bulkeley 2005.

96. Betsill and Bulkeley 2006; Rabe 2007; Selin and VanDeveer 2009.

97. Clapp and Swanston 2009.

98. For background on the movement toward voluntary simplicity, see Maniates 2002, 199–236.

99. The literature on the Fair Trade movement is vast. One recent example is Bacon et al. 2008.

100. For analyses of various movements against the automobile, see McKay 1996; Wall 1999; Robinson 2000; Paterson 2007.

Chapter 4

1. All figures are from World Bank World Development Indicators, <http://www.worldbank.org/data> (accessed July 19, 2010).

2. UNDP 2001, 9.

3. See Human Development Reports: The National and Regional Reports at <http://hdr.undp.org/en/reports/nhdr> (accessed July 19, 2010).

4. See, for example, Tidsell 2005; Tietenberg and Lewis 2008, 2009.

5. Stern 2006.

6. See, for example, Nordhaus 2007.

7. Simon 1996, 54, 67.

8. The Kuznets curve in economics depicts the relationship between growth and inequality over time, as hypothesized by Simon Kuznets (1955). Kuznets argued that in the early stages of economic growth, income inequality would initially worsen, but that over time, once it passed a turning point, it would improve as growth continued. Kuznets based this hypothesis on time-series data for England, Germany, and the United States.

9. Grossman and Krueger 1991, 1995; World Bank 1992. For an overview of this literature, see Dinda 2004.

10. Pempel 1998, 46.

11. Schreurs 2002, 36, 47. Also see McKean 1981; Broadbent 1998; Imura and Schreurs 2005.

12. Bhattarai and Hammig 2001.

13. World Bank 1992, 41.

14. Grossman and Krueger 1995; Dinda 2004.

15. Thomas and Belt 1997. Also see Tietenberg and Lewis 2008, 2009.

16. Munasinghe 1999. Also see Munasinghe et al. 2006.

17. World Bank 1992, 25–32. For an analysis linking poverty and desertification, see Johnson, Mayrand, and Paquin 2006.

18. Millennium Ecosystem Assessment 2005, 13; see also Comim et al. 2008.

19. UNDP 1998, 57; see also Scherr 2000.

20. Cleaver and Schreiber 1992.

21. Barber and Schweithelm 2000, 17.

22. Quoted in Keane 1998.

23. Mink 1993; Comim et al. 2008.

24. World Bank 1994, 161–162.

25. World Bank 1992, 30–31; Comim et al. 2008.

26. World Bank 2006b.

27. UNDP and UNEP 2008.

28. Roodman 1998; Myers and Kent 2001; see also IISD Global Subsidies Initiative at <http://www.iisd.org/subsidies> (accessed July 19, 2010).

29. Elliott 2009; Dauvergne and Neville 2009, 2010.

30. Schmidheiny 1992; Schmidheiny and Zorraquín 1996; DeSimone and Popoff 1997; Holliday, Schmidheiny, and Watts 2002.

31. See the OECD web site on Consumption, Production and the Environment at <http://www.oecd.org/topic/0,3373,en_2649_34289_1_1_1_1_37465,00.html> (accessed July 19, 2010).

32. See the web site for the Marrakech Process at <http://esa.un.org/marrakech process> (accessed July 19, 2010).

33. Keohane and Levy 1996; Hicks et al. 2008.

34. Waring 1999.

35. Ekins, Hillman, and Hutchinson 1992, 36; Daly 1996, 40–42. Also see Daly 1999, 2002.

36. Miranda, Dieperink, and Glasbergen 2006; Lipper et al. 2009.

37. See, for example, Repetto et al. 1989. This study accounts for deforestation, soil erosion, and depletion of oil reserves in the national income and product accounts. It then recalculates Indonesia's annual GDP growth rate from 7.1 to 4 percent.

38. Waring 1999.

39. Ophuls 1973, 112–113. Also, see Daly 1977; Meadows, Randers, and Meadows 2004; Princen 2005. See Princen 2010 for his vision of an economy grounded in natural systems.

40. Wackernagel et al. (2002, 9266–9271) conclude that humanity passed the biosphere's regenerative capacity in the late 1970s, and has been steadily moving beyond it since the mid-1980s. James Lovelock, who in the 1970s developed the "Gaia hypothesis" of the earth as a living, self-regulating, and adaptive super-

organism, issued "A Final Warning" in his 2009 book *The Vanishing Face of Gaia*.

41. Lawn 2004.

42. Ayres 1998, 190.

43. Chapter 5 of Daly 1996 presents his parting suggestions for the World Bank on his resignation.

44. For example, Stern, Common, and Barbier 1996; Ezzati, Singer, and Kammen 2001; Bagliani, Bravo, and Dalmazzone 2008.

45. World Bank 1992, 41; Ansuategi and Escapa 2002.

46. Arrow et al. 1995.

47. Cameron 1996; Hall 2002.

48. Stern, Common, and Barbier 1996, 1156; Suri and Chapman 1998; Clapp 1998, 2001; Tidsell 2001; Bagliani, Bravo, and Dalmazzone 2008.

49. Stern, Common, and Barbier 1996.

50. Arrow et al. 1995; Tidsell 2001, 187.

51. Reardon and Vosti 1995.

52. Mander and Goldsmith 2001.

53. *The Ecologist* 1993, 95.

54. Shiva 1993c; Kimbrell 2002; Weis 2007.

55. Mabogunje 2002.

56. Leach and Mearns 1996; Bryant and Bailey 1997; Broad and Cavanagh 2008.

57. Ostrom 1990. Elinor Ostrom, who won the 2009 Nobel Prize in Economics, is a professor of political science at Indiana University. She is one of the most influential theorists of the relationship between institutional arrangements and the management of common-pool resources. For Ostrom's recent research on the commons, see Dolšak and Ostrom 2003; Ostrom and Hess 2006. For her recent research on institutional analysis, see Ostrom 2005.

58. Bryant and Bailey 1997, 162. Also see Hardin 1968.

59. See, for example, the case studies in Peluso and Watts 2001.

60. Shiva 1997; Miller 2001; Sell 2009.

61. Shiva 1993c, 1997, 2000, 2008.

62. Richards 1985; Fairhead and Leach 1995.

63. Broad 1994.

64. Shiva 1997.

65. Gedicks 2001; McAteer and Pulver 2009; Widener 2009.

66. Broad 1994.

67. *The Ecologist* 1993, 93.

68. Sachs 1999, 168.

69. World Bank World Development Indicators, <http://data.worldbank.org/> (accessed July 19, 2010).

70. See Alcott 2008.

71. World Bank 2005, 236; World Bank World Development Indicators, <http://data.worldbank.org/> (accessed July 19, 2010).

72. Wackernagel and Rees 1996. Mathis Wackernagel is now Executive Director of Global Footprint Network, which aims to "accelerate" the use of ecological footprint analysis. See Global Footprint Network: Advancing the Science of Sustainability at <http://www.footprintnetwork.org> (accessed July 19, 2010).

73. The WWF produces a biannual report, the *Living Planet Report*, which keeps track of trends in ecological footprints. WWF 2008.

74. Wilson 2002, 23. See also Wackernagel and Rees 1996; <http://www.footprintnetwork.org> (accessed July 19, 2010).

75. Wade 2001, 2004.

76. Kates 2000.

77. Myers 1997, 34.

78. Daly 1993b, 27–28.

79. WWF 2008, 28–29.

80. *The Ecologist* 1993, 140. Also see Mies and Shiva 1993, 278.

81. Shiva 1994; Sen 1994, 1999.

82. Schor 1998; Robbins 2002; Schor 2004; Dauvergne 2008.

83. See Dauvergne 2008, chaps. 3–6.

84. Sachs 1999; Princen, Maniates, and Conca 2002.

85. Dauvergne 2010.

86. Sachs 1999, 167. Also, see Princen, Maniates, and Conca 2002; Dauvergne 2008.

87. MacNeill, Winsemius, and Yakushiji 1991; Dauvergne 1997b, 2008.

88. Princen 1997, 2002; Clapp 2002a.

89. See the Basel Action Network web site at <http://www.ban.org> (accessed July 19, 2010).

90. *The Ecologist* 1993; Mander and Goldsmith 2001.

91. O'Connor 1998, 191–196; Dalby 2004.

92. See Maniates 2001 for a critical analysis of the individualization of environmental responsibility. See Maniates and Meyer 2010 for a collection of articles calling on scholars and activists to give more attention to the value of the concept of "sacrifice" as a policy option.

93. See, for example, the Global Green Charter at <http://www.greens.org> (accessed July 19, 2010).

Chapter 5

1. For an overview of the history of this debate, see Esty 1994; Williams 1994, 2001; Sampson and Whalley 2005.

2. WTO and UNEP 2009; WTO statistics online, <http://www.wto.org/english/res_e/statis_e/statis_e.htm> (accessed July 19, 2010).

3. UNCTAD 2009.

4. WTO data; see <http://www.wto.org> (accessed July 19, 2010).

5. World Bank World Development Indicators, <http://www.worldbank.org/data> (accessed July 19, 2010). See also WTO 2008.

6. WTO 2008.

7. WTO 2008, 30.

8. Bhagwati 1993, 2004.

9. Neumayer 2001, 103.

10. WTO 1999, 29.

11. WTO 1999, 4.

12. Neumayer 2001, 104. Also see Lomborg 2009.

13. World Bank 1992, 67.

14. WTO and UNEP 2009.

15. Vogel 1995.

16. Vogel 2002, 366.

17. Logsdon and Husted 2000.

18. Bhagwati 1993, 48.

19. Weinstein and Charnovitz 2001, 150.

20. Dauvergne 2001; Ross 2001.

21. Bhagwati 1993, 44.

22. WTO 1999, 44.

23. Sachs 1999, 183–184. See also Princen, Maniates, and Conca 2002; Dauvergne 2008.

24. Daly and Cobb 1989, 213–218.

25. Daly 2002, 13.

26. Goldsmith 1997, 3. For coffee, see Wild 2004, Talbot 2004, Bacon et al. 2008; for bananas, Striffler 2002, Striffler and Moberg 2003, Bucheli 2005, Soluri 2006; for sugar, Schmitz et al. 2002; for tea, Moxham 2003, MacFarlane and MacFarlane 2004; for oil palm, Casson 2000, McCarthy and Cramb 2009, McCarthy and Zen 2010. Crops like sugar and oil palm are also supporting a growing biofuel industry in the developing world (see Dauvergne and Neville 2009, 2010).

27. Ross 2001; Gedicks 2001; Dauvergne 2001; Bhattacharya 2002; Clover 2004; Dauvergne and Lister 2011.

28. Karliner 1997; Smith, Sonnenfeld, and Pellow 2006.

29. Dauvergne 1997b, 2008.

30. Princen 2002.

31. Ruitenbeek and Cartier 1998, 5.

32. Dauvergne 1997a discusses the theoretical and practical hurdles that the global community would need to overcome to create a genuinely sustainable trading system for tropical timber.

33. Daly 1993a, 25; Conca 2000, 485; Neumayer 2001, 108.

34. WTO and UNEP 2009, 57.

35. Esty 1994.

36. Porter 1999.

37. Goldsmith 1997, 7.

38. Ansell and Vogel 2006.

39. Karliner 1997, 144; Korten 2001; Friends of the Earth International 2003; ActionAid 2006.

40. ActionAid 2006.

41. Khor 2009.

42. Daly 1996, 157.

43. International Forum on Globalization 2002, 76–77. See also Whiteside 2006.

44. For an overview, see Andree 2007.

45. For a critique of research claiming that current amounts of perfluorochemicals in the environment pose a threat to human health, see Weiser 2005.

46. UN 1992, 19.

47. See, for example, Esty 1994.

48. See Perkins and Neumayer 2009.

49. Neumayer 2001, 71–72.

50. Charnovitz 1995.

51. WTO and UNEP 2009.

52. Neumayer 2001, 115–116.

53. See the FSC web site at <http://www.fscoax.org> (accessed July 19, 2010). For a review of the prospects and limits of certification and eco-consumerism for global forest governance, see Dauvergne and Lister 2010b. Also see Cashore 2002; Gulbrandsen 2010.

54. Cashore, Auld, and Newsom 2004; Pattberg 2007.

55. Neumayer 2001, 153–155.

56. Pearce and Warford 1993, 27.

57. Cosbey 2008.

58. Arden-Clarke 1992, 130–131.

59. See Bohne 2010 for an overview of the WTO's institutional history and reforms.

60. The GATT did have dispute panels to make rulings in the case of trade disputes, but any country could veto the resolution of a dispute, so rulings were often ignored. The WTO dispute panels cannot be vetoed (although rulings can be overruled with a unanimous vote by member states, this is unlikely to happen). WTO dispute panels do authorize the use of trade sanctions by contracting parties as a way to enforce its decisions.

61. This round of trade talks was launched nearly two years after the failed attempt to launch a "Seattle Round" of trade talks in Seattle in the fall of 1999. Large protests at that meeting led to the failure to initiate those talks.

62. Clapp 2006; Lee and Wilkinson 2007.

63. As of July 23, 2008; see WTO 2008 and the WTO web site at <http://www.wto.org> (accessed July 19, 2010).

64. See Neumayer 2001, 24–25; Charnovitz 2007.

65. For a listing of these disputes, see <http://www.wto.org/english/tratop_e/envir_e/edis00_e.htm> (accessed July 19, 2010).

66. Esty 1994, 30–31.

67. Neumayer 2001, 134.

68. Vogel 2002, 354.

69. Neumayer 2001, 126–127; also see the WTO web site on environmental disputes at <http://www.wto.org/english/tratop_e/envir_e/edis00_e.htm> (accessed July 19, 2010).

70. For a discussion, see Perkins 1998.

71. DeSombre and Barkin 2002; Neumayer 2004.

72. Charnovitz 2007.

73. Each of these other agreements is explained in full on the WTO web site at <http://www.wto.org> (accessed July 19, 2010).

74. Neumayer 2001, 30, 128.

75. In June 2003 these countries were joined in their complaint by Australia, Brazil, Chile, Colombia, India, Mexico, Peru, and New Zealand.

76. Currie 2006.

77. See Neumayer 2001, 132–133.

78. Vogel 2002, 359–360.

79. See the WTO web site at <http://www.wto.org/english/tratop_e/dda_e/dohaexplained_e.htm#environment> (accessed July 19, 2010).

80. Stilwell and Tarasofsky 2001, 7–8.

81. Stilwell and Tarasofsky 2001, 10–11.

82. Eckersley 2004.

83. Houseman and Zaelke 1995, 315–328.

84. IISD and UNEP 2000, 62.

85. Johannesburg Plan of Implementation, paragraph 92, <http://www.johannesburgsummit.org/html/documents/summit_docs/2309_planfinal.htm> (accessed July 19, 2010).

86. See the WTO web site at <http://www.wto.org/english/tratop_e/envir_e/envir_neg_mea_e.htm> (accessed July 19, 2010) for updates.

87. Weinstein and Charnovitz 2001.

88. Esty 2000.

89. Esty 1994, 41; 2001, 125–126.

90. Biermann 2000, 2001; Biermann and Bauer 2005.

91. Shiva 1993a; Wallach and Sforza 1999.

92. Conca 2000, 487, 489–492.

93. Goldsmith 1997, 7.

94. Hines 2000, 130–131.

95. Starbucks, Ethical Sourcing: Starbucks and Fair Trade, <http://www.starbucks.com/SHAREDPLANET/ethicalInternal.aspx?story=fairTrade> (accessed July 19, 2010).

96. Raynolds 2000, 298; Dauvergne 2008. Organizations that support fair trade include: the International Fair Trade Association (IFAT); the Network of European Worldshops (NEWS!); the Fair Trade Federation; and the Fairtrade Labeling Organizations International (FLO).

97. IISD and UNEP 2000, 75.

98. See Hufbauer et al. 2000; IISD and UNEP, 2000.

99. Hufbauer et al. 2000.

100. On this debate, see Hogenboom 1998.

101. Hufbauer et al. 2000, 50.

102. Logsdon and Husted 2000, 373–374.

103. Vogel 2002, 363.

104. Gallagher 2004, 2.

105. Hufbauer et al. 2000, 50.

106. Jacott, Reed, and Winfield 2001, 44, 52.

107. Prior to May 1, 2004, the membership of the European Union was as follows: Austria, Belgium, Denmark, Finland, France, Germany, Greece, Ireland, Italy, Luxembourg, the Netherlands, Portugal, Spain, Sweden, and the United Kingdom. On May 1, 2004, the following joined the European Union: Cyprus, Czech Republic, Estonia, Hungary Latvia, Lithuania, Malta, Poland, Slovakia, and Slovenia. On January 1, 2007, Bulgaria and Romania joined the European Union.

108. See Geradin 2002.

109. Stevis and Mumme 2000, 31.

110. See the European Union's web page on environmental activities at <http://europa.eu/pol/env/index_en.htm> (accessed July 19, 2010).

111. Geradin 2002, 128–129.

Chapter 6

1. Forbes ranked all of these firms in the top twenty-five in 2008 in terms of sales, profits, assets, and market value. General Electric was number 1, Royal Dutch Shell number 2, and Toyota number 3. Royal Dutch Shell was number 1 in terms of sales, followed by ExxonMobil. See <http://www.forbes.com> (accessed July 19, 2010).

2. In 2007 China was the largest developing-country recipient of foreign direct investment, receiving some US$83.5 billion in investments (UNCTAD 2008, 255).

3. Willetts 2001, 362–363.

4. UNCTAD 2008, xvi.

5. These are China, Hong Kong, Singapore, India, Mexico, and Brazil.

6. All figures in this paragraph are reported in UNCTAD 2008.

7. Calculated from figures provided in UNCTAD 2008, 13.

8. UNCTAD 2008, 10.

9. ETC Group 2008, 7.

10. UNCTAD 2008, 27.

11. ETC Group 2008.

12. For an overview, see Thompson and Strohm 1996; Eskeland and Harrison 2003; Brunnermeier and Levinson 2004; Copeland and Taylor 2004; Ederington 2007.

13. Wheeler 2002, 1.

14. Porter 1999.

15. Leonard 1988; Pearson 1987a; Low 1993.

16. Ferrantino 1997, 52.

17. Neumayer 2001, 55–56.

18. Wheeler 2001.

19. Mani and Wheeler 1998, 215.

20. Repetto 1994, 23; Dean, Lovely, and Wang 2009.

21. Elliott and Shimamoto 2008.

22. Bhagwati 1993, 44; WTO 1999, 44.

23. Internal World Bank memo, December 12, 1991, under the signature of Larry H. Summers. This leaked memo created a furor within the environmental community. Summers defended it, claiming it was intentionally ironic and provocative. In 1998 former World Bank economist Lant Pritchett told John Cassidy

of the *New Yorker* that he was the actual author of the memo. Summers read, signed, and authorized its distribution within the World Bank, and therefore took responsibility for the contents. Pritchett (now a lecturer in public policy at Harvard University) claims the "leaked memo" was a condensed version of his longer memo. He feels the leak was intended to discredit and embarrass Summers.

24. Ruud 2002.

25. See Zeng and Eastin 2007.

26. GEMI 1999, 6, 12.

27. World Bank 1992, 67; Ferrantino 1997, 55–56.

28. Dean, Lovely, and Wang (2009), for example, did not find evidence of investors based in high-income countries seeking pollution havens in China, but did find some evidence of investors in highly polluting industries from other developing countries seeking weaker standards in China.

29. Kent 1991, 10.

30. World Bank 1999, 114.

31. World Bank 1999, 130–135.

32. Dam and Scholtens 2008.

33. Castleman 1985; United Nations Transnational Corporations and Management Division 1992, 223; Fustukian, Sethi, and Zwi 2002, 215.

34. ESCAP and UNCTC 1990, 61; UNCTAD 1993, 60.

35. See Rajan 2001; Jasanoff 2007.

36. Gladwin 1987; MacKenzie 2002.

37. MacKenzie 2002.

38. Morehouse 1994, 164.

39. MacKenzie 2002.

40. Williams 1996, 777–779.

41. Sklair 1993, 79–80.

42. Molina 1993, 232.

43. Bryant and Bailey 1997, 109; Korten 2001; Frey 2003.

44. See also Castleman 1985; Ofreneo 1993; Karliner 1997; Frey 1998; Cavanagh and Mander 2004; Korten 2001, 2009.

45. Pearson 1987b, 121; O'Neill 2001.

46. See, for example, Brunnermeier and Levinson 2004; Levinson and Taylor 2008.

47. Hall 2002, 2009.

48. Clapp 2002b, 14–15.

49. Porter 1999, 136. Also see Zarsky 2006.

50. Neumayer 2001, 70–71.

51. Wheeler 2002, 7.

52. For logging, see Filer 1997; Dauvergne 2001; Ross 2001; Tacconi 2007. For oil, see Gedicks 2001. For mining, see Jackson and Banks 2003; Ali 2004; Tienhaara 2006. For industrial waste, see Clapp 2001; Gallagher and Zarsky 2007; Pellow 2007.

53. See, for example, Global Witness at <http://www.globalwitness.org> (accessed July 19, 2010); World Resources Institute at <http://www.wri.org> (accessed July 19, 2010); Rainforest Action Network at <http://www.ran.org> (accessed July 19, 2010); Mining Watch Canada at <http://www.miningwatch.org> (accessed July 19, 2010); Oil Watch at <http://www.oilwatch.org> (accessed July 19, 2010).

54. Broad (with Cavanagh) (1993) title their book *Plundering Paradise*; Shiva (1997) titles her book *Biopiracy: The Plunder of Nature and Knowledge.*

55. Tokar 1997.

56. See Korten 2001. Korten (1999) calls for a "Post-Corporate World."

57. Karliner 1997.

58. See, for example, Potter 1993; Arentz 1996; Filer 1997; Dauvergne 2001; Humphreys 2006.

59. Japan imported over half of total log production at the height of the log export booms in the Philippines (1964–1973), Indonesia (1970–1980), and Sabah (1972–1987); see Dauvergne 1997b, 2.

60. Frontier forests, as defined by the World Resources Institute (1997), are areas large and pristine enough to still retain full biodiversity.

61. Dauvergne 2001.

62. Gedicks 2001, 29.

63. See Hutchinson 1998; OECD 2002; see also Mining Watch Canada at <http://www.miningwatch.ca> (accessed July 19, 2010).

64. Gedicks 2001; Magno 2003; see also Mining Watch Canada at <http://www.miningwatch.ca> (accessed July 19, 2010).

65. Kennedy (with Chatterjee and Moody) 1998; Perlez and Bonner 2005.

66. McAteer and Pulver 2009; Widener 2009.

67. Mitchell 1994 provides an in-depth analysis of treaty compliance and oil pollution at sea.

68. Rowell 1996; Obi 1997; Hunt 2005.

69. Amnesty International 2009.

70. Smith, Sonnenfeld, and Pellow 2006; Gallagher and Zarsky 2007.

71. Leighton, Roht-Arriaza, and Zarsky 2002, 96–98.

72. See Basel Action Network and Silicon Valley Toxics Coalition 2002.

73. Prakash 2000; Utting 2002.

74. Schmidheiny 1992.

75. Sklair 2001, 204.

76. For a more detailed explanation and history, see Utting and Clapp 2008; Clapp and Thistlethwaite, forthcoming.

77. The UN Global Compact is available at <http://www.unglobalcompact.org> (accessed July 19, 2010). See also Utting and Zammit 2006; Therien and Pouliot 2006.

78. See Gupta 2008.

79. Lamberton 2005.

80. Brown, de Jong, and Teodorina 2009. Also see the GRI web site at <http://www.globalreporting.org> (accessed July 19, 2010).

81. Kolk, Levy, and Pinske 2008.

82. Cashore 2002.

83. See the CCX web site at <http://www.chicagoclimatex.com> (accessed July 19, 2010).

84. See Kollman and Prakash 2001; Prakash and Potoski 2006a; Esty and Winston 2009.

85. Prakash and Potoski 2006b.

86. DeSimone and Popoff 1997, 25.

87. Mol 2002.

88. See Prakash 2000, 3; Dryzek 2005.

89. Levy and Newell 2000; Rowlands 2000.

90. Garcia-Johnson 2000; Prakash and Potoski 2006b.

91. Greer and Bruno 1997; Beder 1997. The nongovernmental organization CorpWatch, for example, gives bimonthly "Greenwash Awards" to offending companies (see <http://www.corpwatch.org/section.php?id=102> [accessed July 19, 2010]).

92. Greer and Bruno 1997; Greenpeace Greenwash Detection kit, available at <http://archive.greenpeace.org/comms/97/summit/greenwash.html> (accessed July 19, 2010). See also Greenpeace's Stop Greenwash campaign at <http://stopgreenwash.org> (accessed July 19, 2010).

93. Rowell 1996; Welford 1997; Parr 2009.

94. Later revised and published as Greer and Bruno 1997.

95. See <http://www.stopgreenwash.org/oil> (accessed July 19, 2010).

96. Greer and Bruno 1997.

97. Sklair 2001, 200.

98. Bruno 2003.

99. CorpWatch 2002.

100. Chatterjee and Finger 1994, 132.

101. See Clapp and Thistlethwaite, forthcoming.

102. See, for example, Falkner 2003; Fuchs 2005; Clapp 2005; Auld, Bernstein, and Cashore 2008; Vogel 2008; Utting and Clapp 2008; Clapp and Fuchs 2009.

103. See Rutherford 2003; Morgera 2004; Clapp 2005.

104. Chatterjee and Finger 1994.

105. Rutherford 2003, 14.

106. Corporate Europe Observer, 2001; Clapp 2005.

107. Sklair 2001; Levy and Newell 2002; Fuchs 2005.

108. Newell and Paterson 1998; Levy 1997; Levy and Egan 1998; Levy and Newell 2002.

109. Falkner 2008 provides a detailed analysis of these three cases.

110. Roht-Arriaza 1995.

111. See Mann 2001; IISD and WWF 2001.

112. See Mann 2001.

113. Neumayer 2001, 89.

114. IISD and UNEP 2000, 58.

115. For an account, see Kobrin 1998; Roberts 1998; Clarke and Barlow 1997.

116. OECD 2000. For the guidelines and more on the ongoing consultations (a consultation was held, for example, in December 2009), see the OECD, Directorate for Financial and Enterprise Affairs at <http://www.oecd.org/department/0,3355,en_2649_34889_1_1_1_1_1,00.html> (accessed July 19, 2010).

117. See FOE England, Wales, and Northern Ireland, 1998.

118. Leaver and Cavanagh 1996, 2.

119. Greenpeace International 2002; FOEI 2002.

Chapter 7

1. Hardin 1974.

2. The countries that have met the target include Denmark, Luxembourg, the Netherlands, Norway, and Sweden; the United Kingdom, Ireland, and Spain are likely to reach it by 2015.

3. See data at <http://www.oecd.org/dataoecd/27/55/40381862.pdf> (accessed July 19, 2010) and Reality of Aid Management Committee 2008.

4. Reality of Aid Management Committee 2008, 8.

5. Reality of Aid Management Committee 2008, 220. Also see the Bill and Melinda Gates Foundation web site at <http://www.gatesfoundation.org> (accessed July 19, 2010).

6. Figures in this paragraph are from World Bank Development Indicators online.

7. BIS 2007.

8. The World Bank Group today includes a number of agencies. In addition to the IBRD and the IDA, there are the International Finance Corporation (IFC), the Multilateral Investment Guarantee Agency (MIGA), and the International Center for the Settlement of Investment Disputes (ICSID).

9. World Bank web site, <http://www.worldbank.org> (accessed July 19, 2010).

10. Ibid.

11. Even today, however, individual researchers at the World Bank are still fully capable of publishing *classic* market liberal arguments about the connections between the global political economy and global environmental change. The World Bank is a large and complex organization, and inevitably a range of views will "emerge" as "World Bank" views. Importantly, however, these all tend to fall within a range from market liberal to institutionalist.

12. Le Prestre 1989, 19; Wade 1997, 618. The name changed again in the early 1980s to the Office of Environmental and Scientific Affairs. For convenience, we call it the Office of the Environment.

13. See for example Rich 1994; Nelson 1995.

14. Wade 1997, 614.

15. Reed 1997, 229–230.

16. George 1988, 161; Wade 1997, 634–637.

17. Rich 1994, 26–29, 152–153; Wade 1997, 637.

18. Turaga 2000; Khagram 2000, 2004; Wood 2007.

19. Rich 1994, 125; Wade 1997, 658–659.

20. Fox and Brown 1998, 508.

21. Wade 1997, 680.

22. Morse 1992.

23. Caufield 1996, 28.

24. Gutner 2002, 53. Also see Le Prestre 1989; Reed 1997.

25. Haas and Haas 1995, 268.

26. World Bank 1992.

27. World Bank 1995.

28. Fox 2000, 279.

29. Gutner 2002, 51.

30. Nelson 1995, 89–90.

31. FOE et al. 2001.

32. Independent Evaluation Group 2008.

33. World Bank 2008.

34. Webster 2009; Rich 2009.

35. Halifax Initiative 2008.

36. Quoted in Webster 2009.

37. Reed 1997, 236.

38. Independent Evaluation Group 2008. See Babb 2009 for a critique of how Washington politics has shaped problematic multilateral bank policies worldwide.

39. Reed 1992; World Bank 2001, 61.

40. See World Bank 1994; Pearce et al. 1995; Glover 1995.

41. World Bank 2001, 65.

42. World Bank 2001, 64.

43. Pearce et al. 1995, 55.

44. Glover 1995, 288.

45. World Bank 2001, 65.

46. Horta 1991; Hogg 1994. Also see Stone and Wright 2006; Phillips 2009.

47. George 1992, 9–14.

48. Toye 1991, 192.

49. Rich 1994; Devlin and Yap 1994, 67–71.

50. George 1992, 3.

51. Hammond 1999.

52. UNEP 2002b, 17.

53. For an overview of the "greening" of aid, see Hicks et al. 2008.

54. Fairman and Ross 1996, 30–31.

55. International Forum on Globalization 2002, 234.

56. See Jordan 1994; Streck 2001; and the GEF web site at <http://www.gefweb.org> (accessed July 19, 2010).

57. These are the Food and Agriculture Organization of the United Nations (FAO), the Inter-American Development Bank (IDB), the United Nations Industrial Development Organization (UNIDO), the Asian Development Bank (ADB), the African Development Bank (AfDB), the European Bank for Reconstruction and Development (EBRD), and the International Fund for Agricultural Development (IFAD).

58. GEF web site, <http://www.gefweb.org> (accessed July 19, 2010).

59. Rich 1994, 176–177.

60. Fairman 1996, 64.

61. Jordan 1994.

62. Chatterjee and Finger 1994, 153–154.

63. Streck 2001, 76. For an explanation of the voting procedures, see table 3.2 of this book.

64. Horta, Round, and Young 2002, 15.

65. Mee, Dublin, and Eberhard 2008.

66. Young 2003; Ervine 2007.

67. See the World Bank web site at <http://beta.worldbank.org/climatechange/overview> (accessed July 19, 2010) and the Adaptation Fund web site at <http://www.adaptation-fund.org> (accessed July 19, 2010).

68. Porter et al. 2008.

69. Halifax Initiative 2008, 5.

70. See, for example, Halifax Initiative 2008; Reality of Aid Management Committee 2008; Porter et al. 2008.

71. Rich 2000, 34.

72. Goldzimer 2003, 4.

73. ECA Watch, <http://www.eca-watch.org/eca/directory.html> (accessed July 19, 2010).

74. OECD 2007; FERN 2008.

75. Rich 2009, 5.

76. FERN 2008.

77. Goldzimer 2003, 4.

78. Goldzimer 2003, 3.

79. ECA Watch, <http://www.eca-watch.org/eca/ecaflyer-english.pdf> (accessed July 19, 2010).

80. Rich 2000, 35; Rich 2009, 10.

81. Rich 2000, 35–36.

82. Berne Declaration et al., 1999.

83. The Jakarta Declaration is available at <http://www.eca-watch.org/goals/jakartadec.html> (accessed July 19, 2010).

84. OECD 2007.

85. Rich 2009, 6–11.

86. For details on the impact of the crisis, see IMF 2009.

87. Schmidheiny and Zorraquin 1996, 8–10.

88. Stephan Schmidheiny founded the World Business Council for Sustainable Development (WBCSD) in 1991. The WBCSD is "a coalition of 170 international companies united by a shared commitment to sustainable development via the three pillars of economic growth, ecological balance and social progress." See "About the WBCSD" at <http://www.wbcsd.ch> (accessed July 19, 2010).

89. See Schmidheiny and Zorraquin 1996.

90. See the Finance Initiative web site at <http://www.unepfi.org> (accessed July 19, 2010).

91. Ganzi et al. 1998, 12.

92. These statements are available at <http://www.unepfi.org/signatories/statements/index.html> (accessed July 19, 2010).

93. Wright and Rwabizambuga 2006.

94. Equator Principles web site, <http://www.equator-principles.com> (accessed July 19, 2010).

95. Cappon 2008; also see the Equator Principles at <http://www.equator-principles.com/index.shtml> (accessed July 19, 2010).

96. Missbach 2004.

97. PRI web site, <http://www.unpri.org> (accessed July 19, 2010).

98. CDP web site, <https://www.cdproject.net> (accessed July 19, 2010).

99. Kolk, Levy, and Pinkse 2008.

100. Kollmus, Zink, and Polycarp 2008.

101. Newell and Paterson 2010.

102. See the Jubilee UK web site at <http://www.jubilee2000uk.org> (accessed July 19, 2010).

103. See Oxfam, CAFOD Christian Aid, and Eurodad 2002.

Chapter 8

1. WBCSD 2002, 2–3.

2. "Our Durable Planet," 1999.

3. WBCSD 2002, three quotes from p. 3.

4. Lomborg 2001, 5.

5. DeSimone and Popoff 1997, 10–11.

6. See Kollman and Prakash 2001. Also see Henderson 2006; Makower and Pike 2009.

7. See Holme and Watts 2000; Watts and Holme 1999; Zadek 2001; Willard 2002; WBCSD 2008.

8. WBCSD 2002, 10.

9. These visions are not confined to market liberals. Authors like Jeremy Rifkin (2002, 8, 253) see an awe-inspiring future in hydrogen, a fuel that "never runs out" and "emits no carbon dioxide." It is, he claims, "a promissory note for humanity's future on Earth."

10. Blackstock et al. 2009.

11. World Bank 2003, 3. Also see Less and McMillan 2005.

12. For more see WBCSD 2000; Lomborg 2009; Steenblik and Kim 2009.

13. UN 1992.

14. See Young 2002; Vogler 2003; Chambers and Green 2005; Young, Schroeder, and King 2008.

15. See, for example, the chapters in Victor, Raustiala, and Skolnikoff 1998; Young 1999; Rechkemmer 2005.

16. See the International Maritime Organization web site at <http://www.imo.org> (accessed July 19, 2010), under "Conventions," then "Status of Conventions—Summary" (accessed November 13, 2009).

17. See, for example, Victor, Raustiala, and Skolnikoff 1998; Wettestad 1999; Young 1999, 2001; Miles et al. 2001; Hovi, Sprinz, and Underdal 2003; Mitchell 2006; Stokke and Hønneland 2007; Chambers 2008. See the notes to chapter 3 for a more complete list of the growing literature on environmental regimes.

18. UNEP 2002b, 405.

19. Fox and Brown 1998; Gutner 2002, 2005. See Park 2010 for an analysis of how environmentalists have influenced the performance of the World Bank Group.

20. Neumayer 2001. Neumayer (2004) argues that the record of the WTO is better than critics portray, but its likely environmental record in the future nevertheless looks "bleak."

21. Biermann 2000, 2001. See also Whalley and Zissimos 2001; Biermann and Bauer 2005.

22. See, for example, von Moltke 2001; Newell 2001; Najam 2003.

23. Sagar and VanDeveer 2005.

24. VanDeveer and Dabelko 2001.

25. UNEP 2002a, 10.

26. World Bank 2003.

27. Rees 2002.

28. Ehrlich 1968, 166.

29. Hardin 1968, 1974.

30. Daly 1993b, 335–340.

31. Brown, Gardner, and Halweil 1999, 18. Paul Ehrlich also continues to see population growth as a core cause of the global environmental crisis. See, for example, Ehrlich and Ehrlich 2008.

32. WWF 2002, 20.

33. This is a popular theme of both moderate bioenvironmentalists and social greens. One of the best books on "branding" is Klein 2000.

34. Wackernagel and Rees 1996; Ryan and Durning 1997.

35. Daly 1993b, 325. See Daly 1973 for his book-length explanation of a steady-state economy. For a more introductory and updated treatment, see Daly and Farley 2003.

36. Daly 1996, 31–32.

37. See Daly and Cobb 1989 (the index was prepared with the assistance of Clifford Cobb).

38. See Cobb, Glickman, and Cheslog 2001.

39. For some sample graphs comparing GDP growth with ISEW and GPI growth, see Friends of the Earth, International Indicators, at <http://www.foe.co.uk/community/tools/isew/international.html> (accessed July 19, 2010).

40. WWF 2002, 2–3.

41. WWF 2008, 2.

42. Rees 2002, 265. Also see Rees 2006.

43. Princen 2001, 2002. For a comprehensive analysis, see Princen 2005.

44. Maniates 2001, 50.

45. Greenpeace International 2001, 7; International Forum on Globalization 2002, 226.

46. Hines 2000, 130–137. Also see International Forum on Globalization at <http://www.ifg.org> (accessed July 19, 2010).

47. Greenpeace International 2001, 11. Some social green thinkers, however, such as Martin Khor, are against the idea of discriminating based on PPMs on justice grounds. See Khor 2009.

48. Hines 2000, 131. Jaffee, an advocate of fair trade, provides a critical evaluation of the social and environmental consequences of fair trade coffee in Mexico (2007).

49. International Forum on Globalization 2002, 223.

50. International Forum on Globalization 2002, 227, 234.

51. New Economics Foundation 2002, 9–10.

52. See Patomaki 2001; Brassett 2010. Also see the Halifax Initiative on the Tobin Tax at <http://halifaxinitiative.org/content/tobin-tax> (accessed July 15, 2010).

53. See Greenpeace International 2002.

54. For a briefing on corporate accountability, see Friends of the Earth 2005.

55. International Forum on Globalization 2002, 237–238.

56. International Forum on Globalization 2002, 109.

57. Hines 2000, 68–69.

58. Meeker-Lowry 1996; Helleiner 2002; Peacock 2006.

59. Norberg-Hodge 1996, 394.

60. *The Ecologist* 1993; Cavanagh and Mander 2004.

61. Hines 2000, 213.

62. Hildyard 1995, 160.

63. See the Sierra Club web site at <http://www.sierraclub.org> (accessed July 19, 2010).

64. The parallel with the logic of Indian Prime Minister Indira Ghandi during her speech at the 1972 Stockholm Conference is striking: "Thus, we see when it comes to the depletion of natural resources and environmental pollution," she eloquently declared, "the increase of one inhabitant in an affluent country at his level of living is equivalent to an increase of many Asians, Africans or Latin Americans at their current material levels of living." It is reasonable to assume that Zuckerman's reasoning on the need to restrict immigration would horrify Ghandi, nicely demonstrating how similar ecological arguments can serve very different political purposes.

65. Zuckerman 2004.

66. See the Sierra Club of Canada at <http://www.sierraclub.ca/en/national/aboutus/index.html%20> (accessed July 15, 2010).

References

ActionAid International. 2006. *Under the Influence: Exposing Undue Corporate Policy Influence over Policy Making at the WTO*. Johannesburg: ActionAid International. Available at <http://www.actionaid.org.uk/doc_lib/174_6_under_the_influence_final.pdf> (accessed July 15, 2010).

Adams, William. 1990. *Green Development: Environment and Sustainability in the Third World*. London: Routledge.

Agyeman, Julian. 2005. *Sustainable Communities and the Challenge of Environmental Justice*. New York: New York University Press.

Agyeman, Julian, Robert D. Bullard, and Bob Evans, eds. 2003. *Just Sustainabilities: Development in an Unequal World*. Cambridge, Mass.: MIT Press.

Akula, Vikram. 1995. Grassroots Environmental Resistance in India. In *Ecological Resistance Movements: The Global Emergence of Radical and Popular Environmentalism*, ed. Bron Taylor, 127–145. Albany: SUNY Press.

Alcott, Blake. 2008. Historical Overview of the Jevons Paradox in the Literature. In *The Jevons Paradox and the Myth of Resource Efficiency Improvements*, ed. John Polimeni, Kozo Mayumi, Mario Giampietro, and Blake Alcott, 8–78. London: Earthscan.

Ali, Saleem H. 2004. *Mining, the Environment, and Indigenous Development Conflicts*. Tucson, Ariz.: University of Arizona Press.

Amnesty International. 2009. Nigeria: Petroleum, Pollution and Poverty in the Niger Delta. London: Amnesty International. Available at <http://www.amnesty.org/en/library/asset/AFR44/017/2009/en/e2415061-da5c-44f8-a73c-a7a4766ee21d/afr440172009en.pdf> (accessed July 15, 2010).

Andersen, Regine. 2002. The Time Dimension in International Regime Interplay. *Global Environmental Politics* 2 (3): 98–117.

Andree, Peter. 2007. *Genetically Modified Diplomacy: The Global Politics of Agricultural Biotechnology and the Environment*. Vancouver: UBC Press.

Andresen, Steinar. 2000. The Whaling Regime. In *Science and Politics in International Environmental Regimes: Between Integrity and Involvement*, ed. Steinar Andresen, Tora Skodvin, Jörgen Wettestad, and Arild Underdal, 35–70. Manchester: Manchester University Press.

Andresen, Steinar. 2001a. Global Environmental Governance: UN Fragmentation and Co-ordination. In *Yearbook of International Co-operation on Environment and Development*, ed. Olav Schram Stokke and Øystein B. Thommessen, 19–26. London: Earthscan.

Andresen, Steinar. 2001b. The Whaling Regime: "Good" Policy but "Bad" Institutions? In *Towards a Sustainable Whaling Regime?*, ed. Robert L. Friedheim, 235–265. Seattle: University of Washington Press.

Ansell, Christopher, and David Vogel, eds. 2006. *What's the Beef? The Contested Governance of European Food Safety*. Cambridge, Mass.: MIT Press.

Ansuategi, Alberto, and Marta Escapa. 2002. Economic Growth and Greenhouse Gas Emissions. *Ecological Economics* 40 (1): 23–37.

Arden-Clarke, Charles. 1992. South-North Terms of Trade: Environmental Protection and Sustainable Development. *International Environmental Affairs* 4 (2): 122–139.

Arentz, Frans. 1996. Forestry and Politics in Sarawak: The Experience of the Penan. In *Resources, Nations and Indigenous Peoples: Case Studies from Australasia, Melanesia and Southeast Asia*, ed. Richard Howitt, with John Connell and Philip Hirsch, 202–211. Melbourne: Oxford University Press.

Arrow, Kenneth, Bert Bolin, Robert Costanza, Partha Dasgupta, Carl Folk, C. S. Holling, Bengt-Owe Jansson, Simon Levin, Karl-Göran Mäler, Charles Perrings, and David Pimentel. 1995. Economic Growth, Carrying Capacity and the Environment. *Science* 268 (April): 520–521.

Auer, Matthew. 1998. Colleagues or Combatants? Experts as Environmental Diplomats. *International Negotiation* 3 (2): 267–287.

Auld, Graeme, Steven Bernstein, and Benjamin Cashore. 2008. The New Corporate Social Responsibility. *Annual Review of Environment and Resources* 33: 413–435.

Axelrod, Regina S., Stacy D. VanDeveer, and David Leonard Downie, eds. 2010. *The Global Environment: Institutions, Law, and Policy*. 3rd ed. Washington, D.C.: CQ Press.

Ayres, Robert. 1998. Eco-thermodynamics: Economics and the Second Law. *Ecological Economics* 26 (2): 189–209.

Babb, Sarah. 2009. *Behind the Development Banks: Washington Politics, World Poverty, and the Wealth of Nations*. Chicago: University of Chicago Press.

Bäckstrand, Karin, and Eva Lövbrand. 2006. Planting Trees to Mitigate Climate Change: Contested Discourses of Ecological Modernization, Green Governmentality and Civic Environmentalism. *Global Environmental Politics* 6 (1): 50–75.

Bacon, Christopher M., V. Ernesto Méndez, Stephen R. Gliessman, David Goodman, and Jonathan A. Fox, eds. 2008. *Confronting the Coffee Crisis: Fair Trade, Sustainable Livelihoods and Ecosystems in Mexico and Central America*. Cambridge, Mass.: MIT Press.

Bagliani, Marco, Giangiacomo Bravo, and Silvana Dalmazzone. 2008. A Consumption-based Approach to Environmental Kuznets Curves using the Ecological Footprint Indicator. *Ecological Economics* 65 (3): 650–661.

Bank for International Settlements (BIS). 2007. Triennial Central Bank Survey. Basel: Bank for International Settlements. Available at <http://www.bis.org/publ/rpfxf07t.pdf?noframes=1> (accessed July 15, 2010).

Barber, Charles Victor, and James Schweithelm. 2000. *Trial by Fire: Forest Fires and Forestry Policy in Indonesia's Era of Crisis and Reform*. Washington, D.C.: World Resources Institute.

Bargagli, Roberto. 2005. *Antarctic Ecosystems: Environmental Contamination, Climate Change, and Human Impact*. Berlin: Springer.

Barlow, Maude. 2008. *Blue Covenant: The Global Water Crisis and the Coming Battle for the Right to Water*. New York: New Press.

Barnet, Richard J., and John Cavanagh. 1994. *Global Dreams: Imperial Corporations and the New World Order*. New York: Simon & Schuster.

Basel Action Network (BAN) and Silicon Valley Toxics Coalition. 2002. *Exporting Harm: The High-Tech Trashing of Asia*. Available at <http://www.ban.org/E-waste/technotrashfinalcomp.pdf > (accessed July 15, 2010).

Beder, Sharon. 1997. *Global Spin: The Corporate Assault on Environmentalism*. Melbourne: Scribe.

Bello, Walden. 2009. *The Food Wars*. London, New York: Verso.

Berne Declaration, Bioforum, Center for International Environmental Law, Environmental Defense, Eurodad, Friends of the Earth, Pacific Environment and Resources Center, and Urgewald. 1999. *A Race to the Bottom: Creating Risk, Generating Debt, and Guaranteeing Environmental Destruction*. Available at <http://www.eca-watch.org/eca/race_bottom.pdf> (accessed July 15, 2010).

Bernstein, Steven. 2001. *The Compromise of Liberal Environmentalism*. New York: Columbia University Press.

Bernstein, Steven, and Maria Ivanova. 2007. Fragmentation and Compromise in Global Environmental Governance: What Prospects for Re-Embedding? In *Global Liberalism and Political Order: Toward a New Grand Compromise?*, ed. Steven Bernstein and Louis W. Pauly, 161–185. Albany: State University of New York Press.

Bernstein, Steven, Jennifer Clapp, and Matthew Hoffmann. 2009. *Reframing Global Environmental Governance: Results of a CIGI/CIS Collaboration*. CIGI Working Paper No. 45. Available at <http://www.cigionline.org/sites/default/files/Working_Paper%2045.pdf> (accessed July 19, 2010).

Betsill, Michele M., and Harriet Bulkeley. 2006. Cities and the Multilevel Governance of Global Climate Change. *Global Governance* 12 (2): 141–159.

Betsill, Michele M., and Elisabeth Corell. 2001. NGO Influence in International Environmental Negotiations: A Framework for Analysis. *Global Environmental Politics* 1 (3): 65–85.

Betsill, Michelle M., and Elisabeth Corell, eds. 2007. *NGO Diplomacy: The Influence of Nongovernmental Organizations in International Environmental Negotiations*. Cambridge, Mass.: MIT Press.

Bhagwati, Jagdish. 1993. The Case for Free Trade. *Scientific American* 269 (5): 42–49.

Bhagwati, Jagdish. 2002. Coping with Antiglobalization: A Trilogy of Discontents. *Foreign Affairs* (Council on Foreign Relations) 81 (1): 2–7.

Bhagwati, Jagdish. 2004. *In Defense of Globalization*. Oxford: Oxford University Press.

Bhattacharya, Hrishikes. 2002. *Commercial Exploitation of Fisheries: Production, Marketing and Finance Strategies*. New York: Oxford University Press.

Bhattarai, Madhusudan, and Michael Hammig. 2001. Institutions and the Environmental Kuznets Curve for Deforestation: A Cross-Country Analysis for Latin America, Africa and Asia. *World Development* 29 (6): 995–1010.

Biermann, Frank. 2000. The Case for a World Environment Organization. *Environment* 42 (9): 22–31.

Biermann, Frank. 2001. The Emerging Debate on the Need for a World Environment Organization. *Global Environmental Politics* 1 (1): 45–55.

Biermann, Frank, and Steffen Bauer, eds. 2005. *A World Environment Organization: Solution or Threat for Effective International Environmental Governance?* Aldershot, UK: Ashgate.

Biermann, Frank, and Philipp Pattberg. 2008. Global Environmental Governance: Taking Stock, Moving Forward. *Annual Review of Environment and Resources* 33 (November): 277–294.

Biermann, Frank, Philipp Pattberg, Harro van Asselt, and Fariborz Zelli. 2009. The Fragmentation of Global Governance Architectures: A Framework for Analysis. *Global Environmental Politics* 9 (4): 14–40.

Biermann, Frank, and Bernd Siebenhüner, eds. 2009. *Managers of Global Change: The Influence of International Environmental Bureaucracies*. Cambridge, Mass.: MIT Press.

Bill and Melinda Gates Foundation. 2008. *Annual Report*. Available at <http://www.gatesfoundation.org/annualreport/2008/Pages/2008-annual-report.aspx> (accessed July 15, 2010).

Blackstock, Jason J., David S. Battisti, Ken Caldeira, Douglas M. Eardley, Jonathan I. Katz, David W. Keith, Aristides A.N. Patrinos, David P. Schrag, Robert H. Socolow, and Steven E. Koonin. 2009. *Climate Engineering Responses to Climate Emergencies*. Santa Barbara: The Novim Group. Available at <http://arxiv.org/ftp/arxiv/papers/0907/0907.5140.pdf> (accessed July 15, 2010).

Bohne, Eberhard. 2010. *The World Trade Organization: Institutional Development and Reform*. London: Palgrave Macmillan.

Brac de la Perrière, Robert Ali, and Frank Seuret. 2000. *Brave New Seeds*. London: Zed Books.

Brack, Duncan, Fanny Calder, and Müge Dolun. 2001. *From Rio to Johannesburg: The Earth Summit and Rio + 10*. London: Royal Institute of International Affairs Briefing Paper, New Series No. 19.

Brassett, James. 2010. *Cosmopolitanism and Global Financial Reform: A Pragmatic Approach to the Tobin Tax*. London: Routledge.

Breitmeier, Helmut, Oran R. Young, and Michael Zürn. 2006. *Analyzing International Environmental Regimes: From Case Study to Database*. Cambridge, Mass.: MIT Press.

Brenton, Tony. 1994. *The Greening of Machiavelli: The Evolution of International Environmental Politics*. London: Royal Institute of International Affairs/Earthscan.

Bretherton, Charlotte. 2003. Movements, Networks, Hierarchies: A Gender Perspective on Global Environmental Governance. In *Global Environmental Governance for the Twenty-First Century: Theoretical Approaches and Normative Considerations*, ed. David Humphreys, Matthew Paterson, and Lloyd Pettiford. Special issue of *Global Environmental Politics* 3 (2): 103–119.

Broad, Robin. 1994. The Poor and the Environment: Friends or Foes? *World Development* 22 (6): 811–822.

Broad, Robin (with John Cavanagh). 1993. *Plundering Paradise: The Struggle for the Environment in the Philippines*. Berkeley: University of California Press.

Broad, Robin, and John Cavanagh. 2008. *Development Redefined: How the Market Met Its Match*. Boulder: Paradigm Publishers.

Broadbent, Jeffrey. 1998. *Environmental Politics in Japan: Networks of Power and Protest*. Cambridge: Cambridge University Press.

Brown, Halina Szenjwald, Martin de Jong, and Teodorina Lessidrenska. 2009. The Rise of the Global Reporting Initiative: A Case of Institutional Entrepreneurship. *Environmental Politics* 18 (2): 182–200.

Brown, Lester R. 2005. *Outgrowing the Earth: The Food Security Challenge in an Age of Falling Water Tables and Rising Temperatures*. New York: Norton.

Brown, Lester R. 2008. *Plan B 3.0: Mobilizing to Save Civilization*. Rev. edition. New York: Norton.

Brown, Lester R., Gary Gardner, and Brian Halweil. 1999. *Beyond Malthus: Nineteen Dimensions of the Population Challenge*. New York: Norton.

Brunnermeier, Smita B., and Arik Levinson. 2004. Examining the Evidence on Environmental Regulations and Industry Location. *Journal of Environment & Development* 13 (1): 6–41.

Bruno, Kenny. 2003. *Presentation to the Public Eye on Davos*. Corporate PR Strategies Panel. Available at <http://www.evb.ch/cm_data/public/Panelprk-bruno.pdf> (accessed July 15, 2010).

Bryant, Raymond. 1997. *The Political Ecology of Forestry in Burma, 1824–1994*. London: Hurst & Co.

Bryant, Raymond, and Sinéad Bailey. 1997. *Third World Political Ecology*. London: Routledge.

Bryner, Gary C. 2001. *Gaia's Wager: Environmental Movements and the Challenge of Sustainability*. Lanham, Md.: Rowman and Littlefield.

Bucheli, Marcelo. 2005. *Bananas and Business: The United Fruit Company in Colombia, 1899–2000*. New York: New York University Press.

Bulkeley, Harriet. 2005. Reconfiguring Environmental Governance: Towards a Politics of Scales and Networks. *Political Geography* 24 (8): 875–902.

Bulkeley, Harriet, and Michele Betsill. 2003. *Cities and Climate Change: Urban Sustainability and Global Environmental Governance*. London: Routledge.

Bulkeley, Harriet, and Susanne C. Moser, eds. 2007. Responding to Climate Change: Governance and Social Action beyond Kyoto. Special issue of *Global Environmental Politics* 7 (2): 1–10.

Bulkeley, Harriet, and Peter Newell. 2010. *Governing Climate Change*. London: Routledge.

Burg, Jericho. 2003. The World Summit on Sustainable Development: Empty Talk or Call to Action? *Journal of Environment & Development* 12 (1): 111–120.

Burtless, Gary, Robert Z. Lawrence, Robert E. Litan, and Robert J. Shapiro. 1998. *Globaphobia: Confronting Fears about Open Trade*. Washington, D.C.: Brookings Institution Press.

Caldwell, Lynton K. 1996. *International Environmental Policy: From the Twentieth to the Twenty-First Century*. Durham, N.C.: Duke University Press.

Cameron, Owen. 1996. Japan and South-East Asia's Environment. In *Environmental Change in South-East Asia: People, Politics and Sustainable Development*, ed. Michael J. G. Parnell and Raymond L. Bryant, 67–94. London: Routledge.

Capoor, Karan, and Philippe Ambrosi. 2009. *State and Trends of the Carbon Market 2009*. Washington, D.C.: World Bank. Available at <http://siteresources.worldbank.org/EXTCARBONFINANCE/Resources/State_and_Trends_of_the_Carbon_Market_2009-FINALb.pdf> (accessed July 15, 2010).

Cappon, Natasha. 2008. Equator Principles: Promoting Greater Responsibility in Project Financing. *Exportwise* (Spring). Available at <http://www.edc.ca/english/publications_14554.htm> (accessed July 15, 2010).

Carmin, JoAnn, and Elizabeth Bast. 2009. Cross-Movement Activism: A Cognitive Perspective on the Global Justice Activities of U.S. Environmental NGOs. *Environmental Politics* 18 (3): 351–370.

Carson, Rachel. 1962. *Silent Spring*. Boston: Houghton Mifflin.

Cashore, Benjamin. 2002. Legitimacy and the Privatization of Environmental Governance: How Non-State Market-Driven (NDSM) Governance Systems Gain Rule-Making Authority. *Governance* 15 (4): 503–529.

Cashore, Benjamin, Graeme Auld, and Deanna Newsom. 2004. *Governing through Markets: Forest Certification and the Emergence of Non-state Authority*. New Haven, Conn.: Yale University Press.

Cass, Loren. 2005. Norm Entrapment and Preference Change: The Evolution of the European Union Position on International Emissions Trading. *Global Environmental Politics* 4 (2): 38–60.

Casson, Anne. 2000. *The Hesitant Boom: Indonesia's Oil Palm Sub-sector in an Era of Economic Crisis and Political Change*. Occasional Paper No. 29. Jakarta: Center for International Forestry Research.

Castleman, Barry. 1985. Double Standards in Industrial Hazards. In *The Export of Hazard*, ed. Jane Ives, 60–89. London: Routledge.

Caufield, Catherine. 1996. *Masters of Illusion: The World Bank and the Poverty of Nations*. London: Macmillan.

Cavanagh, John, and Jerry Mander, eds. 2004. *Alternatives to Economic Globalization: A Better World is Possible*. 2nd ed. San Francisco: Berrett-Koehler.

Chambers, W. Bradnee. 2008. *Interlinkages and the Effectiveness of Multilateral Environmental Agreements*. Tokyo: United Nations University Press.

Chambers, W. Bradnee, and Jessica F. Green, eds. 2005. *Reforming International Environmental Governance: From Institutional Limits to Innovative Reforms*. New York: United Nations University Press.

Chan, Michelle. 2009. *Subprime Carbon? Re-Thinking the World's Largest New Derivatives Market*. Washington, D.C.: Friends of the Earth.

Charnovitz, Steve. 1995. Improving Environmental and Trade Governance. *International Environmental Affairs* 7 (1): 59–91.

Charnovitz, Steve. 2007. The WTO's Environmental Progress. *Journal of International Economic Law* 10 (3): 685–706.

Chasek, Pamela. 2000. The UN Commission on Sustainable Development: The First Five Years. In *The Global Environment in the Twenty-First Century: Prospects for International Cooperation*, ed. Pamela Chasek, 378–398. New York: United Nations University Press.

Chasek, Pamela S., David Downie, and Janet Welsh Brown. 2006. *Global Environmental Politics*. 4th ed. Boulder, Colo.: Westview Press.

Chatterjee, Pratap, and Matthias Finger. 1994. *The Earth Brokers: Power, Politics and World Development*. London: Routledge.

Chester, Charles C. 2006. *Conservation Across Borders: Biodiversity in an Interdependent World*. Washington, D.C.: Island Press.

Chivian, Eric, and Aaron Bernstein, eds. 2008. *Sustaining Life: How Human Health Depends on Biodiversity*. Oxford: Oxford University Press.

Clapp, Jennifer. 1997. Environmental Threats in an Era of Globalization: An End to State Sovereignty? In *Surviving Globalism*, ed. Ted Schrecker, 123–140. New York: St. Martin's Press.

Clapp, Jennifer. 1998. The Privatization of Global Environmental Governance: ISO 14000 and the Developing World. *Global Governance* 4 (3): 295–316.

Clapp, Jennifer. 2001. *Toxic Exports: The Transfer of Hazardous Wastes from Rich to Poor Countries*. Ithaca, N.Y.: Cornell University Press.

Clapp, Jennifer. 2002a. Distancing of Waste: Overconsumption in a Global Economy. In *Confronting Consumption*, ed. Tom Princen, Ken Conca, and Michael Maniates, 155–177. Cambridge, Mass.: MIT Press.

Clapp, Jennifer. 2002b. What the Pollution Havens Debate Overlooks. *Global Environmental Politics* 2 (2): 11–19.

Clapp, Jennifer. 2003. Transnational Corporate Interests and Global Environmental Governance: Negotiating Rules for Agricultural Biotechnology and Chemicals. *Environmental Politics* 12 (4): 1–23.

Clapp, Jennifer. 2005. Global Environmental Governance for Corporate Responsibility and Accountability. *Global Environmental Politics* 5 (3): 23–34.

Clapp, Jennifer. 2006. WTO Agriculture Negotiations: Implications for the Global South. *Third World Quarterly* 27 (4): 563–577.

Clapp, Jennifer. 2009. Food Price Volatility and Vulnerability in the Global South: Considering the Global Economic Context. *Third World Quarterly* 30 (6): 1183–1196.

Clapp, Jennifer, and Doris Fuchs, eds. 2009. *Corporate Power in Global Agrifood Governance*. Cambridge, Mass.: MIT Press.

Clapp, Jennifer, and Linda Swanston. 2009. Doing Away with Plastic Shopping Bags: International Patterns of Norm Emergence and Policy Implementation. *Environmental Politics* 18 (3): 315–332.

Clapp, Jennifer, and Jason Thistlethwaite. Forthcoming. Private Voluntary Programs in Environmental Governance: Climate Change and the Financial Sector. In *Voluntary Programs and Climate Change*, ed. Karsten Ronit.

Clarke, Tony, and Maude Barlow. 1997. *MAI: The Multilateral Agreement on Investment and the Threat to Canadian Sovereignty*. Toronto: Stoddart.

Cleaver, Kevin, and Gotz Schreiber. 1992. Population, Agriculture and the Environment in Africa. *Finance & Development* 29 (2): 34–35.

Clover, Charles. 2004. *The End of the Line: How Overfishing Is Changing the World and What We Eat*. London: Random House.

Cobb, Clifford, Mark Glickman, and Craig Cheslog. 2001. *Genuine Progress Indicator 2000 Update*. Redefining Progress Issue Brief. Available at <http://www.rprogress.org/publications/2001/2000_gpi_update.pdf> (accessed July 15, 2010).

Comim, Flavio. 2008. *Poverty and Environment Indicators: Prepared for the UNDP-UNEP Poverty and Environment Initiative*. Cambridge: St. Edmonds College. Available at <http://www.st-edmunds.cam.ac.uk/csc/research/UNDP_UNEPengD.pdf> (accessed July 15, 2010).

Committee on World Food Security. 2001. *Assessment of the World Food Security Situation*. Available at <http://www.fao.org/docrep/meeting/003/Y0147E/Y0147E00.htm#P79_3644> (accessed July 15, 2010).

Commoner, Barry. 1990. *Making Peace with the Planet*. New York: Pantheon Books.

Conca, Ken. 2000. The WTO and the Undermining of Global Environmental Governance. *Review of International Political Economy* 7 (3): 484–494.

Conca, Ken. 2001. Consumption and Environment in a Global Economy. *Global Environmental Politics* 1 (3): 56–57.

Conca, Ken. 2006. *Governing Water: Contentious Transnational Politics and Global Institution Building*. Cambridge, Mass.: MIT Press.

Conca, Ken, and Geoffrey D. Dabelko, eds. 2010. *Green Planet Blues: Four Decades of Global Environmental Politics*. 4th ed. Boulder, Colo.: Westview Press.

Connelly, James, and Graham Smith. 1999. *Politics and the Environment: From Theory to Practice*. London: Routledge.

Cooney, Rosie, and Barney Dickson, eds. 2005. *Biodiversity and the Precautionary Principle; Risk and Uncertainty in Conservation and Sustainable Use*. London: Earthscan.

Copeland, Brian, and M. Scott Taylor. 2004. Trade, Growth, and the Environment. *Journal of Economic Literature* 42 (1): 7–71.

Corell, Elisabeth. 1999. Actor Influence in the 1993–97 Negotiations of the Convention to Combat Desertification. *International Negotiation* 4 (2): 197–223.

Corell, Elisabeth, and Michelle M. Betsill. 2001. A Comparative Look at NGO Influence in International Environmental Negotiations: Desertification and Climate Change. *Global Environmental Politics* 1 (4): 86–107.

Corporate Europe Observer. 2001. Industry's Rio + 10 Strategy: Banking on Feelgood PR. *CEO Quarterly Newsletter* 10 (December).

CorpWatch. 2002. *Greenwash + 10: The UN's Global Compact, Corporate Accountability and the Johannesburg Earth Summit*. Available at <http://www.corpwatch.org/article.php?id=1348 > (accessed July 15, 2010).

Cosbey, Aaron. 2008. Border Carbon Adjustments. Winnipeg: IISD. Available at <http://www.iisd.org/pdf/2008/cph_trade_climate_border_carbon.pdf> (accessed July 15, 2010).

Costanza, Robert, John Cumberland, Herman Daly, Robert Goodland, and Richard Norgaard. 1997. *An Introduction to Ecological Economics*. Boca Raton, Fla.: St. Lucie Press.

Crosby, Alfred. 1986. *Ecological Imperialism: The Biological Expansion of Europe, 900–1900*. Cambridge: Cambridge University Press.

Currie, Duncan. 2006. Genetic Engineering and the WTO: An Analysis of the Report in the "EC-Biotech" Case. Amsterdam: Greenpeace International. Available at <http://www.wto.org/english/forums_e/ngo_e/posp66_greenpeace_engi_e.pdf> (accessed July 15, 2010).

Dalby, Simon. 2004. Ecological Politics, Violence, and the Theme of Empire. *Global Environmental Politics* 4 (2): 1–11.

Daly, Herman E. 1973. *Toward a Steady-State Economy*. San Francisco: W. H. Freeman.

Daly, Herman E. 1977. *Steady-State Economics: The Economics of Biophysical Equilibrium and Moral Growth*. San Francisco, Calif.: W. H. Freeman.

Daly, Herman E. 1993a. The Perils of Free Trade. *Scientific American* 269 (5): 50–57.

Daly, Herman E. 1993b. The Steady-State Economy: Toward a Political Economy of Biophysical Equilibrium and Moral Growth. In *Valuing the Earth: Economics, Ecology, Ethics*, ed. Herman E. Daly and Kenneth Townsend, 324–356. Cambridge, Mass.: MIT Press.

Daly, Herman E. 1996. *Beyond Growth: The Economics of Sustainable Development*. Boston: Beacon Press.

Daly, Herman E. 1999. Globalization versus Internationalization—Some Implications. *Ecological Economics* 31 (1): 31–37.

Daly, Herman E. 2002. Uneconomic Growth and Globalization in a Full World. *Natur und Kultur*. Available at <http://www.en.unsw.edu.au/downloads/unecon-growth.pdf> (accessed July 15, 2010).

Daly, Herman E., and John Cobb, Jr. 1989. *For the Common Good: Redirecting the Economy toward Community, the Environment, and a Sustainable Future*. Boston: Beacon Press.

Daly, Herman E., and Joshua Farley. 2003. *Ecological Economics: Principles and Applications*. Washington, D.C.: Island Press.

Dam, Lammertjan, and Bert Scholtens. 2008. Environmental Regulation and MNEs Location: Does CSR Matter? *Ecological Economics* 67 (1): 55–65.

D'Amato, Anthony, and Kirsten Engel. 1997. *International Environmental Law Anthology*. Cincinnati, Ohio: Anderson.

Dauvergne, Catherine. 2003. *Challenges to Sovereignty: Migration Laws for the 21st Century*. UN High Commission for Refugees New Issues in Refugee Research Working Paper Series, June.

Dauvergne, Peter. 1997a. A Model of Sustainable International Trade in Tropical Timber. *International Environmental Affairs* 9 (1): 3–21.

Dauvergne, Peter. 1997b. *Shadows in the Forest: Japan and the Politics of Timber in Southeast Asia*. Cambridge, Mass.: MIT Press.

Dauvergne, Peter. 2001. *Loggers and Degradation in the Asia-Pacific: Corporations and Environmental Management*. Cambridge: Cambridge University Press.

Dauvergne, Peter, ed. 2005. *Handbook of Global Environmental Politics*. Cheltenham, England: Edward Elgar.

Dauvergne, Peter. 2008. *The Shadows of Consumption: Consequences for the Global Environment*. Cambridge, Mass.: MIT Press.

Dauvergne, Peter. 2009. *The A to Z of Environmentalism*. Lanham, Md.: Scarecrow.

Dauvergne, Peter. 2010. The Problem of Consumption. *Global Environmental Politics* 10 (2): 1–10.

Dauvergne, Peter, and Jane Lister. 2010a. The Power of Big Box Retail in Global Environmental Governance: Bringing Commodity Chains Back into IR. *Millennium: Journal of International Studies* 39 (1): 145–160.

Dauvergne, Peter, and Jane Lister. 2010b. The Prospects and Limits of Eco-Consumerism: Shopping Our Way to Less Deforestation? *Organization & Environment* 23 (2): 132–154.

Dauvergne, Peter, and Jane Lister. 2011. *Timber.* Cambridge: Polity Press.

Dauvergne, Peter, and Kate J. Neville. 2009. The Changing North-South and South-South Political Economy of Biofuels. *Third World Quarterly* 30 (6): 1087–1102.

Dauvergne, Peter, and Kate J. Neville. 2010. Forests, Food, and Fuel in the Tropics: The Uneven Social and Ecological Consequences of the Emerging Political Economy of Biofuels. *Journal of Peasant Studies* 37 (3): 631–660.

Davies, James B., Susanna Sandström, Anthony Shorrocks, and Edward N. Wolff. 2008. The World Distribution of Household Wealth. UNU-WIDER Discussion Paper 2008/3. Available at <http://www.wider.unu.edu/publications/working-papers/discussion-papers/2008/en_GB/dp2008-03> (accessed July 15, 2010).

Davis, Devra Lee, and Pamela S. Webster. 2002. The Social Context of Science: Cancer and the Environment. *The Annals of the American Academy* 584 (1): 13–34.

Dean, Judith M., Mary Lovely, and Hua Wang. 2009. Are Foreign Investors Attracted to Weak Environmental Regulations? Evaluating the Evidence from China. *Journal of Development Economics* 90 (1): 1–13.

Deke, Oliver. 2008. *Environmental Policy Instruments for Conserving Global Biodi*versity. Berlin: Springer.

Depledge, Joanna. 2007. A Special Relationship: Chairpersons and the Secretariat in the Climate Change Negotiations. *Global Environmental Politics* 7 (1): 45–68.

DeSimone, Livio, and Frank Popoff. 1997. *Eco-Efficiency: The Business Link to Sustainable Development.* Cambridge, Mass.: MIT Press.

DeSombre, Elizabeth R. 2007. *The Global Environment and World Politics.* 2nd ed. London: Continuum.

DeSombre, Elizabeth R., and J. Samuel Barkin. 2002. Turtles and Trade: The WTO's Acceptance of Environmental Trade Restrictions. *Global Environmental Politics* 2 (1): 12–18.

Devlin, John, and Nonita Yap. 1994. Structural Adjustment Programmes and the UNCED Agenda: Explaining the Contradictions. In *Rio: Unravelling the Consequences,* ed. Caroline Thomas, 65–79. London: Frank Cass.

Dimitrov, Radoslav. 2005. *Science and International Environmental Policy: Regimes and Nonregimes in Global Governance.* Lanham, Md.: Rowman & Littlefield.

Dimitrov, Radoslav. 2010. Inside Copenhagen: The State of Climate Governance. *Global Environmental Politics* 10 (2): 18–24.

Dinda, Soumyananda. 2004. Environmental Kuznets Curve Hypothesis: A Survey. *Ecological Economics* 49 (4): 431–455.

Dolšak, Nives, and Elinor Ostrom, eds. 2003. *The Commons in the New Millennium: Challenges and Adaptation*. Cambridge, Mass.: MIT Press.

Doran, Peter. 2002. *World Summit on Sustainable Development (Johannesburg) and Assessment for IISD*. IISD Briefing Paper, October 3. Available at <http://www.iisd.org/pdf/2002/wssd_assessment.pdf> (accessed July 15, 2010).

Dornbusch, Rudiger, Stanley Fischer, and Gordon Sparks. 1993. *Macroeconomics*. 4th ed. Toronto: McGraw-Hill Ryerson.

Downie, David Leonard, and Terry Fenge, eds. 2003. *Northern Lights against POPs: Combatting Toxic Threats in the Arctic*. Montreal, Kingston: McGill-Queen's University Press.

Downie, David, and Marc Levy. 2000. The UN Environment Programme at a Turning Point: Options for Change. In *The Global Environment in the Twenty-First Century: Prospects for International Cooperation*, ed. Pamela Chasek, 355–377. New York: UN University Press.

Dressler, Andrew E., and Edward A. Parson. 2006. *The Science and Politics of Global Climate Change*. Cambridge: Cambridge University Press.

Dryzek, John. 2005. *The Politics of the Earth: Environmental Discourses*. 2nd ed. Oxford: Oxford University Press.

Dryzek, John, and David Schlosberg, eds. 2005. *Debating the Earth: The Environmental Politics Reader*. 2nd ed. Oxford: Oxford University Press.

Easterbrook, Gregg. 1995. *A Moment on the Earth: The Coming Age of Environmental Optimism*. New York: Penguin.

Eberhart, Dave. 2007. Highway in the Sky Suffers Gridlock. *Newsmax.com*. August 6. Available at <http://archive.newsmax.com/archives/articles/2007/8/6/133309.shtml> (accessed July 15, 2010).

Eckersley, Robyn. 2004. The Big Chill: The WTO and Multilateral Environmental Agreements. *Global Environmental Politics* 4 (2): 24–50.

Ecologist, The. 1993. *Whose Common Future? Reclaiming the Commons*. London: Earthscan.

Economic and Social Commission for the Asia Pacific (ESCAP) and the United Nations Center on Transnational Corporations (UNCTC). 1990. *Environmental Aspects of Transnational Corporation Activities in Pollution-Intensive Industries in Selected Asian and Pacific Developing Countries*. Bangkok: UN/ESCAP.

Ederington, Josh. 2007. NAFTA and the Pollution Haven Hypothesis. *PSJ: Policy Studies Journa*l 35 (2): 239–244.

Ehrlich, Paul R. 1968. *The Population Bomb*. New York: Sierra Club–Ballantine.

Ehrlich, Paul R. 1981. An Economist in Wonderland. *Social Science Quarterly* 62 (1): 44–49.

Ehrlich, Paul R., and Anne H. Ehrlich. 1996. *Betrayal of Science and Reason: How Anti-Environmental Rhetoric Threatens Our Future*. Washington, D.C.: Island Press.

Ehrlich, Paul R., and Anne H. Ehrlich. 2004. *One With Nineveh: Politics, Consumption, and the Human Future*. Washington, D.C.: Island Press.

Ehrlich, Paul R., and Anne H. Ehrlich. 2008. *The Dominant Animal: Human Evolution and the Environment*. Washington, D.C.: Island Press.

Ekins, Paul, Mayer Hillman, and Robert Hutchinson. 1992. *The Gaia Atlas of Green Economics*. New York: Anchor Books.

Elliott, Kimberly. 2009. U.S. Biofuels Policy and the Global Food Price Crisis. In *The Global Food Crisis: Governance Challenges and Opportunities,* ed. Jennifer Clapp and Marc J. Cohen, 59–75. Waterloo, Ont.: Wilfrid Laurier University Press.

Elliot, Lorraine. 1994. *International Environmental Politics: Protecting the Antarctic*. London: Macmillan.

Elliot, Lorraine. 2004. *The Global Politics of the Environment*. 2nd ed. New York: NYU Press.

Elliott, Robert, and Kenichi Shimamoto. 2008. Are ASEAN Countries Havens for Japanese Pollution-Intensive Industry? *World Economy* 31 (2): 236–254.

Epstein, Charlotte. 2008. *The Power of Words in International Relations: Birth of an Anti-Whaling Discourse*. Cambridge, Mass.: MIT Press.

Ervine, Kate. 2007. The Greying of Green Governance: Power Politics and the Global Environment Facility. *Capitalism, Nature, Socialism* 18 (4): 125–142.

Eskeland, Gunnar S., and Ann E. Harrison. 2003. Moving to Greener Pastures? Multinationals and the Pollution Haven Hypothesis. *Journal of Development Economics* 70 (1): 1–23.

Esty, Daniel C. 1994. *Greening the GATT: Trade, Environment and the Future*. Washington, D.C.: Institute for International Economics.

Esty, Daniel C. 2000. Environment and the Trading System: Picking up the Post-Seattle Pieces. In *The WTO After Seattle*, ed. J. J. Schott, 243–253. Washington, D.C.: Institute for International Economics.

Esty, Daniel C., and Andrew S. Winston. 2009. *Green to Gold: How Smart Companies Use Environmental Strategy to Innovate, Create Value, and Build Competitive Advantage*. Rev. ed. Hoboken, N.J.: John Wiley.

ETC Group. 2008. *Who Owns Nature? Corporate Power and the Final Frontier in the Commodification of Life*. November. Available at <http://www.etcgroup.org/en/materials/publications.html?pub_id=707> (accessed July 15, 2010).

Ezzati, Majid, Burton Singer, and Daniel Kammen. 2001. Towards an Integrated Framework for Development and Environment Policy: The Dynamics of Environmental Kuznets Curves. *World Development* 29 (8): 1421–1434.

Fairhead, James, and Melissa Leach. 1995. False Forest History, Complicit Social Analysis: Rethinking Some West African Environmental Narratives. *World Development* 23 (6): 1023–1053.

Fairman, David. 1996. The Global Environment Facility: Haunted by the Shadow of the Future. In *Institutions for the Earth*, ed. Robert Keohane and Marc Levy, 55–87. Cambridge, Mass.: MIT Press.

Fairman, David, and Michael Ross. 1996. Old Fads, New Lessons: Learning from Economic Development Assistance. In *Institutions for the Earth*, ed. Robert Keohane and Marc Levy, 55–87. Cambridge, Mass.: MIT Press.

Falkner, Robert. 2003. Private Environmental Governance and International Relations: Exploring the Links. In *Global Environmental Governance for the Twenty-First Century: Theoretical Approaches and Normative Considerations*, ed. David Humphreys, Matthew Paterson, and Lloyd Pettiford. Special issue of *Global Environmental Politics* 3 (2): 72–87.

Falkner, Robert. 2008. *Business Power and Conflict in International Environmental Politics. Houndmills*. Basingstoke: Palgrave Macmillan.

FERN. 2008. *Trade and Investment—Export Credit Agencies: The Need for Binding Guidelines*. Forests and the European Union Resource Network. Available at <http://www.fern.org/campaign_area_extension.html? clid=3&id=2783> (accessed July 15, 2010).

Ferrantino, Michael. 1997. International Trade, Environmental Quality and Public Policy. *World Economy* 20 (1): 43–72.

Filer, Colin, ed. 1997. *The Political Economy of Forest Management in Papua New Guinea*. London and Boroko: International Institute for Environment and Development and National Research Institute.

Finger, Matthias, and James Kilcoyne. 1997. Why Transnational Corporations Are Organizing to "Save the Global Environment." *The Ecologist* 27 (4): 138–142.

Finger, Matthias, and Ludivine Tamiotti. 1999. The Emerging Linkage between the WTO and the ISO: Implications for Developing Countries. Special issue on Globalisation and the Governance of the Environment. *IDS Bulletin* 30 (3): 8–16.

Fisher, Dana. 2010. COP-15 in Copenhagen: How the Merging of Movements Left Civil Society Out in the Cold. *Global Environmental Politics* 10 (2): 11–17.

Food and Agriculture Organization (FAO). 2008. *The State of Food Insecurity in the World 2008*. Rome: FAO.

Food and Agriculture Organization (FAO). 2009. More People Than Ever Are Victims of Hunger. News Release. Available at <http://www.fao.org/fileadmin/user_upload/newsroom/docs/Press%20release%20june-en.pdf> (accessed July 15, 2010).

Food and Agriculture Organization (FAO). 2010. *Global Forest Resources Assessment 2010: Key Findings*. Rome: FAO.

Ford, Lucy H. 2003. Challenging Global Environmental Governance: Social Agency and Global Civil Society. In *Global Environmental Governance for the Twenty-First* Century: Theoretical Approaches and Normative Considerations, ed. David Humphreys, Matthew Paterson, and Lloyd Pettiford. Special issue of *Global Environmental Politics* 3 (2): 120–134.

Founex Report on Development and Environment. 1971. Available at <http://www.stakeholderforum.org/fileadmin/files/Earth%20Summit%202012new/Publications%20and%20Reports/founex%20report%201972.pdf> (accessed July 15, 2010).

Fox, Jonathan. 2000. The World Bank Inspection Panel: Lessons from the First Five Years. *Global Governance* 6 (3): 279–318.

Fox, Jonathan, and L. David Brown, eds. 1998. *The Struggle for Accountability: The World Bank, NGOs, and Grassroots Movements*. Cambridge, Mass.: MIT Press.

Frank, André Gunder. 1967. *Capitalism and Underdevelopment in Latin America*. New York: Monthly Review Press.

Frey, Scott. 1998. The Export of Hazardous Industries to the Peripheral Zones of the World System. *Journal of Developing Societies* 14 (1): 66–81.

Frey, Scott. 2003. The Transfer of Core-Based Hazardous Production Processes to the Export Processing Zones of the Periphery: The Maquiladora Centers of Northern Mexico. *Journal of World-Systems Research* 9 (2): 317–354.

Friedman, Thomas. 2002. Techno Logic. In States of Discord: A Debate between Thomas Friedman and Robert Kaplan. *Foreign Policy* 29 (March–April): 64–71.

Friends of the Earth (FOE), England, Wales, and Northern Ireland. 1998. *A History of Attempts to Regulate the Activities of Transnational Corporations: What Lessons Can Be Learned?* Available at <www.corporate-accountability.org/.../a_history_of_attempts_to_regulate_activities_of_tncs.doc> (accessed July 15, 2010).

Friends of the Earth (FOE), England, Wales and Northern Ireland. 2005. *Briefing: Corporate Accountability*. Available at <http://www.foe.co.uk/resource/briefings/corporate_accountability1.pdf> (accessed July 15, 2010).

Friends of the Earth (FOE), Environmental Defense, Sierra Club, International Rivers Network, and Rainforest Action Network. 2001. *Not in the Public Interest: The World Bank's Environmental Record*. Washington, D.C.: FOE.

Friends of the Earth International (FOEI). 1998. *Benchmarks for Mainstreaming the Environment: Environmental Reform Recommendations for the World Bank Group*. Washington, D.C.: FOE.

Friends of the Earth International (FOEI). 1999. *The IMF: Selling the Environment Short*. Washington, D.C.: FOE. Available at <http://action.foe.org/content.jsp?key=2960> (accessed July 15, 2010).

Friends of the Earth International (FOEI). 2002. *Towards Binding Corporate Accountability*. Available at http://www.foe.co.uk/resource/briefings/corporate_accountability.pdf (accessed July 15, 2010).

Fuchs, Doris. 2005. *Understanding Business Power in Global Governance*. Baden-Baden: Nomos.

Fustukian, Suzanne, Dinesh Sethi, and Anthony B. Zwi. 2002. Workers Health and Safety in a Globalising World. In *Health Policy in a Globalising World*, ed. Kelley Lee, Kent Buse, and Suzanne Fustukian. Cambridge: Cambridge University Press.

Gallagher, Kevin. 2004. *Free Trade and the Environment: Mexico, NAFTA, and Beyond*. Stanford: Stanford University Press.

Gallagher, Kevin, and Lyuba Zarsky. 2007. *The Enclave Economy: Foreign Investment and Sustainable Development in Mexico's Silicon Valley*. Cambridge, Mass.: MIT Press.

Ganzi, John, Frances Seymour, and Sandy Buffett (with Navroz Dubash). 1998. *Leverage for the Environment: A Guide to the Private Financial Services Industry*. Washington, D.C.: World Resources Institute.

Garcia-Johnson, Ronie. 2000. *Exporting Environmentalism: U.S. Multinational Chemical Corporations in Brazil and Mexico*. Cambridge, Mass.: MIT Press.

Gedicks, Al. 2001. *Resource Rebels: Native Challenges to Mining and Oil Corporations*. Boston: South End Press.

Geist, Helmut. 2005. *The Causes and Progression of Desertification*. Aldershot, UK: Ashgate.

George, Susan. 1988. *A Fate Worse Than Debt*. London: Penguin.

George, Susan. 1992. *The Debt Boomerang*. London: Pluto.

Georgescu-Roegen, Nicholas. 1971. *The Entropy Law and the Economic Process*. Cambridge, Mass.: Harvard University Press.

Georgescu-Roegen, Nicholas. 1993. The Entropy Law and the Economic Problem. In *Valuing the Earth: Economics, Ecology, Ethics*, ed. Herman Daly and Kenneth Townsend, 37–49. Cambridge, Mass.: MIT Press.

Geradin, Damien. 2002. The European Community: Environmental Issues in an Integrated Market. In *The Greening of Trade Law: International Trade Organizations and Environmental Issues*, ed. Richard Steinberg, 117–154. Lanham, Md.: Rowman and Littlefield.

Gladwin, Thomas. 1987. A Case Study of the Bhopal Tragedy. In *Multinational Corporations, the Environment, and the Third World*, ed. Charles Pearson, 223–239. Durham, N.C.: Duke University Press.

Gleckman, Harris. 1995. Transnational Corporations' Strategic Responses to "Sustainable Development.". In *Helge Ole Bergesen and Georg Parmann*, ed. Green Globe Yearbook. Oxford: Oxford University Press, 93–106.

Gleick, Peter, William Burns, and Elizabeth Chalecki. 2002. *The World's Water 2002–2003: The Biennial Report on Freshwater Resources*. Washington, D.C.: Island Press.

Global Environment Facility (GEF). 2002. *GEF Replenishment*. Available at <http://www.thegef.org/gef/replenishment> (accessed July 15, 2010).

Global Environment Management Initiative (GEMI). 1999. *Fostering Environmental Prosperity: Multinationals in Developing Countries*. Washington, D.C.: GEMI. Available at <http://www.gemi.org/resources/MNC_101.pdf> (accessed July 15, 2010).

Glover, David. 1995. Structural Adjustment and the Environment. *Journal of International Development* 7 (2): 285–289.

Goldsmith, Edward. 1992. *The Way: An Ecological World-View*. London: Rider.

Goldsmith, Edward. 1997. Can the Environment Survive the Global Economy? *Ecologist* 27 (6): 242–248.

Goldsmith, Edward, and Nicholas Hildyard. 1984. *The Social and Environmental Effects of Large Dams*. Wadebridge, UK: Wadebridge Ecological Centre.

Goldsmith, Edward, and Nicholas Hildyard. 1986. *Green Britain or Industrial Wasteland?* Cambridge: Polity Press.

Goldzimer, Aaron. 2003. Worse Than the World Bank? Export Credit Agencies—The Secret Engine of Globalization. *Food First Backgrounder* 9 (1). Available at <http://www.foodfirst.org/node/50> (accessed July 15, 2010).

Green, Rhys E., Stephen J. Cornell, J. Jörn, P. W. Scharlemann, and Andrew Balmford. 2005. Farming and the Fate of Wild Nature. *Science* 307 (5709): 550–555.

Greenpeace International. 2001. *Safe Trade in the 21st Century: The Doha Edition*. Available at <http://archive.greenpeace.org/politics/wto/doha_report.pdf> (accessed July 15, 2010).

Greenpeace International. 2002. *Corporate Crimes: The Need for an International Instrument on Corporate Accountability and Liability*. Amsterdam: Greenpeace International. Available at <http://archive.greenpeace.org/earthsummit/docs/corpcrimes_1of3.pdf> (accessed July 15, 2010).

Greer, Jed, and Kenny Bruno. 1997. *Greenwash: The Reality Behind Corporate Environmentalism. Penang: Third World Network*. New York: Apex Press.

Grossman, Gene, and Alan Krueger. 1991. *Environmental Impact of a North American Free Trade Agreement*. Working Paper 3914. Cambridge, Mass.: National Bureau of Economic Research.

Grossman, Gene, and Alan Krueger. 1995. Economic Growth and the Environment. *Quarterly Journal of Economics* 110 (2): 353–377.

Grove, Richard. 1995. *Green Imperialism: Colonial Expansion, Tropical Island Edens and the Origins of Environmentalism, 1600–1860*. Cambridge: Cambridge University Press.

Grundmann, Reiner. 2001. *Transnational Environmental Policy: Reconstructing Ozone*. London: Routledge.

Guha, Ramachandra. 2000. *The Unquiet Woods: Ecological Change and Peasant Resistance in the Himalaya*. Expanded ed. Berkeley, Calif.: University of California Press.

Gulbrandsen, Lars H. 2010. *Transnational Environmental Governance: The Origins and Effects of the Certification of Forests and Fisheries*. Cheltenham, UK: Edward Elgar Publishing.

Gunter, Michael M. 2004. *Building the Next Ark: How NGOs Protect Biodiversity*. Dartmouth: Dartmouth College Press and University Press of New England.

Gupta, Aarti. 2008. Transparency Under Scrutiny: Information Disclosure in Global Environmental Governance. *Global Environmental Politics* 8 (2): 1–7.

Gutner, Tamar L. 2002. *Banking on the Environment: Multilateral Development Banks and Their Environmental Performance in Central and Eastern Europe*. Cambridge, Mass.: MIT Press.

Gutner, Tamar L. 2005. Explaining the Gaps between Mandate and Performance: Agency Theory and World Bank Environmental Reform. *Global Environmental Politics* 5 (2): 10–37.

Hamilton, Katherine, Ricardo Bayon, Guy Turner, and Douglas Higgins. 2007. *State of the Voluntary Carbon Market 2007: Picking Up Steam*. London: New Carbon Finance and The Ecosystem Marketplace. Available at <http://www.carbon.sref.info/an-example/market-news> (accessed July 15, 2010).

Haas, Peter. 1999. Social Constructivism and the Evolution of Multilateral Environmental Governance. In *Globalization and Governance*, ed. Aseem Prakash and Jeffrey Hart, 103–133. London: Routledge.

Haas, Peter. 2004. Addressing the Global Governance Deficit. *Global Environmental Politics* 4 (4): 1–15.

Haas, Peter, and Ernst B. Haas. 1995. Learning to Learn: Improving International Governance. *Global Governance* 1 (3): 255–285.

Haas, Peter, Robert Keohane, and Marc Levy, eds. 1993. *Institutions for the Earth: Sources of Effective International Environmental Protection*. Cambridge, Mass.: MIT Press.

Halifax Initiative. 2008. The World Bank, Climate Change and Energy. Issue Brief. October. Available at <http://halifaxinitiative.org/content/issue-brief-world-bank-climate-change-and-energy-october-2008#attachments> (accessed July 15, 2010).

Hall, Derek. 2002. Environmental Change, Protest and Havens of Environmental Degradation: Evidence from Asia. *Global Environmental Politics* 2 (2): 20–28.

Hall, Derek. 2009. Pollution Export as State and Corporate Strategy: Japan in the 1970s. *Review of International Political Economy* 16 (2): 260–283.

Halweil, Brian. 2009. Meat Production Continues to Grow. In *Vital Signs 2009*, ed. Linda Starke, 15–17. Washington, D.C.: Worldwatch Institute.

Hammond, Ross. 1999. The Impact of IMF Structural Adjustment Policies on Tanzanian Agriculture. In the Development Gap and Friends of the Earth, eds., *The All-Too-Visible Hand: A Five-Country Look at the Long and Destructive Reach of the IMF*. Available at <http://www.developmentgap.org/worldbank_imf/index.html> (accessed July 15, 2010).

Hardin, Garrett. 1968. The Tragedy of the Commons. *Science* 162: 1243–1248.

Hardin, Garrett. 1972. *Exploring New Ethics for Survival: The Voyage of the Spaceship Beagle*. Baltimore: Pelican.

Hardin, Garrett. 1974. Living on a Lifeboat. *Bioscience* 24 (10): 561–568.

Hardin, Garrett. 1985. *Filters against Folly: How to Survive Despite Economists, Ecologists, and the Merely Eloquent*. New York: Viking.

Hardin, Garrett. 1995. *Creative Altruism: An Ecologist Questions Motives*. Petoskey, Mich.: Social Contract Press.

Hardin, Garrett. 2000. *Living within Limits: Ecology, Economics, and Population Taboos*. Oxford: Oxford University Press.

Held, David, Anthony McGrew, David Goldblatt, and Jonathan Perraton. 1999. *Global Transformations: Politics, Economics, and Culture*. Stanford, Calif.: Stanford University Press.

Helleiner, Eric. 1994. *States and the Reemergence of Global Finance*. Ithaca, N.Y.: Cornell University Press.

Helleiner, Eric. 2000. New Voices in the Globalization Debate: Green Perspectives on the World Economy. In *Political Economy and the Changing Global Order*, 2nd ed., ed. Richard Stubbs and Geoffrey Underhill. Oxford: Oxford University Press, 60–69.

Helleiner, Eric. 2002. Think Globally, Transact Locally: The Local Currency Movement and Green Political Economy. In *Confronting Consumption*, ed. Thomas Princen, Michael Maniates, and Ken Conca, 255–275. Cambridge, Mass.: MIT Press.

Helliwell, John F. 2002. *Globalization and Well-Being*. Vancouver, B.C.: UBC Press.

Hempel, Lamont C. 1996. *Environmental Governance: The Global Challenge*. Washington, D.C.: Island Press.

Henderson, H. 2006. *Ethical Markets: Growing the Green Economy*. White River, Vt.: Chelsea Green Publishing.

Herber, Bernard P. 2007. *Protecting the Antarctic Commons: Problems of Economic Efficiency*. Tucson, Ariz.: Udall Center for Studies in Public Policy, University of Arizona.

Hicks, Robert L., Bradley C. Parks, J. Timmons Roberts, and Michael J. Tierney. 2008. *Greening Aid? Understanding the Environmental Impact of Development Assistance*. Oxford: Oxford University Press.

Hildyard, Nicholas. 1995. Liberation Ecology. In *The Future of Progress: Reflections on Environment and Development*, ed. Helena Norberg-Hodge, Peter Goering, and Steven Gorelick, 154–160. Foxhole, UK: Green Books.

Hill, Gladwin. 1969. Environment May Eclipse Vietnam as College Issue. *New York Times*, November 30: 1–2.

Hines, Colin. 2000. *Localization: A Global Manifesto*. London: Earthscan.

Hines, Colin. 2003. Time to Replace Globalization with Localization. *Global Environmental Politics* 3 (3): 1–7.

Hirst, Paul, and Grahame Thompson. 2000. *Globalization in Question: The International Economy and the Possibilities of Governance*. 2nd ed. Cambridge: Polity Press.

Hochstetler, Kathryn. 2002. After the Boomerang: Environmental Movements and Politics in the La Plata River Basin. *Global Environmental Politics* 2 (4): 35–57.

Hogenboom, Barbara. 1998. *Mexico and the NAFTA Environment Debate: The Transnational Politics of Economic Integration*. Utrecht: International Books.

Hogg, Dominic. 1994. *The SAP in the Forest*. London: Friends of the Earth.

Holliday, Charles O., Jr., Stephan Schmidheiny, and Philip Watts. 2002. *Walking the Talk: The Business Case for Sustainable Development*. Sheffield, UK: Greenleaf Publishing.

Holme, Richard, and Phil Watts. 2000. *Corporate Social Responsibility: Making Good Business Sense*. Geneva: World Business Council for Sustainable Development. Available at <http://www.wbcsd.org/DocRoot/5mbU1sfWpqAgPpPpUqUe/csr2000.pdf> (accessed July 15, 2010).

Hoogvelt, Ankie. 1982. *The Third World in Global Development*. London: Macmillan.

Horta, Korinna. 1991. Multilateral Development Institutions Bear Major Responsibility for Solving Africa's Environmental Problems. In *Environmental Policies for Sustainable Growth in Africa*. Proceedings of the Fifth International Conference, Montclair State University, Upper Montelair, N.J., May 6, 21–28.

Horta, Korinna, Robin Round, and Zoe Young. 2002. *The Global Environmental Facility: The First Ten Years—Growing Pains or Inherent Flaws?* Washington, D.C.: Environmental Defense, and Halifax: The Halifax Initiative, August. Available at <http://www.environmentaldefense.org/documents/2265_First10YearsFinal.pdf> (accessed July 15, 2010).

Houseman, Robert, and Durwood Zaelke. 1995. Mechanisms for Integration. In *The Use of Trade Measures in Select Multilateral Environmental Agreements*, ed. Robert Housman, Donald Goldberg, Brennan Van Dyke, and Durwood Zaelke, 315–328. Geneva: United Nations Environment Programme.

Hovi, Jon, Detlef F. Sprinz, and Arild Underdal. 2003. The Oslo-Potsdam Solution to Measuring Regime Effectiveness: Critique, Response, and the Road Ahead. *Global Environmental Politics* 3 (3): 74–96.

Hufbauer, Gary, Daniel Esty, Diana Orejas, Luis Rubio, and Jeffrey Schott. 2000. *NAFTA and the Environment: Seven Years Later*. Washington, D.C.: Institute for International Economics.

Humphreys, David. 1996. Regime Theory and Non-Governmental Organizations: The Case of Forest Conservation. In *NGOs and Environmental Policies: Asia and Africa*, ed. David Potter, 90–115. London: Frank Cass.

Humphreys, David. 2006. *Logjam: Deforestation and the Crisis of Global Governance*. London: Earthscan.

Hunt, J. Timothy. 2005. *The Politics of Bones*. Toronto: McClelland & Stewart.

Hurrell, Andrew, and Benedict Kingsbury. 1992. *The International Politics of the Environment*. Oxford: Oxford University Press.

Hutchinson, Moira. 1998. Beyond Best Practice: The Mining Sector. In *Canadian Development Report 1998: Canadian Corporations and Social Responsibility*, ed. Michelle Hibler and Rowena Beamish, 74–90. Ottawa: North South Institute.

Iles, Alastair. 2004. Mapping Environmental Justice in Technology Flows: Computer Waste Impacts in Asia. *Global Environmental Politics* 4 (4): 76–107.

Imura, Hidefumi, and Miranda A. Schreurs. 2005. *Environmental Policy in Japan*. Washington, D.C.: Edward Elgar & the World Bank.

Independent Evaluation Group (World Bank). 2008. *Environmental Sustainability: An Evaluation of World Bank Group Support*. Washington, D.C.: World Bank.

International Air Transport Association (IATA). 2007. Passenger Numbers to Reach 2.75 Billion by 2011. Press Release. 24 October. Available at <http://www.iata.org/pressroom/pr/2007-24-10-01.htm> (accessed July 15, 2010).

International Forum on Globalization (IFG). 2002. *Alternatives to Economic Globalization: A Better World Is Possible*. San Francisco: Berrett-Koehler.

International Institute for Sustainable Development (IISD) and United Nations Environment Programme (UNEP). 2000. *Environment and Trade: A Handbook*. Winnipeg: International Institute for Sustainable Development.

International Institute for Sustainable Development (IISD) and WWF Network. 2001. *Private Rights: Public Problems: A Guide to NAFTA's Controversial Chapter on Investor Rights*. Winnipeg: International Institute for Sustainable Development.

International Monetary Fund (IMF). 2009. World Economic Outlook: October 2009. Washington, D.C.: World Bank. Available at <http://www.imf.org/external/pubs/ft/weo/2009/02/pdf/text.pdf> (accessed July 15, 2010).

International Organization for Standardization (ISO). 2008. ISO Survey 2007. Available at <http://www.iso.org/iso/survey2007.pdf> (accessed July 15, 2010).

International Telecommunications Union (ITU). 2009. *Key Global Telecom Indicators for the World Telecommunication Service Sector*. Available at <http://www.itu.int/ITU-D/ict/statistics/at_glance/KeyTelecom.html > (accessed July 15, 2010).

International Union for Conservation of Nature and Natural Resources (IUCN). 1980. *World Conservation Strategy*. Geneva: IUCN, UNEP, WWF.

Ivanova, Maria. 2007. Designing the United Nations Environment Programme: A Story of Compromise and Confrontation. *International Environmental Agreements* 7 (4): 337–361.

Ivanova, Maria. 2010. UNEP in Global Environmental Governance: Design, Leadership, Location. *Global Environmental Politics* 10 (1): 30–59.

Ivanova, Maria, David Gordon, and Jennifer Roy. 2007. Towards Institutional Symbiosis: Business and the United Nations in Environmental Governance.

Review of European Community and International Environmental Law RECIEL 16 (2): 123–134.

Jackson, Richard, and Glenn Banks. 2003. *In Search of the Serpent's Skin: The Story of the Porgera Gold Mine*. Brisbane: Boolorong Press.

Jacott, Marisa, Cyrus Reed, and Mark Winfield. 2001. *The Generation and Management of Hazardous Wastes and Transboundary Hazardous Waste Shipments between Mexico, Canada, and the United States, 1990–2000*. Austin: Texas Center for Policy Studies. Available at <http://www.texascenter.org/bordertrade/haznafta.htm> (accessed July 15, 2010).

Jaffee, Daniel. 2007. *Brewing Justice: Fair Trade Coffee, Sustainability, and Survival*. Berkeley: University of California Press.

Jasanoff, Sheila. 1997. NGOs and the Environment: From Knowledge to Action. *Third World Quarterly* 18 (3): 579–594.

Jasanoff, Sheila. 2001. Image and Imagination: The Formation of Global Environmental Consciousness. In *Changing the Atmosphere: Expert Knowledge and Environmental Governance*, ed. Clark A. Miller and Paul N. Edwards, 309–337. Cambridge, Mass.: MIT Press.

Jasanoff, Sheila. 2007. Bhopal's Trails of Knowledge and Ignorance. *ISIS: An International Review Devoted to the History of Science and Its Cultural Influences* 98 (2): 344–350.

Jeffery, Michael I., Jeremy Firestone, and Karen Bubna-Litic, eds. 2008. *Biodiversity Conservation, Law and Livelihoods: Bridging the North-South Divide*. New York: Cambridge University Press.

Jha, Raghbendra, and K. V. Bhanu Murthy. 2006. *Environmental Sustainability: A Consumption Approach*. London: Routledge.

Johnson, Pierre Marc, Karel Mayrand, and Marc Paquin, eds. 2006. *Governing Global Desertification: Linking Environmental Degradation, Poverty, and Participation*. Aldershot, UK: Ashgate.

Jordan, Andrew. 1994. Paying the Incremental Costs of Global Environmental Protection: The Evolving Role of GEF. *Environment* 36 (6): 13–20, 31–36.

Jordan, Andrew, Rüdiger K. W. Wurzel, and Anthony R. Zito. 2003. "New" Instruments of Environmental Governance: Patterns and Pathways of Change. *Environmental Politics* 12 (1): 3–24.

Joyner, Christopher C. 1998. *Governing the Frozen Commons: The Antarctic Regime and Environmental Protection*. Columbia, S.C.: University of South Carolina Press.

Karliner, Joshua. 1997. *The Corporate Planet, Ecology, and Politics in the Age of Globalization*. San Francisco: Sierra Club.

Kassa, Stine Madland. 2007. The UN Commission on Sustainable Development: Which Mechanisms Explain its Accomplishments? *Global Environmental Politics* 7 (3): 107–129.

Kates, Robert. 2000. Population and Consumption: What We Know, What We Need to Know. *Environment* 42 (3): 10–19.

Keane, Fergal. 1998. Rain Forest Disaster Ignited by the Hand of Man. *Sunday Telegraph*, April 12.

Keck, Margaret, and Kathryn Sikkink. 1998. *Activists Beyond Borders: Advocacy Networks in International Politics*. Ithaca, N.Y.: Cornell University Press.

Kennedy, Danny (with Pratap Chatterjee and Roger Moody). 1998. *Risky Business: The Grasberg Gold Mine*. Berkeley, Calif.: Project Underground. Available at <http://www.freewestpapua.org/docs/risky_business.pdf> (accessed July 15, 2010).

Kent, Lawrence. 1991. *The Relationship between Small Enterprises and Environmental Degradation in the Developing World, with Special Emphasis on Asia*. Washington, D.C.: Development Alternatives, Inc.

Keohane, Robert, and Marc Levy, eds. 1996. *Institutions for Environmental Aid*. Cambridge, Mass.: MIT Press.

Khagram, Sanjeev. 2000. Toward Democratic Governance for Sustainable Development: Transnational Civil Society Organizing around Big Dams. In *The Third Force: The Rise of Transnational Civil Society*, ed. Ann Florini, 83–114. Washington, D.C.: Carnegie Endowment.

Khagram, Sanjeev. 2004. *Dams and Development: Transnational Struggles for Water and Power*. Ithaca, N.Y.: Cornell University Press.

Khor, Martin. 2009. Unfair to Tax South's Imports on Climate Grounds. Available at <http://www.twnside.org.sg/title2/gtrends/gtrends257.htm> (accessed July 15, 2010).

Kimbrell, Andrew, ed. 2002. *The Fatal Harvest Reader: The Tragedy of Industrial Agriculture*. Sausalito, Calif.: Foundation for Deep Ecology and Island Press.

Kimerling, Judith. 2001. Corporate Ethics in the Era of Globalization: The Promise and Peril of International Environmental Standards. *Journal of Agricultural & Environmental Ethics* 14 (4): 425–455.

Klein, Naomi. 2000. *No Logo: Taking Aim at the Brand Bullies*. Toronto: Random House.

Kneen, Brewster. 1999. *Farmageddon*. Gabriola Island, B.C.: New Society.

Kobrin, Stephen. 1998. The MAI and the Clash of Globalizations. *Foreign Policy* 112 (Fall): 97–109.

Kolk, Ans. 1996. *Forests in International Environmental Politics: International Organisations, NGOs and the Brazilian Amazon*. Utrecht: International Books.

Kolk, Ans, David Levy, and Jonatan Pinkse. 2008. Corporate Responses in an Emerging Climate Regime: The Institutionalization and Commensuration of Carbon Disclosure. *European Accounting Review* 17 (4): 719–745.

Kollman, Kelly, and Aseem Prakash. 2001. Green by Choice? Cross National Variations in Firms' Responses to EMS-Based Environmental Regimes. *World Politics* 53 (3): 399–430.

Kollmuss, Anji, Helge Zink, and Clifford Polycarp. 2008. Making Sense of the Voluntary Market: A Comparison of Carbon Offset Standards. *Stockholm*

Environment Institution and Tricorona. Available at: <http://assets.panda.org/downloads/vcm_report_final.pdf> (accessed July 15, 2010).

Korten, David C. 1999. *The Post-Corporate World: Life After Capitalism.* San Francisco: Berrett-Koehler.

Korten, David C. 2001. *When Corporations Rule the World.* 2nd ed. San Francisco: Berrett-Koehler.

Korten, David C. 2006. *The Great Turning: From Empire to Earth Community.* San Francisco: Berrett-Koehler.

Korten, David C. 2009. *Agenda for a New Economy: From Phantom Wealth to Real Wealth.* San Francisco: Berrett-Koehler.

Krasner, Stephen D. 1983. *International Regimes.* Ithaca: Cornell University Press.

Krueger, Jonathan. 1999. *International Trade and the Basel Convention.* London: Earthscan.

Krut, Riva, and Harris Gleckman. 1998. *ISO 14001: A Missed Opportunity for Global Sustainable Industrial Development.* London: Earthscan.

Kütting, Gabriela. 2000. *Environment, Society and International Relations: Towards More Effective International Environmental Agreements.* London: Routledge.

Kütting, Gabriela, ed. 2010. *Global Environmental Politics: Concepts, Theories and Case Studies.* London: Routledge.

Kumar, Satish. 1996. Ghandi's Swadeshi: The Economics of Permanence. In *The Case against the Global Economy and for a Turn Toward the Local*, ed. Jerry Mander and Edward Goldsmith, 418–424. San Francisco: Sierra Club.

Kuznets, Simon. 1955. Economic Growth and Income Inequality. *American Economic Review* 45 (1): 1–28.

Laferrière, Eric. 2001. International Political Economy and the Environment: A Radical Ecological Perspective. In *The International Political Economy of the Environment: Critical Perspectives, IPE Yearbook*, vol. 12, ed. Dimitris Stevis and Valerie Assetto, 199–216. Boulder, Colo.: Lynne Rienner.

Lallas, Peter L. 2000–2001. The Role of Process and Participation in the Development of Effective International Environmental Agreements: A Study of the Global Treaty on Persistent Organic Pollutants. *UCLA Journal of Environmental Law and Policy* 19 (1): 83–152.

Lamberton, Geoff. 2005. Sustainability Accounting—A Brief History and Conceptual Framework. *Accounting Forum* 29 (1): 7–26.

Lawn, Philip. 2004. How Well Are Resource Prices Likely to Serve as Indicators of Natural Resource Scarcity? *International Journal of Sustainable Development* 7 (4): 369–397.

Layzer, Judith A. 2002. Science, Politics and International Environmental Policy. *Global Environmental Politics* 2 (4): 118–123.

Leach, Melissa, and Robin Mearns. 1996. *The Lie of the Land: Challenging Received Wisdom on the African Environment.* Westport, Conn.: Greenwood.

Lear, Linda. 1997. *Rachel Carson: The Life of the Author of Silent Spring*. New York: Holt.

Leaver, Erik, and John Cavanagh. 1996. Controlling Transnational Corporations. *Foreign Policy in Focus* 1 (6): 1–3.

Lee, Donna, and Rorden Wilkinson, eds. 2007. *The WTO after Hong Kong: Progress in, and Prospects for, the Doha Development Agenda*. London: Routledge.

Leighton, Michelle, Naomi Roht-Arriaza, and Lyuba Zarsky. 2002. *Beyond Good Deeds: Case Studies and a New Policy Agenda for Corporate Accountability*. Berkeley, CA: Nautilus Institute. Available at <http://www.eldis.org/assets/Docs/11134.html> (accessed July 15, 2010).

Lemonick, Michael D., and David Bjerklie. 2004. How We Grew So BIG. *Time* 163 (23): 58–69.

Leonard, H. Jeffrey. 1988. *Pollution and the Struggle for the World Product*. Cambridge: Cambridge University Press.

Le Prestre, Philippe. 1989. *The World Bank and the Environmental Challenge*. Selinsgrove, Pa.: Susquehanna University Press.

Less, Christina Tébar, and Steven McMillan. 2005. *Achieving the Successful Transfer of Environmentally Sound Technologies: Trade-related Aspects*. OECD, Trade Directorate, OECD Trade and Environment Working Papers, no. 2005/2. Paris: OECD.

Levinson, Arik, and M. Scott Taylor. 2008. Unmasking the Pollution Haven Effect. *International Economic Review* 49 (1): 223–254.

Levy, David. 1997. Business and International Environmental Treaties: Ozone Depletion and Climate Change. *California Management Review* 39 (3): 54–71.

Levy, David, and Peter Newell, eds. 2005. *The Business of Global Environmental Governance*. Cambridge, Mass.: MIT Press.

Levy, David, and Daniel Egan. 1998. Capital Contests: National and Transnational Channels of Corporate Influence on the Climate Change Negotiations. *Politics & Society* 26 (3): 337–361.

Levy, David, and Peter Newell. 2000. Oceans Apart? Business Responses to Global Environmental Issues in Europe and the United States. *Environment* 42 (9): 8–20.

Levy, David, and Peter Newell. 2002. Business Strategy and International Environmental Governance: Toward a Neo-Gramscian Synthesis. *Global Environmental Politics* 2 (4): 84–101.

Lipper, Leslie, Takumi Sakuyama, Randy Stringer, and David Zilberman. 2009. *Payments for Environmental Services in Agricultural Landscapes*. New York: Springer.

Lipschutz, Ronnie D. 1999. From Local Knowledge and Practice to Global Environmental Governance. In *Approaches to Global Governance Theory*, ed. Martin Hewson and Timothy J. Sinclair, 259–283. Albany: SUNY Press.

Lipschutz, Ronnie D. 2003. *Global Environmental Politics: Power, Perspectives, and Practice*. Washington, D.C.: Congressional Quarterly Press.

Litfin, Karen, ed. 1998. *The Greening of Sovereignty in World Politics*. Cambridge, Mass.: MIT Press.

Logsdon, Jeanne, and Bryan Husted. 2000. Mexico's Environmental Performance under NAFTA: The First Five Years. *Journal of Environment & Development* 9 (4): 370–383.

Lohmann, Larry. 1993. Resisting Green Globalism. In *Global Ecology: A New Arena of Political Conflict*, ed. Wolfgang Sachs, 159–169. London: Zed Books.

Lomborg, Bjørn, ed. 2009. *Global Crises, Global Solutions: Costs and Benefits*. 2nd ed. Cambridge: Cambridge University Press.

Lomborg, Bjørn. 2001. *The Skeptical Environmentalist*. Cambridge: Cambridge University Press.

Lomborg, Bjørn. 2007. *Cool It: The Skeptical Environmentalist's Guide to Global Warming*. New York: Alfred A. Knopf.

Lovelock, James. 1979. *Gaia: A New Look at Life on Earth*. Oxford: Oxford University Press.

Lovelock, James. 1995. *The Ages of Gaia: A Biography of Our Living Earth*. Rev. ed. New York: Norton.

Lovelock, James. 2009. *The Vanishing Face of Gaia: A Final Warning*. New York: Basic Books.

Low, Patrick. 1993. The International Location of Polluting Industries and the Harmonization of Environmental Standards. In *Difficult Liaison: Trade and the Environment in the Americas*, ed. H. Munoz and R. Rosenberg, 21–50. London: Transaction.

Mabogunje, Akin. 2002. Poverty and Environmental Degradation: Challenges within the Global Economy. *Environment* 44 (1): 9–18.

MacFarlane, Alan, and Iris MacFarlane. 2004. *The Empire of Tea: The Remarkable History of the Plant that Took Over the World*. New York: Overlook Press.

MacKenzie, Deborah. 2002. Fresh Evidence on Bhopal Disaster. *New Scientist* 176 (December 4): 6–7.

MacNeill, Jim, Pieter Winsemius, and Taizo Yakushiji. 1991. *Beyond Interdependence: The Meshing of the World's Economy and the Earth's Ecology*. New York: Oxford University Press.

Maddox, John. 1972. *The Doomsday Syndrome*. London: Macmillan.

Magno, Francisco. 2003. Human and Ecological Security: The Anatomy of Mining Disputes in the Philippines. In *Development and Security in Southeast Asia, Volume 1: The Environment*, ed. David Dewitt and Carolina Hernandez, 115–136. Aldershot, UK: Ashgate.

Makower, J., and C. Pike. 2009. *Strategies for the Green Economy: Opportunities and Challenges in the New World of Business*. New York: McGraw-Hill.

Malthus, Thomas Robert. 1798/1826. *Essay on the Principle of Population.* Available at <http://www.econlib.org/library/Malthus/malPlong.html> (accessed July 15, 2010).

Mander, Jerry, and Edward Goldsmith, eds. 2001. *The Case against the Global Economy: And for a Turn towards Localization.* Rev. ed. London: Earthscan.

Mani, Muthukumara, and David Wheeler. 1998. In Search of Pollution Havens? Dirty Industry in the World Economy, 1960–1993. *Journal of Environment & Development* 7 (3): 215–247.

Maniates, Michael F. 2001. Individualization: Plant a Tree, Buy a Bike, Save the World? *Global Environmental Politics* 1 (3): 31–52.

Maniates, Michael F. 2002. In Search of Consumptive Resistance: The Voluntary Simplicity Movement. In *Confronting Consumption*, ed. Tom Princen, Ken Conca, and Michael Maniates, 199–235. Cambridge, Mass.: MIT Press.

Maniates, Michael F., ed. 2003. *Encountering Global Environmental Politics: Teaching, Learning, and Empowering Knowledge.* Lanham, Md.: Rowman and Littlefield.

Maniates, Michael F., and John M. Meyer, eds. 2010. *The Environmental Politics of Sacrifice.* Cambridge, Mass.: MIT Press.

Mann, Howard. 2001. *Private Rights, Public Problems: A Guide to NAFTA's Chapter on Investor Rights.* Winnipeg: IISD. Available at <http://www.iisd.org/pdf/trade_citizensguide.pdf> (accessed July 15, 2010).

McAteer, Emily, and Simone Pulver. 2009. The Corporate Boomerang: Shareholder Transnational Advocacy Networks Targeting Oil Companies in the Ecuadorian Amazon. *Global Environmental Politics* 9 (1): 1–30.

McCarthy, John F., and R. A. Cramb. 2009. Policy Narratives, Landholder Engagement, and Oil Palm Expansion on the Malaysian and Indonesian Frontiers. *Geographical Journal* 175 (2): 112–123.

McCarthy, John F., and Zahari Zen. 2010. Regulating the Oil Palm Boom: Assessing the Effectiveness of Environmental Governance Approaches to Agro-industrial Pollution in Indonesia. *Law & Policy* 32 (1): 153–179.

McKay, George. 1996. *Senseless Acts of Beauty: Cultures of Resistance Since the Sixties.* London: Verso.

McKean, Margaret A. 1981. *Environmental Protest and Citizen Politics in Japan.* Berkeley: University of California Press.

McKibbon, Bill. 2007. *Deep Economy: The Wealth of Communities and the Durable Future.* New York: Times Books.

McKibbon, Bill. 2010. *Eaarth: Making a Life on a Tough New Planet.* New York: Times Books.

McManis, Charles R., ed. 2007. *Biodiversity and the Law: Intellectual Property, Biotechnology and Traditional Knowledge.* London: Earthscan.

McMurtry, John. 1999. *The Cancer Stage of Capitalism.* London: Pluto.

Meadows, Donella H., Dennis L. Meadows, William W. Behrens, and Jørgen Randers. 1972. *The Limits to Growth.* New York: Club of Rome.

Meadows, Donella H., Dennis L. Meadows, and Jørgen Randers. 1992. *Beyond the Limits: Confronting Global Collapse, Envisioning a Sustainable Future*. White River Junction, Vt.: Chelsea Green.

Meadows, Donella, Jørgen Randers, and Dennis L. Meadows. 2004. *The Limits to Growth: The 30-Year Update*. White River Junction, Vt.: Chelsea Green.

Mee, Laurence, Holly Dublin, and Anton Eberhard. 2008. Evaluating the Global Environment Facility: A Goodwill Gesture or a Serious Attempt to Deliver Global Benefits? *Global Environmental Change* 18 (4): 800–810.

Meeker-Lowry, Susan. 1996. Community Money: The Potential of Local Currency. In *The Case against the Global Economy and for a Turn Toward the Local*, ed. Jerry Mander and Edward Goldsmith, 446–459. San Francisco: Sierra Club Books.

Mehta, Sandeep. 2003. The Johannesburg Summit from the Depths. *Journal of Environment & Development* 12 (1): 121–128.

Mies, Maria, and Vandana Shiva. 1993. *Ecofeminism*. London: Zed Books.

Miles, Edward L., Arild Underdal, Steinar Andresen, Jørgen Wettestad, Jon Birger Skjærseth, and Elaine M. Carlin. 2001. *Environmental Regime Effectiveness: Confronting Theory with Evidence*. Cambridge, Mass.: MIT Press.

Mill, John Stuart. 1848. *Principles of Political Economy*.

Millennium Ecosystem Assessment. 2005. *Ecosystems and Human Well-being: Synthesis*. Available at <http://www.millenniumassessment.org/documents/document.356.aspx.pdf> (accessed July 15, 2010).

Miller, Marian. 2001. Tragedy for the Commons: The Enclosure and Commodification of Knowledge. In *The International Political Economy of the Environment*, ed. Dimitris Stevis and Valerie Assetto, 111–134. Boulder, Colo.: Lynne Rienner.

Mink, Stephen. 1993. *Poverty, Population and the Environment*. World Bank Discussion Paper 189. Washington, D.C.: World Bank.

Miranda, Miriam, Carel Dieperink, and Pieter Glasbergen. 2006. Costa Rican Environmental Service Payments: The Use of a Financial Instrument in Participatory Forest Management. *Environmental Management* 38 (4): 562–571.

Missbach, Andreas. 2004. The Equator Principles: Drawing the Line for Socially Responsible Banks? An Interim Review from an NGO Perspective. *Development* 47 (3): 78–84.

Mitchell, Ronald B. 1994. *Intentional Oil Pollution at Sea: Environmental Policy and Treaty Compliance*. Cambridge, Mass.: MIT Press.

Mitchell, Ronald B. 2002. A Quantitative Approach to Evaluating International Environmental Regimes. *Global Environmental Politics* 2 (4): 58–83.

Mitchell, Ronald B. 2006. Problem Structure, Institutional Design, and the Relative Effectiveness of International Environmental Agreements. *Global Environmental Politics* 6 (3): 72–89.

Mitchell, Ronald B. 2009. *International Politics and the Environment*. London and Thousand Oaks, Calif.: Sage Publications.

Mitchell, Ronald B., William C. Clark, David W. Cash, and Nancy M. Dickson, eds. 2006. *Global Environmental Assessments: Information and Influence*. Cambridge, Mass.: MIT Press.

Mol, Arthur. 2002. Ecological Modernization and the Global Economy. *Global Environmental Politics* 2 (2): 92–115.

Molina, David. 1993. A Comment on Whether Maquiladoras Are in Mexico for Low Wages or to Avoid Pollution Abatement Costs. *Journal of Environment & Development* 2 (1): 221–241.

Moran, Emilio F., and Elinor Ostrom, eds. 2005. *Seeing the Forest and the Trees: Human-Environment Interactions in Forest Ecosystems*. Cambridge, Mass.: MIT Press.

Morehouse, Ward. 1994. Unfinished Business: Bhopal Ten Years After. *Ecologist* 24 (5): 164–168.

Morgera, Elisa. 2004. From Stockholm to Johannesburg: From Corporate Responsibility to Corporate Accountability for the Global Protection of the Environment? *Review of European Community and International Environmental Law RECIEL* 13 (2): 214–222.

Morse, Bradford. 1992. *Sardar Sarovar: Report of the Independent Review*. Ottawa: Resource Futures International.

Moxham, Roy. 2003. *Tea: Addiction, Exploitation and Empire*. London: Constable and Robinson.

Munasinghe, Mohan. 1999. Is Environmental Degradation an Inevitable Consequence of Economic Growth? Tunneling through the Environmental Kuznets Curve. *Ecological Economics* 29 (1): 89–109.

Munasinghe, Mohan, Raoul O'Ryan, Ronaldo Seroa Da Motta, and Carlos De Migu. 2006. *Macroeconomic Policies for Sustainable Growth: Analytical Framework and Policy Studies of Brazil and Chile*. Cheltenham: Edward Elgar.

Murphy, Craig. 1983. What the Third World Wants: An Interpretation of the Development and Meaning of the New International Economic Order. *International Studies Quarterly* 27: 55–76.

Mushita, Andrew T., and Carol B. Thompson. 2002. Patenting Biodiversity? Rejecting WTO/TRIPS in Southern Africa. *Global Environmental Politics* 2 (1): 65–82.

Myers, Norman. 1979. *The Sinking Ark: A New Look at the Problem of Disappearing Species*. New York: Pergamon Press.

Myers, Norman. 1997. Consumption in Relation to Population, Environment and Development. *Environmentalist* 17 (1): 33–44.

Myers, Norman, and Jennifer Kent. 2001. *Perverse Subsidies: How Tax Dollars Can Undercut the Environment and the Economy*. Washington, D.C.: Island Press.

Myers, Norman, and Richard P. Tucker. 1987. Deforestation in Central America: Spanish Legacy and North American Consumers. *Environmental Review* 11 (1): 55–71.

Najam, Adil. 1998. Searching for NGO Effectiveness. *Development Policy Review* 16 (3): 305–310.

Najam, Adil. 2003. The Case against a New International Environmental Organization. *Global Governance* 9 (3): 367–384.

Najam, Adil, Ioli Christopoulou, and William R. Moomaw. 2004. The Emergent "System" of Global Environmental Governance. *Global Environmental Politics* 4 (4): 23–35.

Najam, Adil, Mihaela Papa, and Nadaa Taiyab. 2006. *Global Environmental Governance: A Reform Agenda*. Winnipeg: International Institute for Sustainable Development.

Nash, Jennifer, and John Ehrenfeld. 1996. Code Green. *Environment* 37 (1): 16–45.

Nelson, Paul J. 1995. *The World Bank and Non-Governmental Organizations: The Limits of Apolitical Development*. London: Macmillan.

Neumayer, Eric. 2001. *Greening Trade and Investment: Environmental Protection without Protectionism*. London: Earthscan.

Neumayer, Eric. 2004. The WTO and the Environment: Its Past Record is Better than Critics Believe, but the Future Outlook is Bleak. *Global Environmental Politics* 4 (3): 1–8.

New Economics Foundation (NEF) (Ann Pettifor, Deborah Doane, Romilly Greenhill, and Julian Oram, and Andrew Simms). 2002. Chasing Shadows: Re-Imagining Finance for Development. Available at <http://www.i-r-e.org/fiche-analyse-33_en.html> (accessed July 15, 2010).

Newell, Peter. 2000. *Climate for Change: Non-State Actors and the Global Politics of the Greenhouse*. Cambridge: Cambridge University Press.

Newell, Peter. 2001. New Environmental Architectures and the Search for Effectiveness. *Global Environmental Politics* 1 (1): 35–44.

Newell, Peter. 2008. The Political Economy of Global Environmental Governance. *Review of International Studies* 34 (3): 507–529.

Newell, Peter, and Matthew Paterson. 1998. A Climate for Business: Global Warming, the State, and Capital. *Review of International Political Economy* 5 (4): 679–703.

Newell, Peter, and Matthew Paterson. 2010. *Climate Capitalism: Global Warming and the Transformation of the Global Economy*. Cambridge: Cambridge University Press.

Norberg-Hodge, Helena. 1996. Shifting Direction: From Global Dependence to Local Interdependence. In *The Case Against the Global Economy and For a Turn toward the Local*, ed. Jerry Mander and Edward Goldsmith, 393–406. San Francisco: Sierra Club Books.

Nordhaus, William. 2007. A Review of the Stern Review on the Economics of Climate Change. *Journal of Economic Literature* 45 (3): 686–702.

O'Connor, James. 1998. *Natural Causes: Essays in Ecological Marxism*. New York: Guilford.

O'Neill, Kate. 2000. *Waste Trading among Rich Nations: Building a New Theory of Environmental Regulation*. Cambridge, Mass.: MIT Press.

O'Neill, Kate. 2001. The Changing Nature of Global Waste Management for the 21st Century. *Global Environmental Politics* 1 (1): 77–98.

O'Neill, Kate. 2009. *The Environment and International Relations*. Cambridge: Cambridge University Press.

Obi, Cyril. 1997. Globalization and Local Resistance: The Case of Ogoni versus Shell. *New Political Economy* 2 (1): 137–148.

Ofreneo, Rene. 1993. Japan and the Environmental Degradation of the Philippines. In *Asia's Environmental Crisis*, ed. Michael Howard, 201–219. Boulder, Colo.: Westview Press.

Okereke, Chukwumerije, Harriet Bulkeley, and Heike Schroeder. 2009. Conceptualizing Climate Governance Beyond the International Regime. *Global Environmental Politics* 9 (1): 58–78.

Ophuls, William. 1973. *Ecology and the Politics of Scarcity: Prologue to a Political Theory of the Steady State*. San Francisco: Freeman.

Organization for Economic Cooperation and Development (OECD). 2000. *The OECD Guidelines for Multinational Enterprises*. Paris: OECD.

Organization for Economic Cooperation and Development (OECD). 2002. *Foreign Direct Investment and the Environment: Lessons for the Mining Sector*. Paris: OECD.

Organization for Economic Cooperation and Development (OECD). 2007. *About Environment and Export Credits*, Paris: OECD. Available at <http://www.oecd.org/document/26/0,3343,en_2649_34181_39960154_1_1_1_1,00.html> (accessed July 15, 2010).

Osborn, Derek, and Tom Bigg. 1998. *Earth Summit II: Outcomes and Analysis*. London: Earthscan.

Ostrom, Elinor, and Charlotte Hess, eds. 2006. *Understanding Knowledge as a Commons: From Theory to Practice*. Cambridge, Mass.: MIT Press.

Ostrom, Elinor. 1990. *Governing the Commons: The Evolution of Institutions for Collective Action*. Cambridge: Cambridge University Press.

Ostrom, Elinor. 2005. *Understanding Institutional Diversity*. Princeton, N.J.: Princeton University Press.

Our Durable Planet. 1999. (A Survey of the 20th Century). *The Economist* 352 (September 11): 29–31.

Oxfam, CAFOD Christian Aid, and Eurodad. 2002. *A Joint Submission to the World Bank and IMF Review of HIPC and Debt Sustainability*. Available at <http://www.eurodad.org/uploadedFiles/Whats_New/Reports/eurodad_debtsustainability_hipcreview.pdf> (accessed July 15, 2010).

Paarlberg, Robert. 2000. The Global Food Fight. *Foreign Affairs (Council on Foreign Relations)* 79 (3): 24–38.

Paehlke, Robert. 2003. *Democracy's Dilemma: Environment, Social Equity, and the Global Economy*. Cambridge, Mass.: MIT Press.

Park, Susan. 2005. How Transnational Environmental Advocacy Networks Socialize International Financial Institutions: A Case Study of the International Finance Corporation. *Global Environmental Politics* 5 (4): 95–119.

Park, Susan. 2010. *World Bank Group Interactions with Environmentalists: Changing International Organisation Identities*. Manchester: Manchester University Press.

Parr, Adrian. 2009. *Hijacking Sustainability*. Cambridge, Mass.: MIT Press.

Parson, Edward A. 2002. *Protecting the Ozone Layer: Science, Strategy, and Negotiation in the Shaping of a Global Environmental Regime*. Oxford: Oxford University Press.

Paterson, Matthew, David Humphreys, and Lloyd Pettiford. 2003. Conceptualizing Global Environmental Governance: From Interstate Regimes to Counter-Hegemonic Struggles. In *Global Environmental Governance for the Twenty-First Century: Theoretical Approaches and Normative Considerations*, ed. David Humphreys, Matthew Paterson, and Lloyd Pettiford. Special issue of *Global Environmental Politics* 3 (2): 1–10.

Paterson, Matthew. 1996. *Global Warming and Global Politics*. London: Routledge.

Paterson, Matthew. 2001a. Climate Policy as Accumulation Strategy: The Failure of COP 6 and Emerging Trends in Climate Politics. *Global Environmental Politics* 1 (2): 10–17.

Paterson, Matthew. 2001b. Risky Business: Insurance Companies in Global Warming Politics. *Global Environmental Politics* 1 (3): 18–42.

Paterson, Matthew. 2001c. *Understanding Global Environmental Politics: Domination, Accumulation, Resistance*. London: Palgrave Macmillan.

Paterson, Matthew. 2007. *Automobile Politics: Ecology and Cultural Political Economy*. Cambridge: Cambridge University Press.

Patomaki, Heikki. 2001. *Democratizing Globalization: The Leverage of the Tobin Tax*. London: Zed Books.

Pattberg, Philipp. 2007. *Private Institutions and Global Governance: The New Politics of Environmental Sustainability*. Cheltenham, UK: Edward Elgar.

Peacock, Mark S. 2006. The Moral Economy of Parallel Currencies: An Analysis of Local Exchange Trading Systems. *American Journal of Economics and Sociology* 65 (5): 1059–1083.

Pearce, David, and Jeremy Warford. 1993. *World without End: Economics, Environment and Sustainable Development*. New York: Oxford University Press.

Pearce, David, Neil Adger, David Maddison, and Dominic Moran. 1995. Debt and the Environment. *Scientific American* 272 (June): 52–56.

Pearson, Charles, ed. 1987a. *Multinational Corporations, the Environment, and the Third World*. Durham, N.C.: Duke University Press.

Pearson, Charles. 1987b. Environmental Standards, Industrial Relocation, and Pollution Havens. In *Multinational Corporations, the Environment, and the*

Third World, ed. Charles Pearson, 113–128. Durham, N.C.: Duke University Press.

Pellow, David Naguib. 2007. *Resisting Global Toxics Transnational Movements for Environmental Justice*. Cambridge, Mass.: MIT Press.

Peluso, Nancy Lee, and Michael Watts, eds. 2001. *Violent Environments*. Ithaca: Cornell University Press.

Pempel, T. J. 1998. *Regime Shift: Comparative Dynamics of the Japanese Political Economy*. Ithaca: Cornell University Press.

Perkins, Nancy. 1998. *World Trade Organizations: United States—Import Prohibition of Certain Shrimp and Shrimp Products*. Available at American Society of International Law web site: <http://www.asil.org> (accessed July 15, 2010).

Perkins, Richard, and Eric Neumayer. 2009. Transnational Linkages and Spillover of Environment-Efficiency into Developing Countries. *Global Environmental Change* 19 (3): 375–383.

Perlez, Jane, and Raymond Bonner. 2005. Below a Mountain of Wealth, A River of Waste. *New York Times*. 27 December. Available at: <http://www.nytimes.com/2005/12/27/international/asia/27gold.html> (accessed July 15, 2010).

Peterson, M. J. 1992. Whalers, Cetologists, Environmentalists and the International Management of Whaling. *International Organization* 46 (1): 147–186.

Phillips, David A. 2009. *Reforming the World Bank: Twenty Years of Trial—and Error*. Cambridge: Cambridge University Press.

Pojman, Louis P. 2000. *Global Environmental Ethics*. Mountain View, Calif.: Mayfield.

Population Division of the Department of Economic and Social Affairs of the UN Secretariat. 2001. *World Population Prospects: The 2000 Revision* (Highlights). Available at <http://www.un.org/esa/population/publications/wpp2000/highlights.pdf> (accessed July 15, 2010).

Porter, Gareth. 1999. Trade Competition and Pollution Standards: "Race to the Bottom" or "Stuck at the Bottom"? *Journal of Environment & Development* 8 (2): 133–151.

Porter, Gareth, Neil Bird, Nanki Kaur, and Leo Peskett. 2008. *New Finance for Climate Change and the Environment*. Washington, D.C.: WWF and Heinrich Böll Foundation. Available at <http://assets.panda.org/downloads/ifa_report.pdf> (accessed July 15, 2010).

Potter, Lesley. 1993. The Onslaught on the Forests in South-East Asia. In *South-East Asia's Environmental Future: The Search for Sustainability*, ed. Harold Brookfield and Yvonne Byron, 103–123. Tokyo: United Nations University Press; Melbourne: Oxford University Press.

Pounds, J. Alan, and Robert Puschendorf. 2004. Clouded Futures. *Nature* 427 (January 8): 107–108.

Prakash, Aseem. 2000. *Greening the Firm: The Politics of Corporate Environmentalism*. Cambridge: Cambridge University Press.

Prakash, Aseem, and Matthew Potoski. 2006a. *The Voluntary Environmentalists: Green Clubs, ISO 14001 and Voluntary Environmental Regulations*. Cambridge: Cambridge University Press.

Prakash, Aseem, and Matthew Potoski. 2006b. Racing to the Bottom? Trade, Environmental Governance and ISO 14001. *American Journal of Political Science* 50 (2): 350–364.

Princen, Thomas, Michael Maniates, and Ken Conca, eds. 2002. *Confronting Consumption*. Cambridge, Mass.: MIT Press.

Princen, Thomas. 1994. NGOs: Creating a Niche in Environmental Diplomacy. In *Environmental NGOs in World Politics: Linking the Local and the Global*, ed. Thomas Princen and Matthias Finger, 29–47. London: Routledge.

Princen, Thomas. 1997. The Shading and Distancing of Commerce: When Internalization Is Not Enough. *Ecological Economics* 20 (3): 235–253.

Princen, Thomas. 2001. Consumption and Its Externalities: Where Economy Meets Ecology. *Global Environmental Politics* 1 (3): 11–30.

Princen, Thomas. 2002. Distancing: Consumption and the Severing of Feedback. In *Confronting Consumption*, ed. Thomas Princen, Michael Maniates, and Ken Conca, 103–131. Cambridge, Mass.: MIT Press.

Princen, Thomas. 2005. *The Logic of Sufficiency*. Cambridge, Mass.: MIT Press.

Princen, Thomas. 2010. *Treading Softly: Paths to Ecological Order*. Cambridge, Mass.: MIT Press.

Rabe, Barry. 2007. Beyond Kyoto: Climate Change Policy in Multilevel Governance Systems. *Governance* 20 (3): 423–444.

Rajan, S. Ravi. 2001. Toward a Metaphysic of Environmental Violence: The Case of the Bhopal Gas Disaster. In *Violent Environments*, ed. Nancy Lee Peluso and Michael Watts, 380–398. Ithaca: Cornell University Press.

Rangan, Haripriya. 1996. From Chipko to Uttaranchal: Development, Environment and Social Protest in the Gharhwal Himalayas, India. In *Liberation Ecologies: Environment, Development, Social Movements*, ed. Richard Peet and Michael Watts, 205–226. New York: Routledge.

Rapley, John. 2004. *Globalization and Inequality: Neoliberalism's Downward Spiral*. Boulder, Colo.: Lynne Rienner.

Rapley, John. 2007. *Understanding Development: Theory and Practice in the Third World*. 3rd ed. Boulder, Colo.: Lynne Rienner.

Raynolds, Laura. 2000. Re-embedding Global Agriculture: The International Organic and Fair Trade Movements. *Agriculture and Human Values* 17 (1): 297–309.

Reality of Aid Management Committee. 2008. *The Reality of Aid 2008*. Quezon City: Ibon Books.

Reardon, Thomas, and Stephen Vosti. 1995. Links between Rural Poverty and the Environment in Developing Countries: Asset Categories and Investment Poverty. *World Development* 23 (9): 1495–1506.

Rechkemmer, Andreas, ed. 2005. *UNEO—Towards an International Environment Organization: Approaches to a Sustainable Reform of Global Environmental Governance*. Baden-Baden: Nomos Publishers.

Reed, David, ed. 1992. *Structural Adjustment and the Environment*. Boulder, Colo.: Westview Press.

Reed, David. 1997. The Environmental Legacy of the Bretton Woods: The World Bank. In *Global Governance: Drawing Insights from the Environmental Experience*, ed. Oran K. Young, 227–245. Cambridge, Mass.: MIT Press.

Rees, William E. 2002. Globalization and Sustainability: Conflict or Convergence? *Bulletin of Science, Technology & Society* 22 (4): 249–268.

Rees, William E. 2006. Globalization, Trade and Migration: Undermining Sustainability. *Ecological Economics* 59 (2): 220–225.

Rees, William E., and Laura Westra. 2003. When Consumption Does Violence: Can There Be Sustainability and Environmental Justice in a Resource-Limited World? In *Just Sustainabilities: Development in an Unequal World*, ed. Julian Agyeman, Robert Bullard, and Bob Evans, 99–124. London: Earthscan.

Repetto, Robert. 1994. *Trade and Sustainable Development. UNEP Environment and Trade Series No. 1*. Geneva: United Nations Environment Programme.

Repetto, Robert, William McGrath, Michael Wells, Christine Beer, and Fabrizio Rossini. 1989. *Wasting Assets: Natural Resources in the National Income and Product Accounts*. Washington, D.C.: World Resources Institute.

Ricardo, David. 1817. *On the Principles of Political Economy and Taxation*. Available at <http://www.econlib.org/library/Ricardo/ricP.html> (accessed July 15, 2010).

Rich, Bruce. 1994. *Mortgaging the Earth: The World Bank, Environmental Impoverishment, and the Crisis of Development*. London: Earthscan.

Rich, Bruce. 2000. Exporting Destruction. *The Environmental Forum* (September–October): 32–40.

Rich, Bruce. 2009. *Foreclosing the Future: Coal, Climate and Public International Finance*. Washington, D.C.: Environmental Defense. Available at <http://www.edf.org/documents/9593_coal-plants-report.pdf> (accessed July 15, 2010).

Richards, Paul. 1985. *Indigenous Agricultural Revolution*. London: Unwin Hyman.

Rifkin, Jeremy. 2002. *The Hydrogen Economy: The Creation of the Worldwide Energy Web and the Redistribution of Power on Earth*. New York: Tarcher/Putnam.

Robbins, Richard H. 2002. *Global Problems and the Culture of Capitalism*. 2nd ed. Boston: Allyn and Bacon.

Roberts, Joseph. 1998. Multilateral Agreement on Investment. *Monthly Review* 50 (5): 23–32.

Robinson, Nick. 2000. *The Politics of Agenda Setting: The Car and the Shaping of Public Policy*. Aldershot, UK: Ashgate.

Rodney, Walter. 1972. *How Europe Underdeveloped Africa*. Dar es Salaam: Tanzania Publishing House.

Rogers, Adam. 1993. *The Earth Summit: A Planetary Reckoning*. Los Angeles: Global View Press.

Roht-Arriaza, Naomi. 1995. Shifting the Point of Regulation: The International Organization for Standardization and Global Law-Making on Trade and the Environment. *Ecology Law Quarterly* 22 (3): 479–539.

Roodman, David Malin. 1998. *The Natural Wealth of Nations: Harnessing the Market for the Environment*. New York: Norton.

Ross, Michael L. 2001. *Timber Booms and Institutional Breakdown in Southeast Asia*. Cambridge, Mass.: Cambridge University Press.

Rosendal, G. Kristin. 2001. Impact of Overlapping International Regimes: The Case of Biodiversity. *Global Governance* 7 (2): 95–117.

Rowell, Andrew. 1996. *Green Backlash: Global Subversion of the Environment Movement*. London: Routledge.

Rowland, Wade. 1973. *The Plot to Save the World*. Toronto: Clarke, Irwin, and Company.

Rowlands, Ian. 1995. *The Politics of Global Atmospheric Change*. Manchester: Manchester University Press; New York: St. Martin's Press.

Rowlands, Ian. 2000. Beauty and the Beast? BP's and Exxon's Position on Global Climate Change. *Environment and Planning C: Government & Policy* 18 (3): 339–354.

Ruitenbeek, Jack, and Cynthia Cartier. 1998. *Rational Exploitations: Economic Criteria and Indicators for Sustainable Management of Tropical Forests*. Occasional Paper No. 17, November. Bogor, Indonesia: Center for International Forestry Research.

Runge, C. Ford, and Benjamin Senauer. 2000. A Removable Feast. *Foreign Affairs (Council on Foreign Relations)* 79 (3): 39–51.

Rutherford, Paul. 2003. "Talking the Talk": Business Discourse at the World Summit on Sustainable Development. *Environmental Politics* 12 (2): 145–150.

Ruud, Audun. 2002. Environmental Management of Transnational Corporations in India: Are TNCs Creating Islands of Environmental Excellence in a Sea of Dirt? *Business Strategy and Environment* 11 (2): 103–118.

Ryan, John, and Alan Durning. 1997. *Stuff: The Secret Lives of Everyday Things*. Seattle: Northwest Environment Watch.

Sachs, Wolfgang, ed. 1993. *Global Ecology: A New Arena of Political Conflict*. London: Zed Books.

Sachs, Wolfgang. 1999. *Planet Dialectics: Explorations in Environment and Development*. London: Zed Books.

Sachs, Wolfgang, Tilman Santarius, et al. 2007. *Fair Future: Limited Resources and Global Justice*. London: Zed Books.

Sagar, Ambuj, and Stacy VanDeveer. 2005. Capacity Development for the Environment: Broadening the Scope. *Global Environmental Politics* 5 (3): 14–22.

Salleh, Ariel, ed. 2009. *Eco-Sufficiency and Global Justice: Women Write Political Ecology*. London: Pluto Press.

Sampson, Gary, and John Whalley, eds. 2005. *The WTO, Trade, and the Environment*. Cheltenham, UK: Edward Elgar.

Sassen, Saskia. 2006. *Territory, Authority, Rights: From Medieval to Global Assemblages*. Princeton, N.J.: Princeton University Press.

Sassen, Saskia. 2007. *A Sociology of Globalization*. New York: Norton.

Schafer, Kristin S. 2002. Ratifying Global Toxics Treaties: The United States Must Provide Leadership. *SAIS Review* 22 (1): 169–176.

Scherr, Sara. 2000. A Downward Spiral? Research Evidence on the Relationship between Poverty and Natural Resource Degradation. *Food Policy* 25 (1): 479–498.

Schmidheiny, Stephan (with the World Business Council for Sustainable Development). 1992. *Changing Course: A Global Business Perspective on Development and the Environment*. Cambridge, Mass.: MIT Press.

Schmidheiny, Stephan, and Federico Zorraquin. 1996. *Financing Change: The Financial Community, Eco-Efficiency and Sustainable Development*. Cambridge, Mass.: MIT Press.

Schmitz, A., T. H. Spreen, W. A. Messina, and C. B. Moss. 2002. *Sugar and Related Sweetener Markets*. Wallingford, UK: CABI.

Scholte, Jan Aart. 1997. The Globalization of World Politics. In *The Globalization of World Politics: An Introduction to International Relations*, ed. John Baylis and Steve Smith, 13–30. Oxford: Oxford University Press.

Scholte, Jan Aart. 2005. *Globalization: A Critical Introduction*. 2nd ed. New York: St. Martin's Press Palgrave Macmillan.

Schor, Juliet B. 1998. *The Overspent American: Upscaling, Downshifting, and the New Consumer*. New York: Basic Books.

Schor, Juliet B. 2004. *Born to Buy: The Commercialized Child and the New Consumer Culture*. New York: Scribner.

Schreurs, Miranda A. 2002. *Environmental Politics in Japan, Germany, and the United States*. Cambridge: Cambridge University Press.

Schumacher, E. F. 1973. *Small Is Beautiful: Economics as if People Mattered*. New York: Harper and Row.

Selin, Henrik. 2010. *Global Governance of Hazardous Chemicals: Challenges of Multilevel Management*. Cambridge, Mass.: MIT Press.

Selin, Henrik, and Noelle Eckley. 2003. Science, Politics, and Persistent Organic Pollutants: Scientific Assessments and Their Role in International Environmental Negotiations. *International Environmental Agreement: Politics, Law and Economics* 3 (1): 17–42.

Selin, Henrik, and Stacy D. VanDeveer. 2003. Mapping Institutional Linkages in European Air Pollution Politics. *Global Environmental Politics* 3 (3): 14–46.

Selin, Henrik, and Stacy D. VanDeveer. 2006. Raising Global Standards: Hazardous Substances and E-Waste Management in the European Union. *Environment* 48 (10): 7–18.

Selin, Henrik, and Stacy D. VanDeveer, eds. 2009. *Changing Climates in North American Politics: Institutions, Policymaking, and Multilevel Governance*. Cambridge, Mass.: MIT Press.

Sell, Susan. 2009. Corporations, Seeds and Intellectual Property Rights Governance. In *Corporate Power in Global Agrifood Governance*, ed. Jennifer Clapp and Doris Fuchs, 187–223. Cambridge, Mass.: MIT Press.

Sen, Amartya. 1981. *Poverty and Famines: An Essay on Entitlement and Deprivation*. Oxford: Oxford University Press.

Sen, Amartya. 1999. *Development as Freedom*. New York: Anchor Books.

Sen, Gita. 1994. Women, Poverty and Population: Issues for the Concerned Environmentalist. In *Feminist Perspectives on Sustainable Development*, ed. Wendy Harcourt, 215–225. London: Zed Books.

Shiva, Mira. 1994. Environmental Degradation and Subversion of Health. In *Close to Home: Women Reconnect Ecology, Health and Development Worldwide*, ed. Vandana Shiva, 60–77. Gabriola Island, B.C.: New Society.

Shiva, Vandana. 1989. *Staying Alive: Women, Ecology and Development*. London: Zed Books.

Shiva, Vandana. 1992. *The Violence of the Green Revolution*. London: Zed Books.

Shiva, Vandana. 1993a. GATT, Agriculture and Third World Women. In *Ecofeminism*, ed. Maria Mies and Vandana Shiva, 231–245. London: Zed Books.

Shiva, Vandana. 1993b. The Greening of the Global Reach. In *Global Ecology: A New Arena of Political Conflict*, ed. Wolfgang Sachs, 149–156. London: Zed Books.

Shiva, Vandana. 1993c. *Monocultures of the Mind: Perspectives on Biodiversity and Biotechnology*. London: Zed Books.

Shiva, Vandana. 1997. *Biopiracy: The Plunder of Nature and Knowledge*. Toronto: Between the Lines.

Shiva, Vandana. 2000. *Stolen Harvest: The Hijacking of the Global Food Supply*. Cambridge, Mass.: South End Press.

Shiva, Vandana. 2008. *Soil not Oil: Environmental Justice in an Age of Climate Justice*. Cambridge, Mass.: South End Press.

Simon, Julian L. 1981. *The Ultimate Resource*. Princeton: Princeton University Press.

Simon, Julian L. 1990. *Population Matters*. New Brunswick, N.J.: Transaction.

Simon, Julian L. 1996. *The Ultimate Resource 2*. Princeton: Princeton University Press.

Skjærseth, Jon Birger, and Tora Skodvin. 2001. Climate Change and the Oil Industry: Common Problems, Different Strategies. *Global Environmental Politics* 1 (3): 18–42.

Skjærseth, Jon Birger, Olav Schram Stokke, and Jørgen Wettestad. 2006. Soft Law, Hard Law, and Effective Implementation of International Environmental Norms. *Global Environmental Politics* 6 (3): 104–120.

Sklair, Leslie. 1993. *Assembling for Development*. San Diego: Center for U.S.–Mexican Studies, University of California.

Sklair, Leslie. 2001. *The Transnational Capitalist Class*. London: Blackwell.

Smith, Adam. 1776. *An Inquiry into the Nature and Causes of the Wealth of Nations*. Available at <http://www.gutenberg.org/etext/3300> (accessed July 15, 2010).

Smith, Ted, David Sonnenfeld, and David Pellow, eds. 2006. *Challenging the Chip: Labor Rights and Environmental Justice in the Global Electronics Industry*. Philadelphia: Temple University Press.

Soluri, John. 2006. *Banana Cultures: Agriculture, Consumption, and Environmental Change in Honduras and the United States*. Austin: University of Texas Press.

Soroos, Marvin S. 1997. *The Endangered Atmosphere*. Columbia, S.C.: University of South Carolina Press.

Soroos, Marvin S. 2001. Global Climate Change and the Futility of the Kyoto Process. *Global Environmental Politics* 1 (2): 1–9.

Speer, Lawrence. 1997. From Command-and-Control to Self-Regulation: The Role of Environmental Management Systems. *International Environment Reporter* 20 (5): 227–228.

Speth, James Gustave. 2008. *The Bridge at the Edge of the World*. New Haven, Conn.: Yale University Press.

Speth, James Gustave, and Peter Haas. 2006. *Global Environmental Governance*. Washington, D.C.: Island Press.

Steenblik, Ronald, and Joy A. Kim. 2009. *Facilitating Trade in Selected Climate Change Mitigation Technologies in the Energy Supply, Buildings, and Industry Sectors*. OECD, Trade Directorate, OECD Trade and Environment Working Papers, no. 2009/2. Paris: OECD.

Steinberg, Paul F. 2001. *Environmental Leadership in Developing Countries: Transnational Relations and Biodiversity Policy in Costa Rica and Bolivia*. Cambridge, Mass.: MIT Press.

Stern, David, Michael Common, and Edward Barbier. 1996. Economic Growth and Environmental Degradation: The Environmental Kuznets Curve and Sustainable Development. *World Development* 24 (7): 1151–1160.

Stern, Nicholas. 2006. *The Stern Review on the Economics of Climate Change*. Cambridge: Cambridge University Press.

Stevis, Dimitris, and Valerie Assetto, eds. 2001. *The International Political Economy of the Environment: Critical Perspectives*. Boulder, Colo.: Lynne Rienner.

Stevis, Dimitris, and Stephen Mumme. 2000. Rules and Politics in International Integration: Environmental Regulation in NAFTA and the EU. *Environmental Politics* 9 (4): 20–42.

Stilwell, Matthew, and Richard Tarasofsky. 2001. *Towards Coherent Environmental and Economic Governance: Legal and Practical Approaches to MEA-WTO Linkages*. Gland, Switzerland: WWF Network. Available at <http://www.ciel.org/Publications/Coherent_EnvirEco_Governance.pdf> (accessed July 15, 2010).

Stoett, Peter J. 1997. *The International Politics of Whaling*. Vancouver: UBC Press.

Stokke, Olav Schram, and Geir Hønneland, eds. 2007. *International Cooperation and Arctic Governance: Regime Effectiveness and Northern Region Building*. London: Routledge.

Stokke, Olav Schram, and Øystein B. Thommessen, eds. 2003. *Yearbook of International Co-operation on Environment and Development*. London: Earthscan.

Stone, Diane, and Christopher Wright, eds. 2006. *The World Bank and Governance: A Decade of Reform and Reaction*. New York: Routledge.

Streck, Charlotte. 2001. The Global Environment Facility—A Role Model for International Governance? *Global Environmental Politics* 1 (2): 71–94.

Streeter, Stephen, John Weaver, and William Coleman, eds. 2009. *Empires and Autonomy: Moments in the History of Globalization*. Vancouver: University of British Columbia Press.

Striffler, Steve. 2002. *In the Shadows of State and Capital: The United Fruit Company, Popular Struggle, and Agrarian Restructuring in Ecuador, 1900–1995*. Durham, N.C.: Duke University Press.

Striffler, Steve, and Mark Moberg, eds. 2003. *Banana Wars: Power, Production, and History in the Americas*. Durham, N.C.: Duke University Press.

Strong, Maurice. 2000. *Where on Earth Are We Going?* Toronto: Knopf.

Suri, Vivek, and Duane Chapman. 1998. Economic Growth, Trade and Energy: Implications for the Environmental Kuznets Curve. *Ecological Economics* 25 (2): 195–208.

Susskind, Lawrence. 1994. *Environmental Diplomacy: Negotiating More Effective Global Agreements*. Oxford: Oxford University Press.

Switzer, Jacqueline Vaughn. 2004. *Environmental Politics: Domestic and Global Dimensions*. Belmont, Calif.: Thompson/Wadsworth.

Tabb, William. 2000. After Seattle: Understanding the Politics of Globalization. *Monthly Review* 51 (10): 1–18.

Tacconi, Luca. 2007. *Illegal Logging: Law Enforcement, Livelihoods and the Timber Trade*. London: Earthscan.

Talbot, John M. 2004. *Grounds for Agreement: The Political Economy of the Coffee Commodity Chain*. Boulder, Colo.: Rowman & Littlefield.

Tamiotti, Ludivine, and Matthias Finger. 2001. Environmental Organizations: Changing Roles and Functions in Global Politics. *Global Environmental Politics* 1 (1): 56–76.

Taylor, Bron Raymond, ed. 1995. *Ecological Resistance Movements: The Global Emergence of Radical and Popular Environmentalism*. Albany: SUNY/Press.

Therien, Jean-Philippe, and Vincent Pouliot. 2006. The Global Compact: Shifting the Politics of International Development? *Global Governance* 12 (1): 55–75.

Thomas, Chris, Alison Cameron, Rhys Green, Michel Bakkenes, Linda Beaumont, Yvonne C. Collingham, et al. 2004. Extinction Risk from Climate Change. *Nature* 427 (January 8): 145–148.

Thomas, Vinod, and Tamara Belt. 1997. Growth and the Environment: Allies or Foes? *Finance & Development* 34 (June): 22–24.

Thompson, Peter, and Laura A. Strohm. 1996. Trade and Environmental Quality: A Review of the Evidence. *Journal of Environment & Development* 5 (4): 365–390.

Tidsell, Clem. 2001. Globalization and Sustainability: Environmental Kuznets Curve and the WTO. *Ecological Economics* 39 (2): 185–196.

Tidsell, Clem. 2005. *Economics of Environmental Conservation*. 2nd ed. Cheltenham, UK: Edward Elgar.

Tienhaara, Kyla. 2006. Mineral Investment and the Regulation of the Environment in Developing Countries: Lessons from Ghana. *International Environmental Agreement: Politics, Law and Economics* 6 (4): 371–394.

Tierney, John. 1990. Betting the Planet. *New York Times Magazine* (December 2): 52–81.

Tietenberg, Thomas, and Lynne Lewis. 2008. *Environmental and Natural Resource Economics*. 8th ed. Reading, Mass.: Addison Wesley Longman.

Tietenberg, Thomas, and Lynne Lewis. 2009. *Environmental Economics and Policy*. 6th ed. Reading, Mass.: Addison Wesley Longman.

Tokar, Brian. 1997. *Earth for Sale: Reclaiming Ecology in the Age of Corporate Greenwash*. Boston: South End Press.

Tolba, Mostafa K., and Iwona Rummel-Bulska. 1998. *Global Environmental Diplomacy: Negotiating Environmental Agreements for the World 1973–1992*. Cambridge, Mass.: MIT Press.

Toye, John. 1991. Ghana. In *Aid and Power*, vol. 2 of *Case Studies*, ed. Paul Mosley, Jane Harrigan, and John Toye, 151–200. London: Routledge.

Triggs, Gillian, and Anna Riddell, eds. 2007. *Antarctica: Legal and Environmental Challenges for the Future*. London: British Institute of International and Comparative Law.

Turaga, Uday. 2000. Damming Waters and Wisdom: Protest in the Narmada River Valley. *Technology in Society* 22: 237–253.

United Nations (UN). 1992. *Agenda 21: The United Nations Programme of Action from Rio*. New York: United Nations.

United Nations (UN). 2002. *Johannesburg Declaration on Sustainable Development*. Available at <http://www.johannesburgsummit.org/html/documents/summit_docs/1009wssd_pol_declaration.doc> (accessed July 15, 2010).

United Nations (UN). 2003. *UNDP Budget Estimates for the* Biennium 2004–2005, DP/2003/28. Available at <http://www.undp.org/execbrd/pdf/dp03–28e.pdf> (accessed July 15, 2010).

United Nations (UN). N.d. *Rules of the Governing Council*. Available at <http://www.unep.org/Documents/Default.asp?DocumentID=77&ArticleID=1157> (accessed July 15, 2010).

United Nations Conference on Trade and Development (UNCTAD), Program on Transnational Corporations. 1993. *Environmental Management in Transnational Corporations: Report on the Benchmark Corporate Environmental Survey*. Environment Series No. 4. New York: United Nations.

United Nations Conference on Trade and Development (UNCTAD). 2000. *The Least Developed Countries 2000 Report*. Geneva: United Nations Conference on Trade and Development.

United Nations Conference on Trade and Development (UNCTAD). 2001. *World Investment Report 2001: Promoting Linkages*. Available at <http://www.unctad.org/en/docs/wir01ove_a4.en.pdf> (accessed July 15, 2010).

United Nations Conference on Trade and Development (UNCTAD). 2002. *World Investment Report 2002: Transnational Corporations and Export Competitiveness*. New York: United Nations. (Also used data from the 1990, 1995, and 2000 *Reports*.)

United Nations Conference on Trade and Development (UNCTAD). 2008. *World Investment Report 2008: Transnational Corporations and the Infrastructure Challenge*. Geneva: United Nations. Available at <http://www.unctad.org/en/docs/wir2008_en.pdf> (accessed July 15, 2010).

United Nations Conference on Trade and Development (UNCTAD). 2009. *Trade and Development Report, 2009*. Geneva: UNCTAD. Available at <http://www.unctad.org/en/docs/tdr2009_en.pdf> (accessed July 15, 2010).

United Nations Development Programme (UNDP). 1998. *Human Development Report 1998: Consumption for Human Development*. Oxford: Oxford University Press.

United Nations Development Programme (UNDP). 1999. *Human Development Report 1999: Globalization with a Human Face*. Oxford: Oxford University Press.

United Nations Development Programme (UNDP). 2001. *Human Development Report 2001: Making New Technologies Work for Human Development*. Oxford: Oxford University Press.

United Nations Development Programme (UNDP) and United Nations Environment Programme (UNEP) Poverty-Environment Initiative. 2008. *Making the Economic Case: A Primer on the Economic Arguments for Mainstreaming*

Poverty-Environment Linkages into National Development Planning. Nairobi: UNDP-UNEP PEI.

United Nations Development Programme (UNDP) and United Nations Fund for Population Activities (UNFPA). 2002. *Executive Board of UNDP and UNFPA*. Available at <http://www.undp.org/execbrd/pdf/eb-overview.PDF> (accessed July 15, 2010).

United Nations Environment Programme (UNEP). 2000. *Action on Ozone*. Available at <http://www.unep.ch/ozone/pdf/ozone-action-en.pdf> (accessed July 15, 2010).

United Nations Environment Programme (UNEP). 2001a. *International Environmental Governance: Report of the Executive Director to the Open-Ended Intergovernmental Group of Ministers or Their Representatives on International Environmental Governance*. First meeting: New York, April 18. Available at <http://www.unep.org/IEG/docs/working%20documents/report fromED/IGM_1_2.E.doc> (accessed July 15, 2010).

United Nations Environment Programme (UNEP). 2001b. News Release 01/40; *World Water Day 2001: Water for Health*. Available at http://www.unep.org/Documents/Default.asp?DocumentID=193&ArticleID=2801 (accessed July 15, 2010).

United Nations Environment Programme (UNEP). 2002a. *Capacity Building for Sustainable Development: An Overview of UNEP Environmental Capacity Development Activities*. Geneva: United Nations Environment Programme.

United Nations Environment Programme (UNEP). 2002b. *Global Environment Outlook 3*. London: Earthscan. United Nations Environment Programme (UNEP). No date. *UNEP Resource Mobilization*. Available at <http://www.unep.org/rmu/> (accessed July 15, 2010).

United Nations Fund for Population Activities (UNFPA). 2001. *The State of the World Population 2001. Footprints and Milestones: Population and Environmental Change*. New York: United Nations Fund for Population Activities. Available at <http://www.unfpa.org/swp/2001/english/index.html> (accessed July 15, 2010).

United Nations Transnational Corporations and Management Division, Department of Economic and Social Development. 1992. *World Investment Report 1992*. New York: United Nations.

Utting, Peter. 2002. *The Greening of Business in Developing Countries: Rhetoric, Reality and Prospects*. London: Zed.

Utting, Peter, and Jennifer Clapp, eds. 2008. *Corporate Accountability and Sustainable Development*. Delhi: Oxford University Press.

Utting, Peter, and Ann Zammit. 2006. *Beyond Pragmatism: Appraising UN-Business Partnerships. Programme Paper on Markets, Business and Regulations, Paper No.1*. Geneva: UNRISD.

VanDeveer, Stacy D., and Geoffrey D. Dabelko. 2001. It's Capacity, Stupid: International Assistance and National Implementation. *Global Environmental Politics* 1 (2): 18–29.

Vezirgiannidou, Sevasti-Eleni. 2009. The Climate Change Regime Post-Kyoto: Why Compliance is Important and How to Achieve it. *Global Environmental Politics* 9 (4): 41–63.

Victor, David G. 2006. Toward Effective International Cooperation on Climate Change: Numbers, Interests and Institutions. *Global Environmental Politics* 6 (3): 90–103.

Victor, David, Kal Raustiala, and Eugene Skolnikoff, eds. 1998. *The Implementation and Effectiveness of International Environmental Commitments: Theory and Practice*. Cambridge, Mass.: MIT Press.

Vig, Norman J., and Michael E. Kraft, eds. 2009. *Environmental Policy: New Directions for the Twenty-First Century*. 7th ed. Washington, D.C.: CQ Press.

Vogel, David. 1995. *Trading Up: Consumer and Environmental Regulation in a Global Economy*. Cambridge, Mass.: Harvard University Press.

Vogel, David. 2002. International Trade and Environmental Regulation. In *Environmental Policy: New Directions for the Twenty-First Century*, ed. Norman J. Vig and Michael E. Kraft, 5th ed., 354–373. Washington, D.C.: CQ Press.

Vogel, David. 2008. Private Global Business Regulation. *Annual Review of Political Science* 11 (June): 261–282.

Vogler, John. 2000. *The Global Commons: Environmental and Technological Governance*. 2nd ed. New York: Wiley.

Vogler, John. 2003. Taking Institutions Seriously: How Regime Analysis Can Be Relevant to Multilevel Environmental Governance. In *Global Environmental Governance for the Twenty-First Century: Theoretical Approaches and Normative Considerations*, ed. David Humphreys, Matthew Paterson, and Lloyd Pettiford. Special issue of *Global Environmental Politics* 3 (2): 25–39.

Volk, Tyler. 2008. *CO_2 Rising: The World's Greatest Environmental Challenge*. Cambridge, Mass.: MIT Press.

von Moltke, Konrad. 2001. The Organization of the Impossible. *Global Environmental Politics* 1 (1): 23–28.

Wackernagel, Mathis, and William Rees. 1996. *Our Ecological Footprint: Reducing Human Impact on the Earth*. Gabriola Island, B.C.: New Society.

Wackernagel, Mathis, Niels B. Schulz, Diana Deumling, Alejandro Callejas Linares, Martin Jenkins, Valerie Kapos, et al. 2002. Tracking the Ecological Overshoot of the Human Economy. *Proceedings of the National Academy of Sciences of the United States of America* 99 (14): 9266–9271.

Wade, Robert. 1997. Greening the Bank: The Struggle over the Environment, 1970–1995. In *The World Bank: Its First Half Century*, vol. 2, ed. John Lewis and Richard Webb, 611–735. Washington, D.C.: Brookings Institution Press.

Wade, Robert. 2001. Global Inequality: Winners and Losers. *The Economist* (April 28): 72–74.

Wade, Robert. 2004. Is Globalization Reducing Poverty and Inequality? *World Development* 32 (4): 567–589.

Wall, Derek. 1999. *Earth First! And the Anti-Roads Movement*. London: Routledge.

Wallach, Lori, and Michelle Sforza. 1999. *Whose Trade Organization? Corporate Globalization and the Erosion of Democracy*. Washington, D.C.: Public Citizen.

Wapner, Paul. 1996. *Environmental Activism and World Civic Politics*. Albany: SUNY Press.

Wapner, Paul. 2002. Horizontal Politics: Transnational Environmental Activism and Global Cultural Change. *Global Environmental Politics* 2 (2): 37–62.

Wapner, Paul. 2003. World Summit on Sustainable Development: Toward a Post-Jo'burg Environmentalism. *Global Environmental Politics* 3 (1): 1–10.

Wapner, Paul. 2010. *Living Through the End of Nature: The Future of American Environmentalism*. Cambridge, Mass.: MIT Press.

Waring, Marilyn. 1999. *Counting for Nothing: What Men Value and What Women Are Worth*. Toronto: University of Toronto Press.

Watts, Phil, and Richard Holme. 1999. *Corporate Social Responsibility: Meeting Changing Expectations*. Geneva: World Business Council for Sustainable Development.

Webster, Ben. 2009. World Bank Spends Billions on Coal-fired Power Stations. *The Times* (London). September 16. Available at <http://www.timesonline.co.uk/tol/news/environment/article6836112.ece> (accessed July 15, 2010).

Webster, D. G. 2008. *Adaptive Governance: The Dynamics of Atlantic Fisheries Management*. Cambridge, Mass.: MIT Press.

Weinstein, Michael, and Steve Charnovitz. 2001. The Greening of the WTO. *Foreign Affairs* (Council on Foreign Relations) 80 (6): 147–156.

Weis, Tony. 2007. *The Global Food Economy: The Battle for the Future of Farming*. London: Zed.

Weiser, Rivka. 2005. *Teflon and Human Health: Do the Charges Stick? Assessing the Safety of PFOA*, ed. Gilbert L. Ross. New York: American Council on Science and Health.

Weisman, Alan. 2007. *The World Without Us*. London: Macmillan.

Weiss, Edith Brown, and Harold K. Jacobson, eds. 1998. *Engaging Countries: Strengthening Compliance with International Environmental Accords*. Cambridge, Mass.: MIT Press.

Welford, Richard. 1997. *Hijacking Environmentalism: Corporate Responses to Sustainable Development*. London: Earthscan.

Wenz, Peter S. 2001. *Environmental Ethics Today*. Oxford: Oxford University Press.

Wettestad, Jörgen. 1999. *Designing Effective Environmental Regimes: The Key Conditions*. Cheltenham, UK: Edward Elgar.

Whalley, John, and Ben Zissimos. 2001. What Could a World Environmental Organization Do? *Global Environmental Politics* 1 (1): 29–34.

Wheeler, David. 2001. Racing to the Bottom? Foreign Investment and Air Pollution in Developing Countries. *Journal of Environment & Development* 10 (3): 225–245.

Wheeler, David. 2002. Beyond Pollution Havens. *Global Environmental Politics* 2 (2): 1–10.

Whiteside, Kerry H. 2006. *Precautionary Politics: Principle and Practice in Confronting Environmental Risk*. Cambridge, Mass.: MIT Press.

Widener, Patricia. 2009. Global Links and Environmental Flows: Oil Disputes in Ecuador. *Global Environmental Politics* 9 (1): 31–57.

Wild, Antony. 2004. *Coffee: A Dark History*. London, New York: Fourth Estate.

Willard, B. 2002. *The Sustainability Advantage*. Gabriola Island, B.C.: New Society Publishers.

Willetts, Peter. 2001. Transnational Actors and International Organizations in Global Politics. In *The Globalization of World Politics: An Introduction to International Relations*, ed. John Baylis and Steve Smith, 356–383. Oxford: Oxford University Press.

Williams, Edward. 1996. The Maquiladora Industry and Environmental Degradation in the United States–Mexico Borderlands. *St. Mary's Law Journal* 27 (4): 777–779.

Williams, Marc. 1994. International Trade and the Environment: Issues, Perspectives, and Challenges. In *Rio: Unravelling the Consequences*, ed. Caroline Thomas, 80–97. Ilford, UK: Frank Cass.

Williams, Marc. 2001. In Search of Global Standards: The Political Economy of Trade and the Environment. In *The International Political Economy of the Environment: Critical Perspectives*, ed. Dimitris Stevis and Valerie Assetto, 39–61. Boulder, Colo.: Lynne Rienner.

Wilson, Edward O. 2002. *The Future of Life*. New York: Vintage Books.

Wood, John R. 2007. *The Politics of Water Resource Development in India: The Case of Narmada*. Los Angeles, Calif.: Sage Publications.

World Bank. 1992. *World Development Report 1992*. New York: Oxford University Press.

World Bank. 1994. *Adjustment in Africa: Reforms, Results and the Road Ahead*. Oxford: Oxford University Press.

World Bank. 1995. *Mainstreaming the Environment: The World Bank Group and the Environment Since the Rio Earth Summit*. Washington, D.C.: World Bank.

World Bank. 1999. *Greening Industry: New Roles for Communities, Markets and Governments*. Oxford: Oxford University Press. Available at <http://www.worldbank.org/nipr/greening> (accessed July 15, 2010).

World Bank. 2001. *Adjustment Lending Retrospective: Final Report, Operations Policy and Country Services*. Available at <http://siteresources.worldbank.org/PROJECTS/Resources/ALR06_20_01.pdf> (accessed July 15, 2010).

World Bank. 2003. *World Development Report 2003. Sustainable Development in a Dynamic World: Transforming Institutions, Growth, and Quality of Life.* Washington, D.C.: World Bank.

World Bank. 2005. *World Development Indicators.* Washington, D.C.: World Bank.

World Bank. 2006a. *Global Economic Prospects 2007: Managing the Next Wave of Globalization.* Washington, D.C.: World Bank.

World Bank. 2006b. *Where is the Wealth of Nations? Measuring Capital for the 21st Century.* Washington, D.C.: World Bank.

World Bank. 2008. *The World Bank Annual Report 2008: Year in Review.* Oxford: Oxford University Press. Available at <http://siteresources.worldbank.org/EXTANNREP2K8/Resources/YR00_Year_in_Review_English.pdf> (accessed July 15, 2010).

World Business Council for Sustainable Development (WBCSD). 2000. *Building a Better Future: Innovation, Technology and Sustainable Development.* Available at <http://www.wbcsd.org/DocRoot/xPV0RGh50wmrkQHs1oik/Building.pdf> (accessed July 15, 2010).

World Business Council for Sustainable Development (WBCSD). 2002. *The Business Case for Sustainable Development: Making a Difference toward the Johannesburg Summit 2002 and Beyond.* Geneva: World Business Council for Sustainable Development.

World Business Council for Sustainable Development (WBCSD). 2008. *Sustainable Consumption Facts and Trends, November 2008.* Geneva: World Business Council for Sustainable Development.

World Commission on Environment and Development (WCED). 1987. *Our Common Future.* Oxford: Oxford University Press.

World Health Organization (WHO). 2008. *The World Health Report 2008—Primary Health Care (Now More Than Ever).* Geneva: WHO. Available at <http://www.who.int/whr/2008/en/index.html> (accessed July 15, 2010).

World Health Organization (WHO). 2009. *World Health Statistics.* Geneva: WHO. Available at <http://www.who.int/whosis/whostat/EN_WHS09_Full.pdf> (accessed July 15, 2010).

World Meteorological Organization (WMO). 1997. *Comprehensive Assessment of the Freshwater Resources of the World.* Geneva: WMO.

World Trade Organization (WTO). 1999. *WTO, Trade and Environment. Special Studies 4.* Geneva: World Trade Organization.

World Trade Organization (WTO). 2001. International Trade Statistics 2001. Available at <http://www.wto.org/english/res_e/statis_e/its2001_e/chp_2_e.pdf> (accessed July 15, 2010).

World Trade Organization (WTO). 2003. *WTO Secretariat Budget for 2003.* Available at <http://www.wto.org/english/thewto_e/secre_e/budget03_e.htm> (accessed July 15, 2010).

World Trade Organization (WTO). 2008. *International Trade Statistics 2008*. Geneva: WTO. Available at <http://www.wto.org/english/res_e/statis_e/its2008_e/its2008_e.pdf> (accessed July 15, 2010).

World Trade Organization (WTO) and United Nations Environment Programme (UNEP). 2009. *Trade and Climate Change, WTO-UNEP Report*. Geneva: WTO. Available at <http://www.wto.org/english/res_e/booksp_e/trade_climate_change_e.pdf> (accessed July 15, 2010).

Worldwatch Institute. 2003. *State of the World 2003*. Washington, D.C.: Worldwatch. Summary available at <http://www.worldwatch.org/press/news/2003/01/09> (accessed July 15, 2010).

Wright, Christopher and A. Rwabizambuga. 2006. Institutional Pressures, Corporate Reputation and Voluntary Codes of Conduct: An Examination of the Equator Principles. *Business and Society Review* 111 (1): 89–117.

WWF. 2002. *Living Planet Report 2002*. Gland, Switzerland: WWF–World Wide Fund for Nature.

WWF. 2008. *Living Planet Report 2008*. Gland, Switzerland: WWF–World Wide Fund for Nature. Available at <http://assets.panda.org/downloads/living_planet_report_2008.pdf> (accessed July 15, 2010).

Yoder, Andrew J. 2003. Lessons from Stockholm: Evaluating the Global Convention on Persistent Organic Pollutants. *Indiana Journal of Global Legal Studies* 10 (Summer): 113–156.

Young, Oran R. 1989. *International Cooperation: Building Regimes for Natural Resources and the Environment*. Ithaca: Cornell University Press.

Young, Oran R. 1994. *International Governance: Protecting the Environment in a Stateless Society*. Ithaca: Cornell University Press.

Young, Oran R., ed. 1999. *The Effectiveness of International Environmental Regimes: Causal Connections and Behavioral Mechanisms*. Cambridge, Mass.: MIT Press.

Young, Oran R. 2001. Evaluating the Effectiveness of International Environmental Regimes. *Global Environmental Politics* 1 (1) :99–121.

Young, Oran R. 2002. *The Institutional Dimensions of Environmental* Change: Fit, Interplay, and Scale. Cambridge, Mass.: MIT Press.

Young, Oran R. 2008. The Architecture of Global Environmental Governance: Bringing Science to Bear on Policy. *Global Environmental Politics* 8 (1): 14–32.

Young, Oran R., Heike Schroeder, and Leslie A. King, eds. 2008. *Institutions and Environmental Change: Principal Findings, Applications, and Research Frontiers*. Cambridge, Mass.: MIT Press.

Young, Zoe. 2003. *A New Green Order: The World Bank and the Politics of the Global Environment Facility*. London: Pluto Press.

Zacher, Mark, ed. 1993. *International Political Economy of Natural Resources*, vol. 2. Cheltenham, UK: Edward Elgar.

Zadek, Simon. 2001. *The Civil Corporation: The New Economy of Corporate Citizenship*. London: Earthscan.

Zarsky, Lyuba. 2006. From Regulatory Chill to Deepfreeze? *International Environmental Agreement: Politics, Law and Economics* 6 (4): 395–399.

Zeng, Ka, and Josh Eastin. 2007. International Economic Integration and Environmental Protection: The Case of China. *International Studies Quarterly* 51 (4): 971–995.

Zuckerman, Ben. 2004. Nothing Racist about It. *Globe and Mail* (January 28): A17.

Index